ISBN 978-0-266-72904-4
PIBN 10976004

1 MONTH OF
FREE
READING

at

www.ForgottenBooks.com

By purchasing this book you are
eligible for one month membership to
ForgottenBooks.com, giving you
unlimited access to our entire
collection of over 1,000,000 titles via
our web site and mobile apps.

To claim your free month visit:

www.forgottenbooks.com/free976004

English
Français
Deutsche
Italiano
Español
Português

www.forgottenbooks.com

Mythology Photography **Fiction**
Fishing Christianity **Art** Cooking
Essays Buddhism Freemasonry
Medicine **Biology** Music **Ancient**
Egypt Evolution Carpentry Physics
Dance Geology **Mathematics** Fitness
Shakespeare **Folklore** Yoga Marketing
Confidence Immortality Biographies
Poetry **Psychology** Witchcraft
Electronics Chemistry History **Law**
Accounting **Philosophy** Anthropology
Alchemy Drama Quantum Mechanics
Atheism Sexual Health **Ancient History**
Entrepreneurship Languages Sport
Paleontology Needlework Islam
Metaphysics Investment Archaeology
Parenting Statistics Criminology
Motivational

FOREST PLANTING SITES
IN NORTH MISSISSIPPI AND WEST TENNESSEE

Walter M. Broadfoot

SOUTHERN FOREST EXPERIMENT STATION
Chas A Connaughton, Director
New Orleans, La
FOREST SERVICE, U S DEPT OF AGRICULTURE

FOREST PLANTING SITES
IN NORTH MISSISSIPPI AND WEST TENNESSEE

Walter M. Broadfoot
Southern Forest Experiment Station

Forest plantations in north central Mississippi and western Tennessee can produce merchantable timber on much land which erosion has made useless for other crops. Forest plantings usually can decrease, stop, or forestall erosion economically.

This paper describes the most important sites in the area, predicts what may be expected if they are planted to trees, and suggests the tree species and planting procedures most likely to succeed. The opinions and conclusions are based on field observations during the past three years. The majority of these observations were made during detailed study of 159 plantations on which establishment records were available.

Soil Conditions Affecting Tree Growth

Plant growth is governed chiefly by the ability of the soil to supply water, air, and plant food to the roots. Any soil condition which restricts these essentials also restricts tree growth.

In the light of these requirements, three important influences can be used to evaluate site quality. They are soil depth, fertility of the surface soil layer, and soil moisture conditions.

Plant roots extend several feet into the soil in search of water and nutrients. When an impenetrable layer of hardpan or poorly drained soil keeps tree roots near the surface, they cannot reach enough food and water, and growth is poor. Except in gullies, depth to an impervious layer ordinarily will not vary greatly within fields. In general, the extent to which fertility and soil moisture affect the survival and growth of planted trees depends on whether an impervious layer lies close enough to the surface to deny roots the necessary growing space. Soils more than 24 inches in depth are considered deep in this publication, and those less than 24 inches in depth are considered shallow.

A serious effect of erosion has been to wash the most fertile part. With trees as with row crop in the surface 6 inches, the poorer the growth. Wh parent material is found on the surface, growth is poor.

In order to learn the nature, depth, and pos soil layers in old fields, some digging may be nece soil has not been eroded or otherwise moved or mixe the surface layer, subsoil the second layer, and pa third layer. The topsoil is normally darker and gr brown subsoil. Parent material, from which the sub not differ in color from the subsoil. In this area at depths of more than three feet below the ground serious erosion has removed the topsoil.

In gullies, the composition, depth, and loca soil layers can be learned by examining the sides a base of gully banks and on the gully bottoms the ki surface layer will depend largely on whether or not place. On areas where erosion has stopped or where posited, the surface layer will most likely be of m soil, and parent materials. When the materials are the composition can be estimated from the color and Impervious layers are easily recognized as compact, the case of hardpan, or as grey, waterlogged zones poorly drained soils.

The moisture condition of a site is governed position of the site on the slope. Sites on lower bottoms are naturally moister than sites on ridges

The natural moisture condition of a site is texture of the soil, the direction in which the slc kind of plants already growing on the site. Fine-soils are wetter than coarse sands. North and east more moist than south and west slopes. Soils cover litter absorb more water than those that are bare.

Types of Planting Sites

These three major influences--depth, fertil: combine to determine planting sites. Table 1 summa mon sites in north central Mississippi and western dicates the general results that may be expected on are best suited to pines. Hardwoods are recommend them.

In the table, the probable success of plant duction and erosion control has been rated on a fi

Table 1.--Pine planting sites in north Mississippi and west Tennessee

Soil and site	Old-field sites		Gully sites	
	Planting objective	Probable success	Planting objective	Probable success
I. Soil 24 inches or more deep				
A. Surface 6 inches mostly topsoil				
1. Moist--lower slopes and bottoms	Timber	Excellent [1]	(2/)	...
2. Dry--ridges and upper slopes	Timber	Good to excellent	(2/)	...
Special cases--very rapid internal drainage	Timber	Fair to excellent	(2/)	...
B. Surface 6 inches mostly subsoil				
1. Moist--lower slopes	Timber;erosion control secondary	Good	Erosion control;timber secondary	Fair to Good
2. Dry--ridges and upper slopes	Timber;erosion control secondary	Fair to good	Erosion control;timber secondary	Fair
C. Surface 6 inches mostly parent material	(2/)	...	Erosion control	Poor
II. Soil less than 24 in. deep				
A. Surface 6 inches mostly topsoil				
1. Moist--lower slopes and bottoms	Timber	Good [1]	(2/)	...
Special cases--very poor internal drainage	Timber	Poor to fair	(2/)	...
2. Dry--ridges and upper slopes	Timber	Fair to good	(2/)	...
B. Surface 6 inches mostly subsoil				
1. Moist--lower slopes	Timber; erosion control secondary	Fair	Erosion control;timber secondary	Poor to fair
2. Dry--ridges and upper slopes	Erosion control;timber secondary	Poor to fair	Erosion control	Poor
C. Surface 6 inches mostly parent material	(2/)	...	Erosion control	Very poor

[1] Good results may be expected from planting hardwoods on site IA1, and fair results on site IIA1. On all other sites, hardwoods may be expected to be poor or very poor and cannot be recommended for planting.
[2] Such sites are not commonly found in this area.

running from excellent through good, fair, a
(virtual failure). Survival, vigor, and gro
pleteness of erosion control by needle litto
than trees, have entered into this classifio
emphasized that all but the very poorest rea
An erosion-control gully planting on a shal]
surface layer is mostly subsoil may be wortl
even if only part of the trees live. These
to slow down erosion and eventually they wil
Frequently they will control it more cheapl;
any other means.

Following are fuller descriptions--nu
with the table--of the chief sites in the re

IA1. Soil deep (impervious layer, if
inches below surface); surface 6 inches most
Such sites occur mostly near or at the botto
found mostly in old fields because gullies h
areas in which topsoil predominates in the s
ture may be abundant because the site is nea
and also because soil texture is fine, the s
east, plant cover or litter is good, or beca
these factors occurs. The soil is deep enoug
wetness or waterlogging. These are the best
pine timber production, and hardwoods may do
pines planted on these sites begin to reproc
10 years after planting, and are invaded (fi
other high-quality hardwoods that require go

IA2. Soil deep; surface 6 inches mos
Except for small patches these sites, like t
occur mostly in old fields, but generally or
Their dryness may be the result of position
or western exposure, of coarse soil texture
tective cover, or of a combination of these
survival and growth on these sites (fig. 2)
the sites described above. Invading hardwoc
blackgum, and sassafras--are of less value t
on the moister sites. Eastern redcedar also
sites.

Figure 1.—Ten-year-old shortleaf pine on moist, deep soil in an old field. The plantation has been heavily invaded by white ash, a hardwood of high site requirements.

Figure 2.—Good 10-year-old shortleaf pine about 18 feet high and 3 inches in d.b.h.; survival and vigor are fair to good. Soil is deep and fertile but dry, and the hardwoods which have come in are less valuable than those that invade the moist sites like the one shown in figure 1.

In special cases the effect of the position on the slope is accentuated by excessive or very rapid internal drainage through loose sand or loamy sand in the subsoil. Then survival the first year after planting is likely to be poor, because 15 or 20 days without rain may leave young, short-rooted seedlings without sufficient water. The trees that survive the first year, however, are likely to grow nearly as well (fig. 3) as those on moist sites. Poor survival in dry years may necessitate more replanting on these sites than on many others.

IB1. <u>Soil deep; surface 6 inches mostly subsoil; moist</u>.-- These sites occur on lower slopes, in both old fields and gullies. Good moisture conditions result from their low position, medium- to fine-textured soils, north- or east-facing slopes, and protective plant cover or litter. In old fields such sites are recognizable from the fact that severe sheet erosion has removed most or all of the topsoil. Similar sites occur in gullies in recently abandoned fields, or in overgrazed pastures. Old-field plantings should be successful and produce satisfactory timber, with effective erosion control as a byproduct (fig. 4). Gully plantings may have only fair to good success, but probably will stabilize sand movement; timber can be produced as a secondary objective.

Figure 5 illustrates a common and important type of large gully with steep banks. Moisture as well as nutrient materials concentrate where the bank meets the gully floor. These spots are especially good planting sites. Soil is being deposited in the lower part of the gully and immediately below the outlet. This material contains enough topsoil to support good growth, and overlies deep parent material. The first planting should be on recent, somewhat stabilized deposition areas near the gully mouth (fig. 6) and in spots of moisture and topsoil concentration on the floor along the bank. Survival is likely to be low, but the trees which live will be vigorous and will gradually choke off the sand flow and cause it to be deposited further back from the gully mouth. If the new deposits of sand are planted every two or three years, the entire gully can eventually be covered and stabilized.

IB2. <u>Soil deep; surface 6 inches mostly subsoil; dry</u>.--These sites, like the preceding, occur in both old fields and gullies, but on the drier upper slopes and ridges. Dryness may be accentuated by coarse-textured soils, south and west slopes, and the lack of protective cover. Planting success is fair to good on these dry sites in old fields, where timber production is primary, but only fair in gullies, where erosion control is most important.

Figure 3.—Poor survival but good growth from 10-year-old loblolly pines planted on deep loamy sand that has rapid internal drainage.

Figure 4.—Good loblolly plantation on an old field that had lost most of its topsoil but had plenty of moisture and good subsoil. Note heavy litter cast by these trees, which are 35 feet tall and 7 inches d.b.h. at 14 years of age.

Figure 5.—This common type of gully must be planted piecemeal over several years. Fairly stable soil deposits, particularly near the mouth, should be planted first. When these trees have taken hold, another strip farther inside the gully should be planted, and so on until the entire gully is in trees. Other good planting sites are on the floor along the bank.

Figure 6.--A live dam of loblolly pines across a gully outlet. Trees are 10 years old. Although the gully is not stabilized, it no longer contributes heavy sediment to the stream channels it drains into. The new soil deposits in the foreground are ready for planting.

A dry-site gully of this class is pictured in figure 7. A small gully such as this can be planted straight across, all at one time, and at regular spacing. In 10 years' time, it should be well stabilized (fig. 8) if fire is kept out and there is no livestock or insect damage.

IIA1. Shallow soil (impervious layer always present within 24 inches of surface); surface 6 inches mostly topsoil; moist.--These sites are usually found on gentle lower slopes and bottoms in old fields. Moisture conditions are best where the soil textures are fine to medium, where there is some protective cover, and on north and east slopes. Planting success is good for pine timber production, and fair for hardwoods. In shallow soils, growth is somewhat dependent upon the weather. In wet weather the compact layer, or hardpan as it is sometimes called, tends to hold water, and the site becomes too wet. In dry periods these sites soon dry out. Shortleaf pine is likely to have tipmoth damage. If sheet erosion is not already controlled by minor vegetation before planting, it will be checked by needle litter and minor vegetation soon afterwards.

Sometimes sites of this category are wet because of very poor internal drainage. These wet sites are found on flat areas and small depressions and sometimes on bottoms. Trees planted on such sites are generally poor. There is ample topsoil but the soil is so waterlogged that aeration is poor and growing space is severely limited.

IIA2. Shallow soil; surface 6 inches mostly topsoil; dry.-- These sites are found on ridges and upper slopes in uneroded or slightly eroded old fields. Dryness is accentuated by coarse-textured soils, south and west slopes, and bareness. Fair to good results may be expected for pine timber production.

IIB1. Shallow soil; surface 6 inches mostly subsoil; moist.-- These sites are found on lower slopes, both in old fields and gullies. Moisture conditions are best with fine- to medium-textured soils, on north and east slopes, or under protective cover. Timber production on old field sites of this kind is only fair, but erosion control is a secondary benefit. In gullies, erosion control is the primary objective. Results may be poor to fair, with some timber production in the most favorable cases.

IIB2. Shallow soil; surface 6 inches mostly subsoil; dry.-- These sites occur on ridges and upper slopes in old fields and in gullies. In such locations coarse sands, south and west slopes, and bare areas are especially dry. Even on old fields, planting of these sites is primarily for erosion control, with timber production--only poor to fair at best--of secondary importance. Erosion control is the only justification for planting such sites in gullies.

Figure 7.—This kind of gully can be planted all at one time, with tree rows going straight across.

Figure 8.—Gully like that in figure 7, but ten years after being planted to loblolly pine. This plantation is rated fair for timber. It has stopped erosion.

Figure 9.--Shortleaf 10 years old in a gully where the surface layer is mostly parent material. The shallow, low-quality soil has severely limited the feeding zone of the roots.

IC and IIC. <u>Both deep and shallow soils; surface 6 inches mostly parent material; moist or dry.</u>--These sites occur occasionally in old fields but usually in big gullies that seem to have gotten about as wide and deep as they can. Shallow, dry sites are most common. Often the surface is hard, cemented sandy clay, resistant to root penetration. This infertile material extends across the gully floor underneath a shallow covering of mixed materials. The roots find very little plant food, water, and air. In summer the soil surface becomes hot, often hot enough to kill tender young seedlings. Any moisture in surface layer is quickly dried out by the sun and wind.

Unless the sites are prepared in some way before planting, erosion-control plantations usually give poor results. Survival, growth, and vigor are low (fig. 9). Little or no litter is produced, ground cover is sparse, and the plantations fail to control erosion. Planting on deep silt loam parent material would probably give better

results than indicated in table 1, but inasmuch as such sites are ex
tremely scarce in this category, there is not sufficient planting
evidence to guarantee this prediction.

These sites must be mulched before planting. Mulch will prevent rapid loss of moisture by evaporation, extreme heating of the
soil surface, and rain-drop splash, and will do much to increase suc
cess in erosion control. (See 15 for suggestions for preparing such
sites.)

Tree Species for Planting

Loblolly pine.--Loblolly in most respects has been found to b
the best pine to plant. Along with shortleaf and slash pine, it gro
under the adverse conditions common to areas where erosion must be
checked. Loblolly makes as good growth as any other species and pro
duces more litter.

Under average old-field conditions loblolly crowns close at 1
years when trees have been planted at 6- by 6-foot spacings. In gen
eral, at this age the pines are reproducing and hardwoods are coming
in under them (fig. 2). Loblolly shows the usual decline in appearance with decline in site quality but less so than other species.
Lack of depth of soil necessary for full root development, particularly when caused by poor internal drainage, appears to be the prin
pal limiting site factor for this species.

In gully plantings, loblolly does best on loose, sandy, eros
soil. The poorest growth is found on exposed hard, cemented, erosi
resistant material.

Shortleaf pine.--Shortleaf seems to do as well on adverse si
as loblolly, but its sparse litter production and slower growth mak
it less successful in stabilizing gullies and bare areas.

Shortleaf pine is very susceptible to tipmoth attack, partic
larly on the poorer sites. This factor certainly has reduced its
growth rate.

Slash pine.--Slash pine probably grows as fast as loblolly
faster than shortleaf, but its survival is lower. Other things bei
equal, slash pines produce less litter than loblolly but more than
shortleaf.

Slash pine planted in pure stands or on shallow soils appear
to suffer more ice damage than other species. This should be considered when planting for timber production, but should not prevent
its use when erosion control is the main objective.

Other pines.--A few small plantings of longleaf, Virginia, and northern white pine have been observed in this region. Although these species may have some future for erosion control, they have not had sufficient trial to warrant much optimism in the case of Virginia and longleaf pine, while northern white pine promises even less.

Eastern redcedar.--The few eastern redcedar (Juniperus virginiana) plantings observed in this region have been on very poor sites. Both survival and growth were poor. On fair or good sites redcedar should do better. Its natural occurrence in the old fields of this region indicates this promise. Redcedar has also been found to be a good soil builder. This fact plus the erosion control obtained from good stands should justify its further trial. Timber, or certainly posts, could be secured from the best plantings, but perhaps would be of secondary importance to erosion control and soil improvement.

Black locust.--Black locust shows considerable variation in growth and, in general, appears to be poorly suited for erosion control planting. When planted in gullies it fails except on deep, well-drained deposition areas, or in areas where moisture and nutrients have concentrated. Anything less than an optimum growing site means a sharp decrease in success both for wood production and erosion control. This generally poor showing, however, appears not to be due entirely to poor site conditions. The locust borer, twig borer, and leaf miner have severely attacked most of the plantations on poor sites. Some of the smaller trees are dead. Apparently they were able to survive the adverse site conditions but could not withstand severe borer and miner attack.

One effect of black locust should not be overlooked. That is the beneficial effect it may have on other plants growing nearby or in the planting mixture. Being a legume it can serve a useful purpose as a source of supply of nitrogen in closely-spaced plantings.

Other hardwoods.--Plantings of yellow-poplar, white ash, mimosa, catalpa, and black walnut show sharp variations in results. For good growth, these species appear to demand better sites than the pines do. The right amount of soil moisture seems especially important--there must be neither too much nor too little. For example, no good hardwood plantings were found on old fields located on ridges and upper slopes. On the same site, hardwoods will always look and grow less well than loblolly, slash, and shortleaf pine.

Recommended Planting Procedure

1. Evaluate the sites to be planted, and decide on the main objective of planting and the likelihood of success. For profitable timber production, plant the best sites described in this publication. On many sites of medium quality, forest planting controls erosion, and eventual timber production is possible. On the poorer sites, plant for

erosion control but do not count on getting [
On the poorest sites, forest plantations are
erosion without additional treatment.

2. Choose suitable species. As a ru[
leaf, or slash pine. Loblolly has proven mo[
hardwoods only on fertile, moist soils. Int[
pines or hardwoods, should be used only wher[
that they will do approximately as well as t[

3. Try to get stock grown from seed [
ably within 100 miles of the planting site.
that plantations grown from loblolly pine se[
miles from the planting site produced only h[
plantations from seed gathered near the plan[

4. Make sure that the seedlings do n[
heated during shipment or during storage bet[
Seedlings are packed for shipping with sphag[
When the seedlings arrive, the roots and the
cool, and seedling roots should have no mold
lings stored for more than a day or two betw[
should be heeled-in--placed in thin upright l[
the roots completely covered but with at lea[
above ground.

5. Space planted trees according to [
individual seedlings where and when they have
and grow. The usual spacing for forest plan[
Mississippi is 6 by 6 feet. This is close e[
on the best sites, where erosion control is [
val is assured. On the poorer sites, where [
likely, and especially where trees must be e[
trol erosion, spacing should be closer--but [
feet. Spacing need be approximate only. On
should always be varied to set each seedling
possible chance to survive. Where possible,
moister soil. Plant on level spots, and avo[
may be washed away from the roots, and hollo[
be buried under new soil. Seedlings plante[
or weeds have a good chance for survival.

Deposits of soil near the mouths of g[
good sites, especially where the deposited s[
stabilized because of its location or of inv[
Trees planted in such places are especially [
sediment (fig. 6). Planting should not, ho[
such spots to places where newly planted tre[
by fresh soil deposits. Planting where depo[
postponed until these new depositions have i[
stabilized.

In gully planting, try to distinguish between gullies that can be stabilized by planting all at once (fig. 7) and those it is more effective and economical to plant piecemeal over several years (fig. 5).

6. <u>Plant each tree carefully and correctly.</u> Use either planting bars or mattocks. Set the root collar (which is marked by the change from greenish stem bark to yellowish root bark) at the surface of the ground or, at the most, ½ inch below it--never above. Be sure that the planting hole is at least as deep as the roots, that the roots are not doubled up, that the hole is firmly closed, and that the soil immediately around the seedling is left level. Keep seedling roots moist in bucket or tray during planting. Remove only one seedling at a time so that roots will be moist when planted.

7. <u>Where necessary, take extra precautions to get the trees established.</u> On bare, eroded areas, especially of the poorer, drier sites, mulch the surface with grass, straw, sawdust, or any available plant litter, up to 1 inch in depth. On deep soils, installing small check dams or soil-collecting trenches several months in advance of planting may increase the number of good planting spots. The use of various grasses and legumes both singly and in mixtures is generally recognized as a means of getting a quick cover on exposed surfaces; these usually need fertilization. Native herbaceous cover may be encouraged by·the use of fertilizer alone. Soil quickly stabilized in this way will aid in getting trees established.

8. <u>Furnish proper protection for the planting.</u> All livestock should be fenced out of tree plantations. Fire must be guarded against.

9. <u>Make reinforcement plantings.</u> Replant where trees have died. Plant newly stabilized areas of deposition each year until the planting objective--erosion control--has been reached.

LPWOOD PRODUCTION COSTS
IN SOUTHEAST ARKANSAS, 1950

R. R. Reynolds

SOUTHERN FOREST EXPERIMENT STATION
Chas A Connaughton, Director
New Orleans, La.
FOREST SERVICE, U S. DEPT OF AGRICULTURE

PULPWOOD PRODUCTION COSTS IN SOUTHEAST ARKANSAS, 1950

By

R. R. Reynolds
Southern Forest Experiment Station

In 1940 the Southern Forest Experiment Station made a study of pulpwood production costs in southeast Arkansas.[1] World War II significantly altered these costs, and the study was repeated in 1945.[2] By 1950, continued inflation, the use of bigger trucks, and the introduction of power saws, pallet loading systems, and tractors had made the 1945 data obsolete, so that a third study was made in the fall of 1950.

The 1950 study was carried out in Ashley, Bradley, and Drew counties of southeast Arkansas, as were the previous studies. In all three studies only dry-weather logging conditions were sampled. Production under winter or wet-weather conditions is bound to be somewhat different and costs higher.

In the 1950 study one pulpwood contractor was selected at random in each of nine designated areas of the three counties. Each was visited in the woods and a record was made of equipment, manpower, and methods used, costs of equipment and manpower, hours worked by each crew or portion of a crew, number of cords of wood produced per day, number of loads hauled per day, loading and unloading time and method, distance of haul, and time required per trip. In short, information was collected on every item that would enter into the cost of producing pulpwood.

---

[1] Reynolds, R. R. Pulpwood and log production costs as affected by type of road. Southern Forest Experiment Station, Occasional Paper 96. 1940.

[2] Reynolds, R. R. Pulpwood and log production costs in 1945 as compared with 1940. Southern Forest Experiment Station, Occasional Paper 107. 1945.

Contractors keep few records and do not have co
ment depreciation, insurance, and crew transportation,
do know how long the average piece of equipment lasts
time it is in use each day or each week. Fixed expens
depreciation, have therefore been computed from the be
perience data.

At one time pulpwood cutters and contractors th
tion unprofitable unless the cut per acre was at least
also considered it generally unprofitable to cut pulpw
sawtimber trees. In 1950, however, the average produc
the nine contractors was approximately 1.0 cord and at
of the cut came from tops of sawlog trees. Most of th
cut on lands belonging to large timber companies.

The selection method of cutting of standing tre
moving defective, very rough, forked, or badly suppres
pulpwood. Usually only trees below 10 inches in diame
pulpwood. Trees of larger sizes were cut into sawlogs
be removed was marked before cutting.

Felling and Bucking

Up until about 1943 all trees were felled and b
wood by hand labor. Two-man crosscut saws were genera
although some one-man bow saws were used where the tim
below 8 inches d.b.h. Chain saws were introduced in a
met with varied success in the early years. Now they
standard equipment of nearly every contractor and two-
saws are rarely used.

Chain saws, however, have not taken over the en
bucking job. Most contractors have found that small s
be felled and bucked into pulpwood lengths most cheapl
bow saws. It is customary for the contractor first to
crews through the stand to cut the small marked trees
8 inches in diameter, and then to have chain saw crews
ately to cut the larger trees. One chain saw crew to
is the usual combination.

Bow saw cutters are paid either by the hour or
wood they cut. When on an hourly basis they generally
per hour and are covered by social security, workmen's
and industrial insurance. Hourly rate employees are f
axes, and other tools, which are maintained by the con
bow saw cutters work on a task or contract basis, they
$0.90 to $1.00 per 6-foot pen of wood. Contract cutte
nish and maintain their own tools, although the contra
security and other benefits for them.

The average production for the bow saw cutters is 8 pens of 5-foot wood per man per day. The timber is small, generally running from four to five 6-foot pens of wood per standard cord, and the average production per 8-hour day is from 1.6 to 2.0 cords. Wood produced by the bow saws, therefore, costs from $4.28 to $5.39 per standard cord, depending on the average size and wage rates.

Although several of the 13 chain saws in use by the 9 pulpwood contractors are supposed to be one-man saws, some of the contractors have found the work heavy enough so that they are using two men on a one-man saw. The two additional men that make up the usual four-man chain saw crew limb the trees felled by the saw crew, and also the tops of the sawlog trees that are to be cut. They help get the saw out of pinches and sometimes aid in felling standing trees.

All chain saws used on the various jobs were of the bow type. All contractors except one reported that the chain saws were expensive to operate. Perhaps because of poor operation procedures, some sprockets were good for only 20 to 40 cords. Chains, which cost $27.00 each, sometimes lasted for only about 100 cords of wood.

The estimated cost of operating the average chain saw was $0.87 per standard cord (appendix table 8). This is exclusive of operating labor. Cost of chain-saw crew labor averaged $0.805 per man-hour-- plus social security, workmen's compensation, and insurance. With a cost of $0.036 per man-hour for social security and industrial insurance, $0.042 for workmen's compensation, $0.070 for supervision, and $0.063 for crew transportation, the total cost per man-hour was $1.016 or $8.128 per 8-hour day. Total cost per 8-hour day for the four-man crew was, therefore, $32.51. Average production for the crews studied was 10.8 cords. Labor, therefore, costs $3.01 per cord. Adding $0.87 per cord for the cost of the saw makes the total net cost $3.88 per cord.

Average net felling and bucking cost for pulpwood produced by all contractors was $4.20 per cord (table 1). If 20 percent is allowed for profit and risk, the average net cost becomes $5.04 per standard cord. In connection with these figures (table 1) it is of interest to note that the two contractors with lowest production costs used chain saws exclusively--but so did the two contractors having the most expensive wood. The differences are accounted for by the amount of wood cut per acre and the amount taken from tops of sawlog trees.

Table 1.--Pulpwood felling and bucking cost per cord, by contractors

Con- trac- tor	Men in crew	Labor	Power saw	Super- vision	Crew trans- porta- tion	Social security, unemploy- ment ins., workmen's comp.	Total	Total cost plus 20 percent	Total cords pro- duced per day
	Number	--------- Dollars ---------							Cords
1	9	1.80	0.87	0.17	0.07	0.18	3.09	3.71	30.0
2	6	2.57	.62	.24	.38	.26	4.07	4.88	14.0
3	15	3.62	.27	.26	.30	.32	4.77	5.72	32.0
4	14	2.97	.59	.24	.10	.28	4.18	5.02	32.7
5	6	3.58	.87	.14	.12	.25	4.96	5.95	24.0
6	4	1.60	.87	.15	.25	.16	3.03	3.64	15.0
7	11	3.33	.25	.27	.26	.32	4.43	5.32	22.5
8	7	3.13	.87	.26	.35	.30	4.91	5.89	15.0
9	7	3.00	.50	.28	.20	.31	4.29	5.15	14.0
Weighted average		2.90	.61	.22	.20	.27	4.20	5.04	

Skidding or Bunching

Under dry-weather conditions during the years up to about 1945, most of the pulpwood was loaded directly from the ground or pen to the truck that hauled it to the mill or railroad car. Any necessary bunching was done with one or two horses and a slide or wagon. In 1950, however, there were almost as many different bunching and loading systems as there were contractors. Two contractors were using a pallet system. One loaded the pallets in the woods by use of a special slide and Caterpillar D-2 and D-4 tractors. The other bunched the wood with a team and wagon and loaded the pallets on the road. Some contractors loaded most of their wood directly by hand onto the trucks (which were driven into the woods) and bunched only the scattered wood. Some used teams and wagons for this bunching, and some used tractors and special heavy iron slides. One contractor did not bunch any of his wood but loaded all of it directly onto the trucks.

One contractor who had very reasonable bunching, loading, and hauling costs bunched all of his wood with an International TD-9 tractor and a special iron slide that held approximately 2.3 cords of wood-- a good 1½-ton truckload. Each slide load was skidded to some woods road or point easily accessible to the trucks. At this spot a specially adapted Loggers Dream loader equipped with cable slings raised the whole load high enough to permit a truck to back directly under. If no truck was at hand the load was picked off the slide and left swinging in the air. As soon as a truck arrived, the supervisor or one of the bunching crew helped the truck driver lower the load onto the truck. After the load was on the truck it was necessary to straighten a few sticks and level the load somewhat, but the whole loading operation was accomplished in less than 15 minutes.

Only the two contractors using the pallet systems and the one using the Loggers Dream method of loading bunched all of their pulpwood prior to hauling. The others bunched only wood--about 10 to 30 percent--that was scattered or in places hard to get to with a truck.

Cost of bunching has been computed for each contractor. The costs are based on the number of men, the number and size of tractors and other equipment, and the number of hours the manpower or equipment was used each day. Use was made of the machine rates shown in appendix tables 9 and 10.

Because only a small proportion of the wood is bunched during dry weather, the average bunching cost was only $1.16 per cord (table 2). This average, of course, includes some producers who bunch all of their wood and some who bunch little or none.

The figures on skidding or bunching costs do give some indications of the additional cost that applies any time that ground conditions are such that all wood must be bunched to a road. The average cost for those contractors who bunch all of their wood was $2.47 per cord. It seems reasonable to expect, therefore, that the cost of the wood during wet weather should be the difference between what the various contractors are spending for bunching in dry weather and $2.47. The average dry-weather bunching cost for those contractors who are not using pallet systems is $0.50 per cord. Wet-weather costs, therefore, probably will be close to $2.00 per cord more than dry-weather costs.

Loading and Hauling

There is a very definite change away from the use of small and light trucks toward the use of larger and more powerful ones for the hauling of pulpwood. At the time of the two previous studies practically all trucks were of the 1½-ton, 85-horsepower variety. Out of a total of twenty-three trucks in use in 1950, however, nine were in the 2-ton class and two were 2½-ton. Largely as a result of this increase in truck size the average load of wood increased in volume from 1.98 standard cords in 1940 and 1945 to 2.47 cords in 1950.

Table 2.—Pulpwood skidding or bunching cost per cord

Contractor	Men in crew (Number)	Hours per day (Number)	Labor cost 1/ (Dollars)	Team or machine (Number)	Team or machine rate per hour (Dollars)	Machine cost	Cable, slides, etc.	Supervision 2/	Crew transportation cost	Total cost	Total cords produced (Cords)	Cost per cord (Dollars)
						... Dollars ...						
1	...	...	...	...	...	...	...	...	...	...	30.0	...
2	1½	8	9.92	8	0.438	3.50	1.00	0.84	1.33	16.59	12.0	1.38
3	4	2	7.66	2	2.76	5.52	.50	.51	.64	14.83	30.0	.49
4	3½	8	27.53	6	2.76	16.56	.50	1.77	.81 }	47.17	27.0	1.75 }
5	3/ 7	8	46.28	3, 5	2.40, 2.76	21.00	1.00	3.54	3.45 }	75.27	22.8	3.30 }
6	4	2	6.61	2	2.76	5.52	.50	.51	.92	14.06	15.0	.94
7	3	1	2.48	1	2.76	2.76	.06	.19	.20	5.69	25.0	.23
4/8										36.00	14.4	2.50
5/9										11.55	14.0	.82
Total			100.48			54.86	3.56	7.36	7.35	221.16	190.2	1.16
Cost per cord other than contracts			.62			.34	.02	.05	.05			1.08

The average distance over which the wood is hauled has also increased in the last few years. Previously, the average one-way haul was about 8 miles. Now it averages just under 15 miles.

In 1940 most truck tires were 32 by 6. In 1945 most trucks had 7.50 by 20 tires on the front and 8.25 by 20 on the rear wheels. Many tires in 1950 were still 8.25 by 20, but some were 9.00 by 20 and larger.

As has been indicated previously, most of the loading of the trucks is still by manpower. Some contractors have tried mechanical loading and have found that hand-loading is the cheapest. Perhaps it is, even though it requires approximately 35 minutes to complete the average load as contrasted to 15 minutes for the pallet and Loggers Dream loading.

Trucks worked an average of 9.11 hours per day in 1950. Of this, 1.71 hours were spent in loading, 3.55 hours in hauling, 1.34 hours in unloading, and 2.51 hours in delay.

Pulpwood loading and hauling cost was computed in the following manner:

Loading labor.--Used exclusively for loading (except the driver). The number of hours worked per day and the rate paid per hour were obtained from the contractors. Total labor cost plus social security, industrial insurance, and workmen's compensation was divided by the total amount of wood produced per day to get the loading labor cost per cord.

Truck costs.--These costs were computed from machine rates (appendix tables 11 and 12) that were themselves computed from average experience data on cost of trucks, average life, cost of repairs, sale values for used trucks, and average operating costs. Truck costs were broken down into fixed cost and operating cost. The former is based on the amount required per year for interest on the investment, license, taxes, and depreciation. This cost was charged against the truck for each day of operation regardless of the miles traveled. In effect, it was charged against the operation whether the truck was standing still, being loaded, or was traveling down the highway. Operating costs, on the other hand, were only charged when the truck was traveling. The amount charged depended upon the size of the truck, the number of miles traveled, and the size of the load.

Truck driver.--This cost was charged separately from other costs. It was arrived at by dividing the total cost of the driver per day, including social security and other extras, by the number of cords hauled.

Supervision.--As will be explained later, supervision was charged at so much per man hour of labor used in loading or hauling on the various jobs.

Crew transportation.--As will also be explained later, this is the cost of picking up the men in the morning, transporting them to the work site, and taking them home at night. The total cost for the men used in the loading and hauling crews divided by the number of cords produced per day gave the cost per cord.

The cost per cord for loading and hauling the wood averaged $3.92 (table 3), exclusive of any charge for profit and risk. The three contractors with mechanical loading devices had unusually low loading and hauling costs, but their advantage here is largely offset by the high cost of bunching the wood.

Table 3.--Pulpwood loading and hauling cost per cord

Con- tractor	Loading labor cost	Other loading cost	Truck fixed cost	Truck hauling cost	Truck driver cost	Super- vision	Crew trans- por- tation	Total cost
- - - - - - - - - - - - Dollars - - - - - - - - - - - -								
1	2.20	...	0.64	0.94	1.10	0.28	0.11	5.27
2	1.10	...	.64	1.96	1.10	.19	.30	5.29
3	1.02	...	.61	1.56	1.02	.15	.17	4.53
4	...	0.33	.29	.74	.69	.04	.02	2.11
5	(1/)	...	.65	1.43	.76	.05	.04	2.93
6	.88	...	.56	1.49	.88	.15	.25	4.21
7	.79	...	.55	1.89	.79	.13	.13	4.28
8	(1/)	...	.55	1.30	.46	...	...	2.31
9	.94	...	.55	1.46	.94	.16	.11	4.16
Weighted average	.82	.05	.56	1.37	.87	.13	.12	3.92

1/ Pallet system used. No labor except driver needed.

Crew Transportation

During and after World War II much of the country labor either left the State or moved to the towns. Woods workers became scarce and those available lived long distances from the woods operations. To get sufficient laborers, contractors in 1950 had to furnish good transportation and sometimes haul the men 20 to 30 miles to the job. Thus, the beginning of the "work truck" and the need for charging crew transportation against the cost of producing pulpwood.

The average contractor drives his work truck 43.7 miles per day in transporting his crew to and from the job. The cost of the truck per mile of operation (appendix table 13) is $0.103. The cost of the driver is $0.045 per mile. The total cost of the crew transportation is $0.505 per man-day of labor transported, or $0.340 per cord of wood produced (table 4).

Table 4.--Crew transportation cost per man-day

Con-tractor	Work truck miles traveled per day	Truck cost per day 1/	Driver hours	Driver cost per day	Total truck and driver cost 2/	Men trans-ported	Cost per man-day	Cost per cord
	Miles	Dollars	Hours	Dollars		Number	Dollars	
1	37	3.81	1.5	1.69	5.50	24	0.229	0.183
2	60	6.18	2.4	2.70	8.88	10	.888	.740
3/3	100	10.30	4.0	4.50	14.80	23	.643	.493
4	25	2.58	1.0	1.12	3.70	16	.231	.137
5	30	3.09	1.2	1.35	4.44	9	.493	.195
6	50	5.15	2.0	2.25	7.40	8	.925	.493
7	60	6.18	2.4	2.70	8.88	17	.522	.355
8	40	4.12	1.6	1.80	5.92	8	.740	.411
9	35	3.60	1.4	1.58	5.18	13	.398	.370
Total	437	45.01		19.69	64.70	128		
Average	43.7	4.50		1.97	6.47	12.8	.505	.340

1/ At $0.103 per mile.
2/ At $1.125 per hour. No social security, workmen's compensation, or unemployment insurance included.
3/ Used two trucks.

Supervision

Contractors with small crews usually supervise the whole job themselves. Those with larger operations hire a man for the purpose but also often spend all or a portion of their own time in super-vision. In this cost study, each man who worked at supervising the various crews and equipment was assumed to be performing a necessary function, and his time was charged against the wood production cost at $8.00 per day. Dividing the total amount charged for supervision for all contractors by the total number of men employed by all con-tractors gave $0.56 as the cost per man-day of labor employed. This is the rate used in all tables.

This report would be incomplete without one
subject of supervision. Some contractors feel tha
crews in the woods, tell them what to do, and then
them for long periods of time. Production always
jobs. By observing the closeness and character of
each job, one was able to forecast very accurately
production costs would be. Contractors who have s
close supervision undoubtedly produce cheaper wood
than many of those having large, poorly supervised

Total Production Costs

Total net costs of producing pulpwood varie
$11.19 per standard cord and averaged $9.28 (table
of net cost were allowed for profit and risk the v
from $9.65 to $13.43 and would average $11.14 per
When computed for a uniform 15-mile haul for all j
producing the wood, plus 20 percent for profit and
$9.79 to $13.39 and averages $11.43 per standard c

Table 5.--Pulpwood production cost by contractors-
conditions, 1950

Con- tractor	Felling and bucking	Skidding or bunching	Loading and hauling	Total net cost	Net For
	Dollars - - -				
1	3.09	...	5.27	8.36	
2	4.07	1.38	5.29	10.74	
3	4.77	.49	4.53	9.79	
4	4.18	1.75	2.11	8.04	
5	4.96	3.30	2.93	11.19	
6	3.03	.94	4.21	8.18	
7	4.43	.23	4.28	8.94	
8	4.91	2.50	2.31	9.72	
9	4.29	.82	4.16	9.27	
Weighted average	4.20	1.16	3.92	9.28	

1/ If all contractors hauled over same road condi
same speed.

Costs for Various Lengths of Haul

In the area of the study, trucks of three different sizes each carried different average loads per trip over different lengths of haul. Table 6 shows what the cost would be for 1½- and 2-ton trucks carrying given average loads of wood over distances varying from 5 to 20 miles.

Table 6.--<u>Pulpwood production cost per cord for various lengths of haul[2] and various loads--1½- and 2-ton trucks. Dry weather only</u>

Length of haul[2] (miles)	Loading and hauling cost				Total cost [3]			
	1½-ton truck		2-ton truck		1½-ton truck		2-ton truck	
	2.25 cords per load	2.50 cords per load	2.75 cords per load	3.00 cords per load	2.25 cords per load	2.50 cords per load	2.75 cords per load	3.00 cords per load
	- - - - - - - - - - - - - Dollars - - - - - - - - - - - -							
5	4.06	3.78	4.00	3.77	9.96	9.68	9.90	9.67
7½	4.51	4.20	4.40	4.15	10.41	10.10	10.30	10.05
10	5.08	4.70	4.91	4.61	10.98	10.60	10.81	10.51
12½	5.58	5.15	5.36	5.02	11.48	11.05	11.26	10.92
15	6.11	5.63	5.82	5.45	12.01	11.53	11.72	11.35
17½	6.61	6.08	6.26	5.84	12.51	11.98	12.16	11.74
20	7.13	6.55	6.70	6.25	13.03	12.45	12.60	12.15

1/ Based on net cost plus 20 percent for profit and risk.
2/ Mainly over first-class roads.
3/ Including $5.04 per cord for felling and bucking and $0.86 per cord for bunching or skidding, excluding pallet systems.

For short hauls there is apparently little to choose between 1½-ton trucks carrying loads averaging 2.25 cords per load and 2-ton trucks carrying 2.75 cords per load, or between 1½-ton trucks carrying 2.50 cords as compared to 2-ton trucks carrying 3 cords. For the longer hauls, however, the 2-ton trucks and the larger loads produce the cheaper wood.

There is little question but that for hauls over 15 miles, still larger trucks would be cheapest, but then bunching all the wood to the road or central loading point would be necessary. Trucks larger than 2-ton are very difficult to maneuver through the woods.

Comparison of 1940, 1945, and 1950 Costs

Who will remember delivering pulpwood to the railroad car or pulp mill for $3.45 per standard cord? It sounds quite impossible. Yet, ten short years ago, that was the cost for wood hauled 8 miles. Even five years ago the cost was $7.23 per standard cord. Today the cost averages $11.14 per standard cord, though the average load is hauled somewhat farther.

Over the last 10 years, felling and bucking cost has gone up 198 percent. The cost of bunching, loading, unloading, and delay has risen 269 percent, and hauling cost is up 227 percent. The total average cost of delivering wood today is 223 percent more than it was ten years ago--in short, it has more than tripled.

Manpower efficiency in 1945 was lower than in 1940, but efficiency in 1950 was again up to the level of 1940. Trucks and other equipment are undoubtedly better than in 1940. The tremendous increase in pulpwood production cost over the last 10 years, therefore, can practically all be charged to greatly increased wages, and to cost of equipment. Only a very small percentage of it can be charged to increased length of haul.

Table 7.--Comparison of 1940, 1945, and 1950 pulpwood production costs per standard cord[1], dry weather conditions

Item	Cost			Percent increase over 1940	
	1940	1945	1950	1945	1950
	- - - Dollars - - -			- Percent -	
Felling and bucking[2]	1.69	3.69	5.04	118	198
Bunching, loading, unloading, delay	.81	1.52	2.99	88	269
Hauling[3]	.95	2.02	3.11	113	227
Total	3.45	7.23	11.14	110	223

[1] Computed at cost plus 20 percent for profit and risk.
[2] Wood cut in 5-foot lengths.
[3] In 1940 and 1945 average length of haul 8 miles, in 1950 average length of haul 15 miles.

Table 8.--Estimated costs: one-man chain saw used to produce pulpwood

	Dollars
Investment	
Saw complete with bow and chain	441.58

	Costs per cord Dollars
Fixed expenses	
Interest on investment or carrying charge at 7 percent[1]	0.01
Depreciation[2]	.22
Maintenance--labor	.20
Maintenance--parts	.25
Risk or insurance	.02
Gasoline, oil, wedges, etc.	.17
Total per cord	0.87

[1] On one-half of original investment.
[2] Life 1 year or 2,000 cords or 200 days.

Table 9.--Estimated costs for Loggers Dream used to load pulpwood

	Dollars
Investment	
Loader complete with cable and pulleys	3,450.00
Less trade-in value	- 600.00
Total amount to be depreciated	2,850.00
Fixed expenses	
Interest on investment[1] or carrying charge at 7 percent	158.38
License and taxes per year	59.00
Insurance or risk	123.00
Total fixed expenses per year	340.38
Fixed expenses per day (200 days per year)	1.70
Depreciation of truck per day--life = 1,200 days	2.38
Total fixed expenses per day (8-hour day)	4.08
Running expenses per day	
Tires	.50
Gasoline	.84
Oil and grease	.45
Repair labor	1.00
Repair supplies, cable, etc.	2.00
Total	4.79

[1] Average investment = $\frac{\text{initial investment + trade-in value}}{2}$ + $\frac{\text{annual depreciation}}{2}$ = $\frac{\$3,450.00 + \$600.00}{2}$ + $\frac{\$475.00}{2}$ = $2,262.50

Table 10.--Machine rate for tractors

Item	:Caterpillar: D-2 or equal	:International TD-9 or equal
	Dollars	Dollars
Investment		
Tractor complete with radiator and oil pan guards, bumper, pull hooks, side panels, and starter motor	4,400.00	5,100.00
Less trade-in	- 600.00	- 800.00
Amount to be depreciated	3,800.00	4,300.00
Fixed expenses		
Interest on investment[1] or carrying charge at 7 percent	214.88	251.63
Taxes	18.00	20.00
Operating overhead and risk	90.00	100.00
Total fixed expenses per year	322.88	371.63
Fixed expenses per 6-hour day (200 days per year)	1.61	1.86
Depreciation of tractor[2] per day	5.70	6.45
Total fixed expenses per day	7.31	8.31
Fixed expenses per hour (6-hour day)	1.22	1.38
Running expenses per hour		
Fuel (at $0.13 per gallon)	.203	.255
Lubricating oil (at $0.60 per gallon)	.042	.047
Grease (0.32 pound at $0.16)	.051	.051
Gas to start engine (0.027 gallon at $0.28)	.008	.008
Service labor	.020	.020
Repair parts and labor	.900	1.000
Total	1.224	1.381
Total hourly operating cost	2.44	2.76

[1]/ Average investment in 1950 = $3,069.72 for D-2 and $3,594.68 for TD-9.
[2]/ Life = 667 days or 4,000 hours (6-hour days).

Table 11.--Estimated costs per truck used for pulpwood hauling--1½- and 2-ton trucks

Item	Truck 1½-ton, 105 h.p.	Truck 2-ton, 105 h.p., 161-in. wheelbase
	Dollars	Dollars
Investment		
Truck complete with cab, dual wheels, and metal pulpwood rack	2,058.00	2,551.00
Minus tires[1]	- 627.28	- 627.28
Net investment	1,430.72	1,923.72
Minus truck trade-in value	- 450.00	- 600.00
Amount to be depreciated	980.72	1,323.72
Fixed expenses		
Interest on investment[2] or carrying charges at 7 percent	85.44	111.50
License and taxes per year	55.25	57.00
Insurance or risk	72.61	91.00
Total fixed expenses per year	213.30	259.50
Fixed expenses per day (200 days per year)	1.07	1.30
Depreciation of truck per day[3]	2.80	3.31
Total fixed expenses per day	3.87	4.61
Fixed expenses per hour (8-hour day), truck only	.48	.58
Running expenses per mile--woods or low-quality road		
Tires--life = 6,000 miles	.105	.105
Gasoline--5 miles per gallon	.056	.056
Oil and grease	.006	.006
Repair labor	.020	.020
Repair supplies	.020	.020
Total	.207	.207
Running expenses per mile--graded dirt or better quality road		
Tires--life = 10,000 miles	.063	.063
Gasoline--9 miles per gallon	.031	.031
Oil and grease	.003	.003
Repair labor	.012	.012
Repair supplies	.012	.012
Total	.121	.121

[1] Cost of tires charged against running expenses.
  Front tires 7:50 x 20 = $93.54 each.
  Rear tires 8:25 x 20 = $110.05 each.
[2] Average investment = $1,220.56 for 1½-ton truck, $1,592.79 for 2-ton.
[3] Life--1½-ton truck = 350 days, 2-ton truck = 400 days.

Table 12.--Estimated costs for truck used for pulpwood hauling--2½-ton,
8-cylinder, 159-inch wheel base, 5-speed transmission

	Dollars
Investment	
Truck complete with cab, dual wheels, and pallet system	5,208.00
Minus tires[1]/	- 726.34
Net investment	4,481.66
Minus truck and pallet trade-in value	-1,200.00
Amount to be depreciated	3,281.66
Fixed expenses	
Interest on investment[2]/ or carrying charge at 7 percent	237.14
License and taxes per year	65.00
Insurance or risk	122.40
Total fixed expenses per year	424.54
Fixed expenses per day (200-day year)	2.12
Depreciation of truck per day--life = 600 days	5.47
Total fixed expenses per day	7.59
Fixed expenses per hour (8-hour day)	.95
Running expenses per mile--woods or low-quality roads	
Tires--life = 6,000 miles	.121
Gasoline--4 miles per gallon	.070
Oil and grease	.007
Repair labor	.025
Repair supplies	.025
Total	.248
Running expenses per mile--graded dirt or better-quality roads	
Tires--life = 10,000 miles	.073
Gasoline--8 miles per gallon	.035
Oil and grease	.004
Repair labor	.015
Repair supplies	.015
Total	.142

1/ Cost of tires charged against running expenses.
   Front tires 8:25 x 20 = $110.05 each.
   Rear tires 9:00 x 20 = $126.56 each.
2/ Average investment = $3,387.77.

Table 13.--Estimated costs for 3/4-ton pick-up truck used to haul
        woods crews

	Dollars
Investment	
Truck complete with covered pick-up body	1,748.70
Minus tires[1]/ 7:00 x 16	- 119.20
Net investment	1,629.50
Minus truck trade-in value	- 300.00
Amount to be depreciated	1,329.50
Fixed expenses	
Interest on investment[2]/ or carrying charges at 7 percent	76.84
License and taxes per year	22.00
Insurance or risk	44.00
Total fixed expenses per year	142.84
Fixed expenses per day (200-day year)	.71
Depreciation of truck per day--life = 1,000 days	1.33
Total fixed expenses per day	2.04
Fixed expenses per mile of operation	.047
Running expenses per mile	
Tires--life = 12,000 miles	.010
Gasoline--10 miles per gallon	.028
Oil and grease	.002
Repair labor	.008
Repair supplies	.008
Total	.056

1/ Cost of tires charged against running expenses.
      Tires 7:00 x 16 all around at $29.80 each.
2/ Average investment = $1,097.70

Southern Forest Experiment
Station Occasional Paper
No. _122_ is out of print.
Please bind this in
its place.

STORING SOUTHERN PINE SEED

Philip C. Wakeley

SOUTHERN FOREST EXPERIMENT STATION
Harold L Mitchell, Director
New Orleans, La
FOREST SERVICE, U S DEPT OF AGRICULTURE

STORING SOUTHERN PINE SEED

Philip C. Wakeley
Southern Forest Experiment Station

This paper gives directions for dry, cold storage of southern pine seed. Such storage, although not infallible, is the best method yet developed for keeping a high percentage of the seed capable of the vigorous germination essential to good results in the nursery.

Dry, cold storage involves refrigerating seed promptly and continuously at a temperature not exceeding 41° F. and preferably between 32° and 5° F., and at a constant moisture content between 6 and 9 percent for longleaf pine and between 9 and 12 percent for other southern pine seeds. The details of the process must, however, be varied to fit local circumstances.

Principles of Successful Storage

Storage problems can best be overcome if three main facts are kept in mind.

1. So long as a seed is alive, it _respires_. That is, it consumes the plant food stored within it; it uses oxygen; it liberates carbon dioxide, water, and heat. Some respiration is essential to keep the seed alive, but too much rapidly uses up the stored food on which seedling growth depends, and injures the seed in other ways. Keeping respiration just above the minimum safe level is therefore basic to successful seed storage. The rate of respiration increases slowly with increases in storage temperature up to about 41° F., and with increases in seed moisture content up to 10 percent with longleaf pine and to somewhere between 10 and 12 percent with other southern pines. Above these levels, the rate of respiration increases tremendously with increases in temperature and seed moisture content. It also increases tremendously with injury to the seed.

2. Seed is _in_ _storage_ from the time the cone ripens until pregermination treatment or sowing--not just while in containers or buildings specifically set aside for storage purposes. Excessive respiration at any time between ripening and sowing may weaken or kill the seed. For example, many lots of southern pine seed properly refrigerated most of the time between extraction and use have deteriorated badly during brief exposure to adverse conditions before refrigeration or between refrigeration and sowing.

3. Storage succeeds only when _all_ influences that materially affect respiration are kept favorable. These influences include not only storage temperature and seed moisture content, but also the initial soundness and vitality of the seed. Thus correct collection, extraction, and dewinging, and correct drying if special drying is necessary, are the foundations of successful storage. Only if these processes are properly carried out will the seed enter storage with the necessary soundness and vitality.

In the light of these three main facts, successful cold storage narrows down to the following jobs:

Locating a dependable cold storage locker or warehouse. The two main things to consider are the level and maintenance of the temperature and humidity.

Any temperature between 5° and 41° F. is satisfactory, but temperatures between 5° and 32°, although they may injure excessively moist seed, preserve properly dried seed better than do temperatures between 32° and 41°. Prolonged or frequent increases of temperature above 41° make a warehouse unsuitable for seed storage. Occasional brief rises in temperature while defrosting the refrigerator will, however, do no harm to dry seed.

The relative humidity of the air in the refrigerator should preferably be low and constant. If these conditions cannot be met, the seed must be dried before storage and stored in airtight containers, as described later.

Avoiding injury to the seed. The chief causes of injury, other than actual storage at too high temperatures and moisture contents, are:

Letting the cones or the extracted seed heat spontaneously, or mold.

Excessive artificial heat during extraction or drying.

Bruising the seed or cracking or scarifying the seed coats during dewinging or cleaning.

Taking too long to extract or dry the seed, letting the seed lie around too long before drying or storage, or letting dried seed reabsorb moisture before or during storage.

Drying the seed to the right moisture content before or in the earliest part of the storage period.

Care During Collection and Extraction

Seed should not be collected until the cones are fully mature. A simple, reliable test for cone maturity is to pick one sample cone apiece from each of 20 or more standing trees and drop each cone immediately into SAE 20 lubricating oil, or into a mixture of 1 part of kerosene with 4 parts of raw linseed oil. When sound cones from 19 out of 20 trees will float in the oil, the crop is ripe enough for picking. Cones that sink are still immature. The test is unreliable if applied to wormy cones, to cones from trees felled more than a few hours previously, or to cones that have been separated from the tree for more than a few minutes.

Cones may be gathered at any time after they pass the oil test, but the longer collection can be postponed without losing the seed through opening of the cones on the tree, the better the seed is likely to keep in storage.

A wetting right after collection may not harm the cones, but it is safer to protect them from rain. Free circulation of air through the piles or around each sack of cones will prevent heating and reduce not only molding but also shipping weight and the length of time needed for extraction. High or fluctuating seed moisture content (even for very brief periods and especially if accompanied by exposure to moderate or high air temperature) between collection and extraction or storage may prevent successful storage even if it does not immediately reduce germinability.

Cones should not be left in sacks more than a week after collection--ten days at most. Preferably they should be spread in curing sheds or on extracting racks or trays within 3 or 4 days after collection. The importance of such spreading cannot be emphasized too strongly, and the necessary space and equipment should always be provided before collection starts.

For final drying to extract the seed, cones should never be spread in layers more than 2 cones deep, even in air-temperature extraction on wire shelves or trays. In kilns, or in air extraction on tight floors, they should never be spread in layers more than one cone deep. Any apparent saving of space or of investment in equipment made by using deeper layers is false economy. Deeper layers delay drying, prolong the extracting period, and necessitate rehandling that more than offsets the saving in equipment costs. They

often reduce the yield and quality of seed, and may prevent altoget[
the functioning of certain types of kilns.

Drying cones in kilns

In most of the southern pine region, kiln-drying the cones r
duces seed more nearly to the right moisture content for storage
than does air-drying. Unlike air-drying, however, kiln-drying involves some danger to the seed from excessive artificial heat. To
extract seed without injury, a kiln must circulate the hot air free
and rapidly among all the cones, keep temperature and humidity belo
levels injurious to the seed, and permit adjustment of temperature
and humidity schedules to meet the requirements of different batche
of cones[1].

Air circulation.--Rapid circulation of the air around every
cone serves two important purposes. One is to get all the cones dr
and open as quickly as possible and at about the same time. The
other is to keep the temperature of the seeds safely below that of
the air in the kiln while the seeds are still very moist, as they
are in unopened cones freshly placed in the kiln. Seeds are most
easily injured by high temperature when their moisture content is
high. As long as the air moves rapidly, cones and the seeds they
contain are cooled by evaporation from the cone surfaces. If the a
moves sluggishly, evaporation slows or ceases and the moist cones a
seeds become as hot as the kiln air. Then ordinarily safe kiln air
temperatures may injure seeds, and slightly higher temperatures may
kill them outright.

Temperatures.--As a general rule, longleaf seed should be ex
tracted at a maximum air temperature of 115° F., and shortleaf,
loblolly, and slash at a maximum of 120° F. Occasionally longleaf
has been extracted at 120° to 130° F., loblolly and slash at 130° F
and shortleaf at 140° F., without excessive injury, but these temper
atures are not recommended.

Kiln temperatures are measured as close as possible to the
place where the incoming hot air first hits the cones. In a forced
draft kiln this usually is high on the wall opposite the air inlet.
In a kiln utilizing upward convection currents it is under the lowe
screen. In one utilizing downward convection currents it is above
one of the topmost trays. No kiln should be without a direct-readi
and preferably also a maximum thermometer at the point described. /

---

[1]. Rietz, R. Kiln design and development of schedules for
extracting seed from cones. U. S. Dept. Agr. Tech. Bul. 773, 70 pp.
illus. 1941.

recording hygrothermograph at the hottest point is essential to safe, efficient operation of a forced-draft kiln.

Humidity.--The kiln should always be run at the lowest humidity attainable without wasting fuel. This not only safeguards the seed but speeds up cone opening. For an hour or more at the start of a run the humidity cannot be brought as low as it can later. Moderately high humidity is least dangerous at the start of the run, however, because the rapid evaporation which keeps the humidity up also cools the cones and keeps the temperature of the seed many degrees below that of the kiln air. Humidity is decreased mainly by increasing the rate at which the hot air with its load of moisture is allowed to escape from the kiln.

Since seed injury at any temperature is likely to increase with duration of exposure, seed should be removed from the kiln as soon as possible after the cones have opened completely.

Dewinging Seed

Seed keeps better if the wings are left on during storage and removed just before sowing time. Where cold storage space is at a premium, however, or where the seed is to be shipped, dewinging seed of all species except longleaf is usually justified to reduce weight and bulk. Seeds may be dewinged by hand rubbing, by machine, or by wetting the wings.

Hand rubbing is least likely to injure the seed. Though slow, it is frequently the most economical method for small lots, and some operators prefer it even for large lots. Generally, however, mechanical dewingers are an economy wherever large amounts of seed have to be handled.

To operate mechanical dewingers at full capacity without injuring the seed requires great care. The brushes must usually be of fiber instead of wire, and neither too soft to be effective nor so stiff as to crack the seed coats, especially of longleaf. They must be replaced before the bristles become so short as to lose their springiness. Care is also necessary in adjusting revolutions per minute and rate of feed. Sometimes the seed must be dried artificially to facilitate mechanical dewinging. Optimum adjustment and procedure must be determined and maintained for each dewinger and species by trial runs and by frequent close examinations of the seed (preferably with a hand lens), and in some cases by special germination tests.

When the wing of any southern pine seed except longleaf is thoroughly moistened, the two curved prongs which attach the wing to the seed straighten out within a few seconds and the seed falls away at a touch. Advantage can be taken of this fact by dipping the hands repeatedly in water during dewinging by hand rubbing, or by spreading the seed on screens in layers about an inch deep, hosing it until thoroughly moist, and stirring it repeatedly until dry.

These wetting methods frequently are cheaper than mechanical dewinging of dry seed. Their disadvantage is that they usually increase seed moisture content enough to cause deterioration or spoilage unless special precautions are taken to dry the seed after dewinging.

Drying Seed

For best results, as already mentioned, longleaf seed must be dried to a moisture content of 6 to 9 percent, and seed of other southern pines to a moisture content of 9 to 12 percent. The drying may be done before or during the earliest days of cold storage. Once dried, seed must be maintained at the lowest moisture content reached. The reasons are the excessively rapid respiration of seed at moisture contents above the levels mentioned, and possible injury to the seed from repeated changes in moisture content at any levels.

There are two ways of getting seed dry enough without over-drying it. The more precise way is to dry each seed lot to a certain weight calculated from the weight before drying and from the moisture content percent of carefully drawn samples dried in an electric oven[2]. An approximate method more suitable for small-scale operations is to expose the seed for several hours or days to some combination of temperature and relative humidity which tests have shown will dry it to the desired level.

The curves in figure 1, although based on a sample of longleaf pine seed, show approximately the moisture contents reached by southern pine seed of any species at the temperatures and humidities shown at the side and bottom, respectively, of the chart.

The chart can be used to check whether any given combination of temperature and humidity within the limits shown will dry seed to the right content. For example, if the temperature of the air

---

2/. Wakeley, Philip C. Planting the southern pines. Southern Forest Experiment Station Occasional Paper 122. (Processed.) 579 pp., illus. 1951.

remains at 75° and its relative humidity at 55 percent, the seed
will eventually dry to slightly below 10 percent moisture content.
If, however, the temperature is 75° and the relative humidity
90 percent, the seed will dry little if any below 16 percent, which
is too high for successful storage.

The higher the temperature, the smaller the seed lot, and
the drier the seed is at the start, the sooner the seed will come
into moisture equilibrium with the air--that is, become as dry as
the particular combination of temperature and humidity will ever
make it. Lots weighing only a few pounds may dry sufficiently at
75° or 80° F. in a day or less; lots weighing several hundred
pounds may take three or four weeks, especially if the temperature
is near freezing.

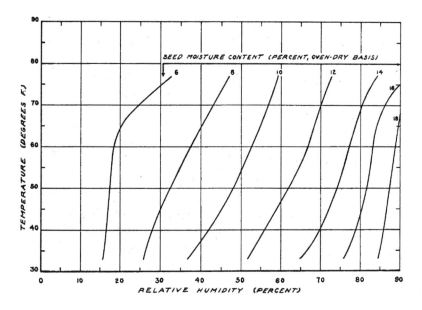

Figure 1.--Moisture content percentages of fresh longleaf pine seed
(1938 crop, Mississippi, extracted at air temperature) in
equilibrium with air at various temperatures and relative
humidities.

Drying by artificial heat.--When seed comes from a cone kiln too moist for storage, it can be dried by placing it in the kiln (in shallow layers in wire trays) for a few hours at temperatures slightly below those recommended for extracting the species concerned. Later batches can be extracted at more nearly correct moisture content by precuring the cones more thoroughly, loading the kiln less heavily, increasing kiln temperature, reducing relative humidity within the kiln, lengthening the kiln run, or substituting "progressive" for "batch" operation of the kiln[2/].

In kiln-drying extracted seed, free movement of air over and among the seeds is essential. Excessive drying at any temperature may injure the seed. Such injury increases with degree or duration of temperature and with the moisture content of the seed when drying starts, and is often intensified by subsequent storage of the seed.

Detection of excessive seed moisture content in the first place, control of kiln-drying of extracted seed, and confirmation of effective drying after kiln operation has been altered, all require oven-drying and repeated weighing of carefully drawn samples, and are not feasible where ovens and balances are lacking.

Sun drying.--For operations which lack either kilns or other artificial drying chambers, or the ovens and balances necessary for controlling drying, sunning the seed in shallow layers in trays for several days seems the most practical method of drying seed before placing it in cold storage.

Direct sunlight reduces seed moisture content to about the optimum for storage, apparently without ever reducing it too much. Over night, and on cloudy or rainy days before drying is complete, the seed should be returned to tightly covered metal cans to reduce reabsorption of moisture from damp air. At one large nursery, seed has for many years been sunned with great success in 20-pound lots in loosely woven cotton sacks large enough to hold about 40 pounds if full. The sacks are shaken up and turned over frequently each day, and always left with the seed spread out as thinly as possible. Over night the sacks are placed in a room completely lined with tin, which keeps out moist air.

Drying in refrigerators.--There is some evidence that southern pine seed, especially longleaf, endures slow drying at low temperatures better than fast drying at high temperatures. Refrigerators operated at 35° F. and 33 percent relative humidity, or at other suitable combinations (fig. 1), effectively reduce the moisture content of southern pine seed in cloth sacks or other containers that allow moisture to escape easily. Seed placed in refrigerators at 13 or 15 percent moisture content and dried in this way to 8 to 10 percent appears to stand storage better than seed dried to the same

level before being refrigerated. Where facilities are available,
such refrigeration may be the best way to complete the drying of
southern pine seed for storage.

How Containers Affect Storage

Although many confusing statements about the influence of
containers on seed storage have been published, the principles are
very simple. Containers influence the keeping quality of southern
pine seed largely, if not entirely, through their effect on seed
moisture content. They should be chosen, therefore, first for their
effect on moisture content and second for cheapness and convenience.

If the humidity of the storage locker or warehouse is low
(fig. 1) and constant, seed can be stored in burlap or cotton bags,
or in metal containers with non-airtight covers, such as garbage or
shortening cans. Cloth and metal containers are equally good if the
seed has already been dried to moisture equilibrium with the air inside the refrigerator. If the seed is to be brought to final
correct moisture content by drying in the refrigerator itself, cloth
containers are better.

If the humidity of the refrigerator is too high, or if it
fluctuates, the seed must be dried to the right moisture content
before storage, and placed immediately in air-tight containers. For
air-tight, sealed storage of commercial lots, gasketed grease drums
and corked glass carboys have proved most satisfactory (except that
longleaf seed will not pour freely through the narrow necks of the
carboys), but any metal container that can be sealed air-tight by
soldering or caulking will serve.

The effects of containers and seed moisture contents may be
illustrated by a brief account of the extensive tests that led to
the foregoing recommendations.

In 1937, lots of freshly collected longleaf seed (with wings
on) were stored at moisture contents ranging from 6 to 18 percent.
Three different kinds of containers were used: sealed glass jars,
slip-top tin cans, and cheesecloth sacks. Half the samples were
stored at 41° F. and half were kept in a normally heated office.

Results are summarized in table 1. All seed stored at air
temperature was dead in two years, and only the lot stored at 6 percent moisture in sealed jars remained viable through the first year.
At 41° the seed kept well for 4 years and fairly well for 5 when
maintained at 6 or 9 percent moisture by sealed jars. But seed
stored in glass at 18 percent moisture deteriorated considerably in

one year and was dead in two, despite refrigeration. In brief, sealed containers, which kept the seed dry, were better than unsealed when the seed was stored at moisture contents below 10 percent. Unsealed containers, because they permitted drying in storage, were better for seed placed in storage at initial moisture contents of 15 and 18 percent.

Table 1.--Five-year storage of 1937 longleaf pine seed at two temperature levels and varying seed moisture contents.

Approximate initial moisture content of seed; storage container	Germination percent at 32 to 38 days						
	Before storage	After air-temperature storage for:		After storage at 41° F. for:			
		1 year	2-5 years	1 yr	2 yrs	4 yrs	5 yrs
6 percent							
Sealed glass jar	77	62	0	88	82	71	47
Slip-top tin can	77	1	0	84	86	52	32
Cheesecloth sack	77	0	0	85	92	60	21
9 percent							
Sealed glass jar	76	1	0	89	92	78	40
Slip-top tin can	76	1	0	88	83	58	14
Cheesecloth sack	76	0	0	82	84	48	28
12 percent							
Sealed glass jar	83	0	0	85	87	32	18
Slip-top tin can	83	1	0	96	84	41	16
Cheesecloth sack	83	1	0	82	76	30	15
15 percent							
Sealed glass jar	83	0	0	69	78	14	13
Slip-top tin can	83	1	0	87	73	41	19
Cheesecloth sack	83	0	0	82	76	42	16
18 percent							
Sealed glass jar	87	0	0	50	0	0	0
Slip-top tin can	87	0	0	86	79	30	5
Cheesecloth sack	87	1	0	77	76	46	21

A test with dewinged slash pine seed gave closely similar results. All lots, even the moistest (18 percent) remained useful after 5 years of refrigeration at 41°, but at air-temperature only dry seed in glass jars lived for 2 years and none stayed alive for 3. Rather surprisingly, the lots originally at 6 and 9 percent moisture content and stored in slip-top tin cans at 41° deteriorated more by the end of the fifth year than did the moister seed in such cans. The explanation is that these lots were below moisture equilibrium with the air of the refrigerator, and the cans permitted the seed to

increase in moisture content during storage. This is a forceful re-
minder of the danger, previously mentioned, of letting seed moisture
content rise or fluctuate during storage.

Slash is less exacting than longleaf in its requirements for
long storage, but numerous studies strongly suggest that it be refrigerated at no more than 12 percent moisture content, and at no less
than 9 percent for open containers and 6 percent for sealed.

Less is known about requirements for storing shortleaf and
loblolly seed, but they clearly resemble those for slash more closely
than those for longleaf. Shortleaf seed, like slash, has been stored
successfully for 19 years at 35° to 38° and approximately 10 percent
moisture content, whereas longleaf has not been kept successfully
under these conditions for more than 10 years.

The foregoing recommendations apply to overwinter storage as
well as to long-term storage. Tests of storage for 1, 2½, and 3½
months very strongly confirm the findings of the long-term studies
in favor of dry, cold storage, with drying during storage as second
choice and dry storage at intermediate temperatures (as little as
possible above 41°, perhaps in an unheated building) as third choice,
even for the short period between seed extraction and spring sowing.
Longleaf seed at 22 percent moisture content, stored in sealed glass
at temperatures higher than 38° F., was completely dead at the end
of 2½ months. These studies also showed that incorrect overwinter
storage, in addition to wasting seed and running up costs, may decrease the reliability of germination tests as guides to sowing
rates.

Special storage techniques.--Many refinements of storage techniques have been recommended in various publications, but have
given inconsistent or harmful results and at best seem unnecessary
with southern pine seed if the main principles of dry cold storage
are followed closely. These refinements include disinfecting the
seed with mercury compounds or formaldehyde before storage, sealing
charcoal in the containers with the seed, and sealing the seed in a
vacuum.

Delays are Dangerous

More important than minor refinements in storage or even very
precise determination of storage temperature and seed moisture is the
need for getting the seed into storage promptly and keeping it there
until it is sown. When extraction is delayed too long, some seed will
die in the cones. The germinability of extracted seed held at air
temperature may decrease seriously in 4 to 8 weeks. Even if immediate
germinability is not affected, rapid respiration before cold storage
depletes the food reserves within the seed.

For these reasons, nurserymen should avoid the common practice of holding seed at air temperature until part of it has been sown in spring. It is much better to put all the seed in cold storage as it is extracted and to withdraw it, as needed, immediately before sowing. A possible alternative, if refrigerator space is at a premium, is to refrigerate immediately all seed to be held a year or more and to keep overwinter at air temperature only the minimum amount likely to be sown the spring after extraction.

Removing the seed from cold storage much before sowing or testing it may do even more harm than holding it at air temperature before it has been sensitized by refrigeration. This has been shown most clearly with longleaf pine, which often deteriorates seriously in two to four weeks, especially if it is at high moisture content, but results with commercial lots of shortleaf, loblolly, and slash seed indicate that no southern pine seed should be removed from cold storage more than a week before sowing or testing.

Deferring removal in this way may be impossible when seed is to be shipped abroad, especially to South America or South Africa, where the sowing season differs by six months from that in the United States. Rather than expose refrigerated seed to possible high temperatures in transit, it is preferable to arrange export well in advance, ship seed immediately after extraction and cleaning, and keep it in cold storage at its destination from receipt until sowing time.

Recommendations

For storage beyond the first spring following extraction.--
Provided seed moisture content can be kept constant after preparation for storage, the seed should:

a. Be collected, extracted, and dewinged (longleaf should be left with wings on) with minimum injury;

b. Be dried to 6 to 9 percent moisture content for longleaf, or 9 to 12 percent moisture content for slash, loblolly, and shortleaf(or brought to these final levels in the refrigerator, as previously noted);

c. Be placed in cold storage within a week or two after extraction, cleaning, and drying;

d. Be stored at a temperature not higher than 41° F., preferably at 5° to 32° F.; and

e. Be removed from cold storage not more than a week before testing or sowing, or before pregermination treatment if such treatment is necessary.

The seed can be maintained at constant low moisture content either by sealing the containers, or by storing it in air-permeable containers in a refrigerator having a constant low relative humidity (fig. 1).

If sealed containers cannot be used and the seed must be stored in a refrigerator too humid to maintain the moisture content at the most favorable level, the seed should be placed in storage at or slightly above the moisture content at which it will come into equilibrium with the air in the refrigerator. Reducing the seed moisture content below this level and letting it rise in storage should be avoided, as should repeated changes in moisture content during storage.

If storage at 41° F. or below is impossible, seed of all species should be kept at 6 percent moisture content in sealed containers at the lowest temperature available.

For overwinter storage only.--Preferably, seed should be stored overwinter precisely as for longer periods; that is, refrigerated at 41° F. or below, at constant moisture content of 6 to 9 percent for longleaf, or 9 to 12 percent for slash, loblolly, and shortleaf, and otherwise as described for long storage.

Second choice, refrigeration at or below 41° F., at constant moisture content not above 15 percent (any species).

Third choice, storage at temperatures as little as possible above 41° F., and at constant moisture content of 6 to 9 percent for longleaf, or 9 to 12 percent for slash, loblolly, and shortleaf.

For shipment abroad, especially to the Southern Hemisphere.-- Preferably, ship immediately after extraction and cleaning, in sealed containers, at moisture content of 6 to 9 percent for longleaf, or 9 to 12 percent for slash, loblolly, and shortleaf. Receiver should refrigerate seed at 41° F. or lower, at the same or lower moisture content (the latter will necessitate unsealing the containers) from receipt until use.

Second choice (especially applicable to seed already refrigerated before shipment), ship at moisture content similar to the above, either in refrigerated holds or by air express with instructions to keep as cool as possible. Refrigerate from receipt until use.

COSTS OF
PRODUCING MINE PROPS

R. M. Osborn

SOUTHERN FOREST EXPERIMENT STATION
Harold L Mitchell, Director
New Orleans, La
FOREST SERVICE, U S DEPT OF AGRICULTURE

CONTENTS

Page

COSTS OF PRODUCING MINE PROPS

R. M. Osborn
Southern Forest Experiment Station

In the heavily industrialized area of north-central Alabama, the forest resource has been depleted to a degree equalled in few other parts of the South. In addition to the usual timber consumption, the best trees of small sizes are continually cut to make props for the region's coal and iron mines. These mines used one-fifth of the wood marketed in the territory in 1945, one-half of it in the form of props.

Today the forests of the region are saddled with large proportions of hardwood stumpage of undesirable species or form, or limby and with considerable butt rot or other defects. Much of this stumpage, though undesirable for other wood products, is still suitable for mine wood, especially props. Such use would reduce the pressure on the more valuable trees--the pines and the hardwoods of good form and species. Although the mines will readily accept hardwood props, cutters are reluctant to produce them as long as pine is available.

Recently the Birmingham Branch of the Southern Forest Experiment Station carried out a study to determine the cost of producing props from low-grade hardwoods and from pine, and to compare the various tools available for logging jobs. This paper presents the results of the study.

Highlights of the Findings

1. For props from the general run of tree sizes, the average direct costs of producing hardwood props were 20 percent greater than the costs of cutting pine props.

2. For felling trees 8 inches d.b.h. and larger, the three-man power chain saw crew is cheaper than the two-man bow saw crew or the two-man crosscut saw crew. This is true, though, only as long as the chain saw is operating steadily; delays quickly reduce its advantage.

3. For bucking logs or bolts larger than 7 inches in diameter, the three-man power chain saw with bow attachment is cheaper (as long as it runs steadily) than any of the tools studied.

4. Splitting the lower bolts from large trees reduces the cost of props.

5. Topping time increases rapidly as diameters larger than 6.5 inches are chopped.

6. Virginia pine is appreciably more expensive to limb than either loblolly or shortleaf pine or hardwoods.

Description of Study

Data were collected in Jefferson County, Alabama, on commercial prop operations and also on the Flat Top Experimental Forest, where props were being produced by the Experiment Station's woods crew. The terrain was moderately rough with few rock outcrops. Ridge tops were generally broad with moderate slopes into drainages. Slopes of 50 to 100 percent were occasionally found.

Although generally quite extensive, the woods road system was poor, most roads being little more than brushed-out rights of way on the ridge tops and gentle slopes. The main county roads were well-graded clay and rock or were paved.

By far the greater portion of the area where the study was made was of the mixed pine-hardwood type--the pines represented primarily by loblolly and shortleaf with some Virginia pine, and the hardwoods mainly by hickory, southern red oak, scarlet oak, black oak, post oak, and white oak. Most of the stands were poorly stocked; they averaged about 100 sound stems per acre (5 inches or larger in d.b.h.), with a volume of about 515 cubic feet per acre ($7\frac{1}{2}$ cords). About one-half of the volume was in trees from 5 to 9 inches in d.b.h., and only 10 percent in trees 15 inches and over.

The heaviest cut removed 85 trees per acre. On the Flat Top Experimental Forest, trees were marked prior to cutting and only about 45 trees per acre were removed.

Nearly all of the mines in the territory are supplied with props by contractors who cut, usually without restriction, on company lands. Most of the contractors observed in the study had had at least five years of experience in cutting props. The Experiment Station woods crew had had about two years' experience in producing props.

Field data were collected during the fall and winter of 1946-47 and during the winter of 1949-50. Temperatures on working days averaged about 60 degrees Fahrenheit. Data for the power saw are based on felling and bucking 829 trees, for the crosscut on 404 trees, and for the bow saw on 396 trees. The range in d.b.h. was from 5 inches through 19 inches, with most trees falling between 5 and 11 inches d.b.h. Skidding data are based on 177 trees and hauling data on 18 loads.

Logging, general.--Prop producers in the territory generally use one of three logging methods, the choice being determined primarily by the terrain and skidding facilities available. Where it is feasible to drive a truck near the felled trees, felling is immediately followed by bucking; the props are then loaded directly onto the truck. Since no skidding is necessary, this procedure is the cheapest but can be employed only on ridge tops and flats. When timber in the hollows or on the lower slopes is to be logged, some skidding is necessary. One method is to fell and buck on the spot, and load the props onto a sled carrying about 20 props. This sled is then pulled by a team to a point accessible to trucks and the props transferred. The third method is to fell and skid tree lengths to a landing at which point the trees are bucked and the props loaded onto the truck. This last procedure is the one applicable to the greatest area and the one to be considered in detail.

Since the present market for props is limited, a producer who wishes to give full time to prop production must usually have one small crew perform all phases of the work. If he has an integrated logging operation, with separate crews for each job, he is likely to find himself overproduced at rather frequent intervals. The contractor or subcontractor usually works as a member of the crew, and for this reason no charge for supervision has been made. This study was concerned with the small single-crew type of operation only.

Method of handling costs.--Two kinds of cost have been recognized, direct and indirect. As used here, direct costs include expenses incurred only when a specific tree is logged; labor is the chief item. Indirect costs are those not affected by the handling of any particular tree, but chargeable to the entire logging job--such as interest and taxes. Some items of expense may include both kinds,

and a proportionate division is not readily determined. I
is partly a matter of time and partly wear through use. F
tools, wear is the more important component; consequently,
cost has been treated as a direct cost. For heavy equipme
a tractor, the decrease in value is largely a function of
therefore, has been treated as an indirect cost.

Felling

Crew organization

Three-man power chain saw, Mall, 24-inch bar.--In f
from 5 to 11 inches d.b.h., one man with an 8-foot push po
marked trees, decided upon direction of fall, and aided ir
fall in the desired direction as the two saw operators cut
were made only on the larger trees that were felled agains

Walking time included necessary swamping and delays
than two minutes' duration; longer delays were separated a
under delay time. The difference between total elapsed ti
sum of cutting and delay time was divided by the number of
felled to give average walking time per tree, in this cas
seconds per tree.

Poor engine performance of the power saw was respo
delays that amounted to 16 percent of the total elapsed t
this test was made, better operation has been obtained by
spare spark plug, which allows frequent alternation, and b
buretor adjustment. In view of this and the results of W
in Tennessee, an estimate of 10 percent for major delays
able.

The cost of power saw operation, $0.58 per hour, i
1,114 hours of use and is classed as a direct cost (see A
details). The $0.028 per hour chargeable to interest, t
like is by definition an indirect cost. However, since h
very minor, it will not be separated from the direct cost
pected life of 2,000 operating hours has been used for de
Other investigators[1] [2] have used a considerably longer
in view of the present condition of the saw and known fac
ing its care and service, the shorter period seems justif

---

[1] Wiesehuegel, E. G. Power chain saws and manua
saws in the production of hardwood logs. South. Lumberma
46-50. 1946.
[2] Campbell, R. A. Pine pulpwood production--a s
and power methods. Southeast. Forest Expt. Sta. Tech. N
19 pp. 1946.

Three-man crosscut saw (5-foot).--As with the chain saw, one man located marked trees, aided when necessary to relieve pinching, and chopped a small undercut on most trees felled. Since all this required less time than the felling and was done while the other crew members were sawing, no time was required in excess of walking time. The men alternated between the different tasks. Walking time per tree was 50 crew-seconds. All charges for tools are considered as direct costs.

Two-man crosscut saw (5-foot), and two-man bow saw (Sandvik 36-inch).--When using either of these tools for prop size trees, one of the two men carried an ax and chopped a small undercut; the two then sawed together. When it was necessary to relieve pinching, one man pushed while the other completed the cut. Here undercutting required time in addition to sawing and walking time, as one man was usually idle during this inter-val. All costs for tools are considered direct.

Performance

As figure 1 and table 1 show, hardwoods required only about 5 percent more felling time per tree than did pine. In most of the tree sizes studied, the 3-man power chain saw crew proved to be faster and cheaper than any of the other felling methods. Trees were felled in about one-third the time re-quired by crosscut or bow saw crews, and costs were appreci-ably lower for all trees 8 inches d.b.h. and larger. De-lays with the power saw are, however, expensive, and the ad-vantage can easily be wiped out by poor engine performance or an inefficient crew. The great-er walking time required by the crosscut saw crew was probably an expression of greater fatigue.

The least efficient fell-ing crew for all size trees studied was the three-man cross-cut. For trees up to 9 inches d.b.h. there was little differ-ence between the two-man crosscut

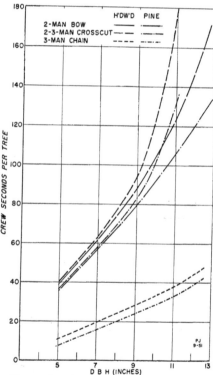

Figure 1.--Time required for felling pine and hardwood trees.

Table 1.--<u>Total direct felling cost per tree</u>[1]

D.b.h. (inches)	Three-man crosscut saw	Two-man crosscut saw	Two-man bow saw	Three-man saw, 24-in
		- - - - - - - Dollars - - - - - - -		
		PINE		
5	0.059	0.046	0.045	0.047
6	.066	.053	.051	.051
7	.073	.059	.058	.054
8	.081	.066	.065	.058
9	.087	.073	.074	.062
10	.096	.080	.083	.065
11	.104	.088	.093	.070
		HARDWOOD		
5	.061	.048	.046	.051
6	.069	.055	.053	.054
7	.077	.061	.061	.05
8	.086	.069	.069	.06
9	.095	.077	.079	.06
10	.106	.087	.089	.07
11	.120	.098	.111	.07

[1] See Appendix for detailed rates.

and the two-man bow saw crews. But for trees above 9 inches,
man crosscut saw crew was superior to the bow saw crew.

Trimming

When skidding was done by tree lengths, the double-bit
used for topping as well as for limbing and for cutting the b
of the stump. Topping time varied directly with diameter of
about 6.5 inches inside bark; for sizes larger than this, top
increased rapidly. However, since most prop trees are topped
smaller than 6.5 inches, the ax method appears satisfactory.
trees that were not completely severed from the stumps in fel
quired only a very small portion of the total trimming time.

Time required for limbing the bole varied directly with d.b.h.
Virginia pine required appreciably longer to limb than did loblolly or
shortleaf pine or the hardwoods.
Trimming of hardwoods, although
confined to a few large limbs,
required on the average about
one-third longer than loblolly
and shortleaf pine. Figure 2
compares total chopping time required for the different species.
Table 2 gives costs for total
trimming time. Average walking
time per tree was 0.02 man-hour.
Here all charges for replacing
and maintaining tools are classed
as direct costs.

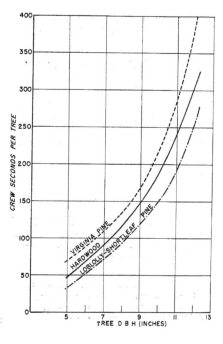

Figure 2.--Total chopping time for
trimming pine and hardwood trees.

Table 2.--Total direct trimming
cost per tree[1]

D.b.h. (inches)	Loblolly and shortleaf pine	Hard- wood	Vir- ginia pine
	- - - Dollars - - -		
5	0.023	0.026	0.032
6	.027	.031	.036
7	.031	.036	.040
8	.036	.042	.046
9	.042	.049	.054
10	.049	.058	.064
11	.059	.069	.079
12	.073	.084	.102

[1] See Appendix for detailed
rates.

Skidding

An Oliver "Cletrac" Model BD, 38 horsepower, crawler-type tractor
with winch was used for bunching and ground skidding. Two men assisted
the operator in dragging cable to trees, in choking, and in unhooking
chains at landing. Trees were bunched within an average radius of 35
feet and skidded an average distance of 100 feet. It was not practicable to determine average load by d.b.h. class. However, the study
average was three trees per load with most trees in the 9-inch d.b.h.
class. Nearly all trees were bunched uphill.

Although skidding time undoubtedly varies with s
cient data were collected to determine the variation by
different slope percents. Ground skidding was confined
15 percent or less; when slopes exceeded 15 percent, tr
by cable. Considerable bunching was done on slopes up
Total direct skidding costs per turn are listed in tabl

Table 3.--Total direct skidding cost per turn[1]

Operation	Minutes
Bunching, three trees in average load	5.00
Ground skid and return, per 100 feet	2.75
Unhooking	.75
Delay	.50
Total cost per turn	9.00

[1] See Appendix for detailed rates. A turn is the com
the tractor from the landing to the woods, bunching
logs, and ground-skidding the load back to the land

It will be seen from tables 8 and 9 that skiddin
expensive operation. Road construction and hauling cos
therefore, be carefully balanced against costs of skidd
determine the cheapest procedure. Taxes and charges fo
and interest on investment are termed indirect costs; a
direct.

Bucking

Crew organization

Three-man power chain saw with bow attachment.--
of trees that were concentrated at a landing, one man m
lengths and assisted the two saw operators by placing t
for bucking and by moving them when the saw pinched. C
about one-third of the total time and walking and placi
position for bucking about one-half. Delays accounted

When piles of trees were large, considerable tim
to move trees and props so that remaining trees could b
the saw operators had to stop cutting and aid the third

piles generally contained about 14 trees, but at times as many as
25 trees were skidded to a single landing before bucking. When piles
were small, placing trees in position and releasing the saw from
pinching could usually be handled by one man without slowing the saw
operators unduly. Small piles usually contained about seven trees
and were the result of skidding one or two tractor loads to the roadside. Placing the trees in position and walking between cuts required
about twice as much time when bucking large piles as when bucking
small ones. Details will be found in table 4. These figures indicate
that by limiting piles to 6 or 8 trees and leaving about ten feet
between piles, bucking costs may be lowered considerably without increasing skidding costs.

Table 4.--Time required for walking between cutting and for placing
trees in position for bucking as affected by size of
pile1/ at landing and size of trees

D.b.h. of trees in pile (inches)	Two-man bow saw		Three-man chain saw	
	Small piles	Large piles	Small piles	Large piles
	- - - - - Crew seconds per prop - - - - -			
5 to 9	16	33	11	22
5 to 14	27	35	15	21
10 to 14	30	47	10	23

1/ The average number of trees in small piles was 7; in large piles,
14.

Two-man bow saw or two-man crosscut saw.--With either of these
tools one man placed the tree in position for bucking and then
assisted the other in completing the cut. At times both men were
needed to place trees in position. With these tools, cutting accounted for about half of the total time; walking and placing trees in
position required a slightly smaller proportion, and delay time was
very small. When small piles were bucked, walking between cuts,
placing trees in position, and relieving pinching took only one-half
to two-thirds as long as for large piles.

Performance

As in felling, the bucking times and costs were only about 5
percent greater for hardwood than for pine. For all diameters, the
power saw with bow attachment proved to be faster; it was also cheaper

for diameters larger than 6 inches (figure 3 and tab
man bow saw competed closely in dollar cost for the
but cost increased rapidly for diameters 8 inches an
about 10 inches and up, the two-man crosscut saw is
bow saw. As in felling, poor operation of the power
sible for considerable costly delay. Proper adjustm
this delay factor to about 10 percent of total time.

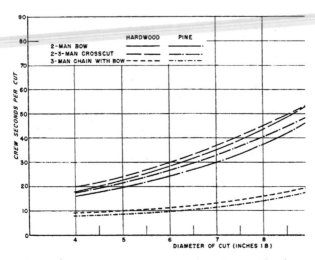

Figure 3.--Cutting time for bucking pine and hardwoo

Data for the one-man bow saw were inadequate
conclusions, but this tool appears to have good poss
man able and willing to stand the strain.

Splitting Large Bolts

Some mines will accept split props that meet
fications. When 3- and 4-foot props are being prod\
cuts of trees 8 inches d.b.h. and larger are usuall;
and 6-foot props, trees must usually be 10 inches in
they can be split. In large timber, several props <
split from one bolt, but in this study the trees we:
no more than two props could be split from any one s

- 10 -

Table 5.--Total direct bucking cost per cut [1]

Diameter of cut inside bark (inches)	Two-man crosscut saw	Two-man bow saw	Three-man chain saw with bow attachment
	- - - - - - - Dollars - - - - - - -		
	PINE		
4	0.019	0.017	0.019
5	.021	.018	.019
6	.023	.020	.020
7	.026	.024	.022
8	.030	.027	.024
9	.034	.032	.028
10	.039	.038	.030
11	.044	.045	.033
	HARDWOOD		
4	.020	.019	.020
5	.022	.020	.020
6	.024	.023	.021
7	.028	.026	.023
8	.032	.030	.026
9	.036	.035	.029
10	.041	.042	.035
11	.047	.050	.041

[1] See Appendix for detailed rates.

Splitting costs were found to vary with length of props being produced. On the average, splitting required 1.1 manhours per 100 3- or 4-foot props and 1.8 manhours for 100 5- or 6-foot props. As tables 8 and 9 show, splitting reduced production costs considerably, both for hardwood and for pine props. If anything, the savings were greater in hardwood.

Loading and Hauling

Three men usually did the loading from small pi
along a woods road or from large piles, either "as buck
stacked. Since stacking required about one man-hour p
imately two cords, and since there was no practical di
loading times by the three procedures, loading from la
is the least efficient.

Two men could readily load and unload props six
or shorter. It was not practicable to determine loadir
sizes of props cut from each d.b.h. class. Likewise, i
props per load given in table 6 are averages for props
of different sizes; the number per load would undoubte
diameter of trees from which they were cut. Loading a
accounted for over 60 percent of the total time, haulir
and return for about 20 percent, and delays for about i
of the delay time accumulated at the mine while the mer
load to be checked.

Table 6.--Size requirements, number in load, and unit p
props

Prop size		Average in load		
Length (feet)	Min. diameter (inches)	Pine	Hardwood	
		- Number -		
3	4	362	338	
4	4	242	225	
5	5	175	170	
6	5	160	154	

Delivering props from the cutting area to the m
average haul of $\frac{1}{2}$ mile on woods roads and 2 miles on i
or asphalt road. Sufficient data will be found in tab
Appendix to determine costs for the longer hauls requi
tractors. The average load per truck was two cords.
preciation, license fees, taxes, insurance, and intere
are indirect costs; all other are direct.

Table 7.--Total direct loading and hauling costs

Prop size		Load-ing (three men)	Un-load-ing (two men)	Delay (two men)	Haul-ing (two men)	Total time for 3-man crew	Cost		
Length (feet)	Av. diam. small end, i.b. (inches)						Tools and truck	Labor	Total
		- - Man-hours per load - -					Dollars per load		
3	4.2	2.00	1.00	0.67	0.67	4.34	0.67	3.52	4.19
6	5.5	1.50	.50	.67	.67	3.34	.67	2.71	3.38
6	8.3	2.00	.67	.67	.67	4.01	.67	3.25	3.92
7	5.5	1.50	.50	.67	.67	3.34	.67	2.71	3.38

Production Costs Per 100 Props

So far, costs have been given only in terms of specific operations, without regard to the total cost of producing a given quantity
of props from trees of different size classes. To complete the comparison, however, it is necessary to compute the cost of producing a given
quantity of props from trees of each d.b.h. class in both hardwood and
pine. This has been done in tables 8 and 9, which give costs for
units of 100 props.

The tables show that the direct costs per 100 props range from
10 to 40 percent greater for hardwood than for pine. The chief cause
of this important differential is that the upland oaks and hickories
in this territory have shorter boles than the pines (or yellow-poplar,
which was not included in this study), a fact that necessitates handling more hardwood trees than pines in order to get a given quantity
of props (see column 3 of tables 8 and 9). On the average, for props
from the general run of tree sizes, the hardwood props will cost about
20 percent more to produce than pine props.

In computing tables 8 and 9, the most efficient tools of those
covered in the study were selected. In table 8 the costs for felling
and bucking are those for the two-man bow saw crew, which is the most
efficient crew for sawing trees 8 inches in d.b.h. and smaller. Five-
and six-foot props are usually cut from trees 8 inches in d.b.h. and
larger, and for this reason the costs of producing them with a bow saw
have not been included. In table 9 the costs for felling are those
for the power chain saw with a 24-inch bar and a 3-man crew, and for
bucking the power chain saw with a bow attachment and a 3-man crew.
The power chain saw is the most efficient tool in all respects for
sawing trees over 9 inches in d.b.h.

Table 8A.--Pine props: direct costs of production per 100 props, using two-man bow saw crew for felling and bucki[ng]

Tree d.b.h. (inches) (1)	Av. utilized volume per tree to minimum top diam., outside bark, of 4 in. (2) Cords	Trees per 100 props (3) Number	Labor costs (Man-hours)						Costs other than labor (Dollars[3])			
			Felling (4)	Trimming (5)	Skidding (6)	Bucking[1] (7)	Hauling[2] (8)	Total stump to mine (9)	Felling bucking & trimming (10)	Skidding (11)	Hauling[2] (12)	t[o
				THREE-FOOT PROPS								
5	0.018	23.1	1.34	0.67	2.60	1.81	1.20	7.75	0.10	1.01	0.19	
6	.030	14.6	.95	.50	1.75	1.86	1.20	6.37	.09	.68	.19	
7	.051	11.5	.86	.46	1.51	1.94	1.20	6.00	.09	.58	.19	
4/8	.079	10.0	.84	.46	1.39	2.11	1.20	5.89	.09	.54	.19	
4/9	.108	9.0	.83	.48	1.35	2.33	1.20	5.94	.10	.52	.19	
5/8	.079	9.1	.76	.42	1.26	2.02	1.20	5.57	.09	.49	.19	
5/9	.108	7.6	.70	.40	1.14	2.15	1.20	5.55	.08	.44	.19	
				FOUR-FOOT PROPS								
5	.018	30.8	1.78	.89	3.46	1.78	1.7	9.86	.11	1.34	.28	
6	.030	19.5	1.27	.66	2.34	1.86	1.7	8.03	.10	.90	.28	
7	.051	15.4	1.15	.60	2.02	1.94	1.7	7.53	.10	.78	.28	
4/8	.079	13.3	1.11	.61	1.85	2.06	1.79	7.36	.10	.72	.28	
4/9	.108	11.9	1.09	.63	1.79	2.22	1.79	7.36	.11	.69	.28	
5/8	.079	11.8	.99	.54	1.64	1.95	1.79	6.85	.09	.63	.28	
5/9	.108	9.6	.88	.51	1.44	2.01	1.79	6.71	.09	.56	.28	

Table 8B.--Hardwood props: Direct costs of production per 100 props, using two-man bow saw crew for felling and [

				THREE-FOOT PROPS								
5	.009	50.0	2.98	1.65	5.62	1.89	1.28	13.53	0.17	2.17	0.20	
6	.024	23.2	1.56	.90	2.78	2.00	1.28	8.52	.12	1.08	.20	
7	.043	16.9	1.31	.76	2.22	2.17	1.28	7.63	.12	.86	.20	
4/8	.064	14.1	1.23	.73	1.96	2.33	1.28	6.99	.11	.76	.20	
4/9	.090	12.3	1.22	.75	1.84	2.69	1.28	7.34	.12	.71	.20	
5/8	.064	12.4	1.08	.64	1.72	2.20	1.28	6.43	.04	.67	.20	
5/9	.090	9.8	.97	.60	1.47	2.38	1.28	6.55	.03	.57	.20	
				FOUR-FOOT PROPS								
5	0.009	66.7	3.98	2.20	7.50	1.72	1. 3	17.61	.19	2.90	.30	
6	.024	31.2	2.10	1.22	3.74	2.00	1. 3	10.99	.14	1.45	.30	
7	.043	22.7	1.76	1.02	2.98	2.19	1.93	9.75	.13	1.15	.30	
4/8	.064	18.5	1.61	.96	2.58	2.33	1.93	9.19	.13	1.00	.30	
4/9	.090	16.4	1.62	1.00	2.46	2.56	1.93	9.23	.14	.95	.30	
5/8	.064	15.6	1.36	.81	2.18	2.14	1.93	8.23	.11	.84	.30	
5/9	.090	14.1	1.39	.86	2.12	2.36	1.93	8.56	.12	.82	.30	

[1]/ Figures in this column were determined in the following manner: From taper tables, the diameters at 3- and[
intervals were determined; from Figure 3 and Table 5 the costs of bucking these diameters were determined a[
averaged for each d.b.h. class. This average figure was multiplied by 100 to get the cost per 100 props.
split props are included (see footnote 5) the average cost by d.b.h. was determined as before but was multi[
by the number of cuts necessary to produce 100 round and split props combined. To this bucking cost was a[
cost of splitting those props that were large enough.

[2]/ Cost per load was multiplied by the factor $\frac{100}{\text{Number of props per average load}}$; each load contained about two[

[3]/ Expressed in 1950 dollars.

[4]/ Data in this line are for round props.

[5]/ Data include split props meeting mine specifications.

Table 9A.--Pine props: Direct costs of production per 100 props, using three-man power chain saw crew for felling and bucking

Tree d.b.h. (inches) (1)	Average utilized volume per tree[1] (2)	Trees per 100 props (3)	Labor costs						Costs other than labor				
			Fell-ing (4)	Trim-ming (5)	Skid-ding (6)	Buck-ing[2] (7)	Haul-ing[3] (8)	Total stump to mine (9)	Felling and trimming (10)	Skid-ding (11)	Buck-ing[2] (12)	Haul-ing[3] (13)	Total stump to mine (14)
	Cords	Number	- - - - - - Man-hours - - - - - -						- - - - - - Dollars[4] - - - - - -				
THREE-FOOT PROPS													
5	0.018	23.1	1.09	0.67	2.60	1.94	1.20	7.50	0.22	1.01	0.38	0.19	1.80
6	.030	14.6	.74	.50	1.75	1.97	1.20	6.16	.15	.68	.38	.19	1.40
7	.051	11.5	.63	.46	1.51	1.97	1.20	5.77	.13	.58	.38	.19	1.28
8[5]	.079	10.0	.58	.46	1.39	2.00	1.20	5.63	.12	.54	.39	.19	1.24
9[5]	.108	9.0	.56	.48	1.35	2.08	1.20	5.67	.12	.52	.40	.19	1.23
8[6]	.079	9.1	.54	.42	1.27	1.92	1.20	5.35	.11	.49	.36	.19	1.15
9[5]	.108	7.6	.47	.40	1.14	2.11	1.20	5.32	.10	.44	.35	.19	1.08
FOUR-FOOT PROPS													
5	0.018	30.8	1.45	.89	3.46	1.94	1.79	9.53	.30	1.34	.38	.28	2.30
6	.030	19.5	.99	.66	2.34	1.97	1.79	7.75	.21	.90	.38	.28	1.77
7	.051	15.4	.84	.60	2.02	1.97	1.79	7.22	.18	.78	.38	.28	1.62
8[5]	.079	13.3	.78	.61	1.85	2.00	1.79	7.03	.16	.72	.39	.28	1.55
9[5]	.108	11.9	.74	.63	1.79	2.06	1.79	7.01	.16	.69	.40	.28	1.53
8[6]	.079	11.8	.69	.54	1.64	1.89	1.79	6.55	.15	.63	.34	.28	1.40
9[5]	.108	9.6	.60	.51	1.44	2.09	1.79	6.43	.13	.56	.32	.28	1.29
FIVE-FOOT PROPS													
5	.011	55.6	2.62	1.61	6.20	1.94	1.90	14.27	.54	2.43	.38	.38	3.73
6	.024	31.2	1.59	1.06	3.74	1.94	1.90	10.23	.33	1.45	.38	.38	2.54
7	.043	22.7	1.24	.89	2.98	1.97	1.90	8.98	.26	1.15	.38	.38	2.17
8	.069	18.5	1.08	.85	2.58	2.00	1.90	8.41	.23	1.00	.39	.38	2.00
9	.096	16.7	1.04	.89	2.51	2.06	1.90	8.40	.22	.97	.40	.38	1.97
10[5]	.125	15.2	1.00	.93	2.42	2.19	1.90	8.44	.21	.94	.43	.38	1.96
11[5]	.152	14.3	1.01	1.04	2.43	2.33	1.90	8.71	.22	.94	.45	.38	1.99
10[6]	.125	13.2	.87	.81	2.10	2.14	1.90	7.82	.19	.81	.37	.38	1.75
11[5]	.152	12.5	.88	.91	2.13	2.32	1.90	8.14	.19	.82	.40	.38	1.79
SIX-FOOT PROPS													
5	.011	66.7	3.15	1.93	7.50	1.94	2.08	16.60	.65	2.90	.38	.42	4.35
6	.024	37.0	1.88	1.26	4.44	1.94	2.08	11.60	.39	1.72	.38	.42	2.91
7	.043	27.0	1.48	1.05	3.54	1.97	2.08	10.12	.31	1.37	.38	.42	2.48
8	.069	22.2	1.29	1.02	3.09	2.00	.208	9.48	.27	1.19	.39	.42	2.27
9	.096	20.0	1.24	1.06	3.00	2.06	2.08	9.44	.26	1.16	.40	.42	2.24
10[5]	.125	18.2	1.20	1.11	2.90	2.17	2.08	9.46	.26	1.12	.42	.42	2.22
11[5]	.152	17.2	1.21	1.26	2.92	2.33	2.08	9.80	.26	1.13	.45	.42	2.26
10[6]	.125	15.4	1.01	.94	2.45	2.12	2.08	8.60	.21	.95	.36	.42	1.94
11[5]	.152	14.7	1.03	1.07	2.50	2.31	2.08	8.99	.22	.97	.39	.42	2.00

[1] To a minimum 4-inch top (inside bark) for 3- and 4-foot props; and to a 5-inch top for 5- and 6-foot props.

[2] Figures in this column were determined in the following manner: From taper tables, the diameters (at 3- and 4-foot intervals for 3- and 4-foot props, and 5- and 6-foot intervals for 5- and 6-foot props) were determined, from Figure 3 and Table 5 the costs of bucking these diameters were determined and averaged for each d.b.h. class. This average figure was multiplied by 100 to get the cost per 100 props. When split props are included (see footnote 6) the average cost by d.b.h. was determined as before but was multiplied by the number of cuts necessary to produce 100 round and split props combined. To this bucking cost was added the cost of splitting those props that were large enough.

[3] Cost per load was multiplied by the factor $\dfrac{100}{\text{Number of props per average load}}$; each load contained about two cords.

[4] Expressed in 1950 dollars.

[5] Data in this line are for round props.

[6] Data include split props meeting mine specifications.

—15—

Table 9B.—Hardwood props: Direct costs of production per 100 props, using three-man power chain saw crew for felling and bucking

Tree d.b.h. (inches) (1)	Average utilized volume per tree[1] (2)	Trees per 100 props (3)	Labor costs						Costs other than labor				
			Felling (4)	Trimming (5)	Skidding (6)	Bucking[2] (7)	Hauling[3] (8)	Total stump to mine[2] (9)	Felling and trimming (10)	Skidding (11)	Bucking[2] (12)	Hauling[3] (13)	Total stump to mine[4] (14)
	Cords	Number	— — — — — — Man-hours — — — — — —						— — — — — Dollars[4] — — — — —				
						THREE-FOOT PROPS							
5	0.009	50.0	2.50	1.65	5.62	2.00	1.28	13.05	0.52	2.17	0.39	0.20	3.28
6	.024	23.2	1.26	.90	2.78	2.00	1.28	8.22	.26	1.08	.39	.20	1.93
7	.043	16.9	.99	.76	2.22	2.06	1.28	7.31	.21	.86	.40	.20	1.67
5/8	.064	14.1	.87	.41	1.96	2.11	1.28	6.63	.18	.76	.41	.20	1.55
5/9	.090	12.3	.81	.75	1.84	2.25	1.28	6.93	.17	.71	.44	.20	1.52
6/8	.064	12.4	.77	.36	1.72	1.99	1.28	6.12	.16	.67	.36	.20	1.39
6/9	.090	9.8	.65	.60	1.47	2.23	1.28	6.23	.14	.57	.37	.20	1.28
						FOUR-FOOT PROPS							
5	.009	66.7	3.34	2.20	7.53	2.00	1.	17.00	.69	2.90	.39	.30	4.28
6	.024	31.2	1.69	1.22	3.74	2.00	1.	10.58	.35	1.45	.39	.30	2.49
7	.043	22.7	1.32	1.02	2.98	2.06	1.	9.31	.28	1.15	.40	.30	2.13
5/8	.064	18.5	1.15	.96	2.58	2.11	1.93	8.73	.24	1.00	.41	.30	1.95
5/9	.090	16.4	1.09	1.00	2.46	2.22	1.93	8.70	.23	.95	.43	.30	1.91
6/8	.064	15.6	.97	.81	2.18	1.95	1.93	7.84	.20	.84	.35	.30	1.69
6/9	.090	14.1	.93	.86	2.12	2.26	1.93	8.10	.20	.82	.38	.30	1.70
						FIVE-FOOT PROPS							
6	.018	50.0	2.71	1.65	6.00	2.00	1.	14.32	.56	2.32	.39	.39	3.66
7	.038	33.3	1.94	1.50	4.37	2.04	1.	11.83	.41	1.69	.40	.39	2.89
8	.060	26.3	1.63	1.37	3.66	2.08	1.	10.70	.34	1.42	.40	.39	2.55
9	.083	22.7	1.50	1.38	3.41	2.19	1.	10.44	.32	1.32	.43	.39	2.46
5/10	.109	20.4	1.43	1.47	3.25	2.31	1.96	10.42	.31	1.26	.45	.39	2.41
5/11	.130	19.2	1.62	1.65	3.26	2.44	1.96	10.93	.35	1.26	.47	.39	2.47
6/10	.109	16.9	1.19	1.22	2.69	2.22	1.96	9.28	.26	1.04	.38	.39	2.07
6/11	.130	16.1	1.36	1.38	2.74	2.41	1.96	9.85	.29	1.06	.41	.39	2.15
						SIX-FOOT PROPS							
6	.018	58.8	3.19	2.29	7.06	2.00	2.16	16.70	.66	2.73	.39	.44	4.22
7	.038	40.0	2.33	1.80	5.25	2.03	2.16	13.57	.49	2.03	.39	.44	3.35
8	.060	31.2	1.93	1.62	4.35	2.08	2.16	12.14	.41	1.68	.40	.44	2.93
9	.083	27.0	1.79	1.65	4.05	2.19	2.16	11.84	.38	1.57	.42	.44	2.81
5/10	.109	24.0	1.69	1.76	3.88	2.25	2.16	11.74	.36	1.50	.44	.44	2.74
5/11	.130	23.3	1.96	2.00	3.96	2.42	2.16	12.50	.42	1.53	.47	.44	2.86
6/10	.109	19.6	1.38	1.41	3.12	2.16	2.16	10.23	.30	1.21	.36	.44	2.31
6/11	.130	18.8	1.58	1.62	3.20	2.40	2.16	10.96	.34	1.24	.39	.44	2.41

1/ To a minimum 4-inch top(inside bark) for 3- and 4-foot props, and to a 5-inch top for 5- and 6-foot props.

2/ Figures in this column were determined in the following manner: From taper tables, the diameters (at 3- and 4-foot intervals for 3- and 4-foot props, and 5- and 6-foot intervals for 5- and 6-foot props) were determined; from Figure 3 and Table 5 the costs of bucking these diameters were determined and averaged for each d.b.h. class. This average figure was multiplied by 100 to get the cost per 100 props. When split props are included (see footnote 4) the average cost by d.b.h. was determined as before but was multiplied by the number of cuts necessary to produce 100 round and split props combined. To this bucking cost was added the cost of splitting those props that were large enough.

3/ Cost per load was multiplied by the factor $\dfrac{100}{\text{Number of props per average load}}$; each load contained about two cords.

4/ Expressed in 1950 dollars.

5/ Data in this line are for round props.

6/ Data include split props meeting mine specifications.

All trees 7 inches in d.b.h. and larger can be felled most efficiently with the power chain saw, but trees under 9 inches in d.b.h. can be bucked at lower cost with a Swedish-type bow saw. A combination of these two tools would, therefore, provide the most efficient means for felling and bucking in the production of mine props.

Although butt cuts of the larger trees are usually split, the costs of producing the round props are included for comparison and for use of producers of other products.

A power chain saw will usually reduce costs when trees 8 inches d.b.h. and larger are logged. However, farmers or part-time loggers will probably find that the high initial cost cannot be depreciated within a reasonable length of time, while hand tools are low in cost and quite efficient in logging. Such operators, too, would find it advisable to skid with mules. Mules would be restricted by steep slopes and rough terrain, but on average operations can skid about 100 lineal feet of prop-size trees in an hour.

The data in tables 8 and 9 may readily be adapted to other products--such as fence posts, pulpwood, or chemical wood--with similar size requirements. For such purposes the figures may be expressed in cords instead of 100 props. Cords per 100 props can be determined by multiplying the appropriate value in column 2 by its corresponding value in column 3. When divided by the resulting product, costs in the table will be converted to a cord basis. For example, the labor cost for felling when producing 100 five-foot props from eight-inch pine trees is 1.08 man-hours. The volume produced is 0.069 cord per tree (column 2) times 18.5 trees per 100 props (column 3) or 1.28 cords per 100 props. Thus 1.08 man-hours is required to produce 1.28 cords-- or 0.85 man-hour per cord.

Indirect costs applicable to the tractor and truck will be found in the Appendix.

Implications of the Study for the Land Manager

In the Birmingham territory, props are sold by the piece, delivered at the mine; the price varies with the size of the prop. Since it costs more to make props from hardwoods than from pine, this procurement policy encourages the contractor to take the best trees and leave the short, low-quality boles. The result is that the prop contractor gets his return partly at the expense of the income that the landowner could realize by managing his forest for high-grade saw timber, either pine or good-quality hardwoods. Since good timber is extremely scarce in the territory, the loss probably is considerable. Much of it falls on the mines themselves, for many of the props that the mines purchase are cut, without restriction as to species or type of tree, from company lands.

Landowners could encourage the cutting of low-quality trees by exacting no stumpage or royalty for them, while demanding fair value for the better hardwoods and the pine. Another course might be to restrict cutting to Virginia pine and hardwoods (except yellow-poplar). Mines could readily control such cutting through the timber checker at the mine. The improvement in its own forests might also justify a mine in paying enough of a premium for hardwood props to offset the contractor's extra expense in producing them. Steps like these would not by themselves rehabilitate the region's forests, but they would be a realistic start and would help to relieve the pressure on trees of high potential value.

Table 10.--Direct tool cost rates per crew for power chain saw

Item	Cost per hour
	Dollars
Depreciation (excluding chains)	0.275
Interest, taxes, etc. (10 percent of average annual investment[1]/)	.028
Repairs (excluding chains)	.127
Chains (wear, repair, and sharpening)	.089
Gasoline (0.189 gallon per hour)	.047
Oil (0.093 quart per hour)	.019
Total	.585

1/ An indirect cost here classified as direct for convenience since it is very small.

Data from which costs were determined.--Work year: 250 eight-hour days or 2,000 hours. Expected life of saw: 2,000 hours of operation; of chain, 300 hours. Cost (1947) of saw with 24-inch bar and 18-inch bow (excluding chains): $550. Cost of 24-inch chain, $19; 36-inch chain, $27. Sharpening chains: by machine, once in 150 operating hours at $3.00. By hand filing, once in 30 operating hours, ½-hour. These data cover 1,114 hours of operation, during which 211 gallons of gasoline (at $0.25) and 104 quarts of oil (at $0.20) were consumed. The costs for repairs, parts, and labor were $141.48.

Table 11.--Direct tool cost rates per crew for crosscut saw, bow and ax

Item	:Crosscut: saw[2]	Bow : saw[3]	Double : bitted
	- - Dollars per hour -		
Depreciation (1/2000 of annual equipment cost)	0.013	0.010	0.00[4]
Maintenance, including labor	.048	.049	.02[3]
Interest, taxes, etc. (10 percent of average annual investment) [1]	.001	.001	...
Total	.062	.060	.02[7]

1/ For crosscut and bow saws, an indirect cost classed as direct convenience inasmuch as it is a very small item. For the ax, indirect costs are negligible.

2/ Crosscut saw. Estimated cost of equipment: two saws at $9.5[0] axes at $3.50, or $26.00 per year. Maintenance charges: one f every four weeks at $0.20; one-half hour labor per day at $0.7

3/ Bow saw. Estimated annual costs of equipment: one saw frame a $4.75; four blades at $1.95; two axes at $3.50, or $19.55 per year. Maintenance charges: one file kit per five years; one f every four weeks at $0.20; 1/2 hour labor per day at $0.75.

4/ Ax. Estimated annual costs of equipment: two axes at $3.50; t handles at $0.75; or $8.50 per year. Maintenance charges: on file every eight weeks at $0.20; ¼ hour labor per day at $0.7[5]

Table 12.--Equipment costs, Ford 1½-ton stake truck

Item	Indirect costs per year	Direct costs per mile	
		Woods road	Improved road
		- - - - - Dollars - - - - - -	
Depreciation	542		
License	23		
Interest, taxes, insurance, etc. (10 percent of av. annual investment)	122		
Total	687		
Tires		0.118	0.063
Gasoline		.050	.028
Oil and grease (estimated)		.003	.003
Repairs, parts, and labor (estimated)		.025	.025
Total direct cost per mile		.196	.119
Total direct cost per hour		1.058	2.689

Data from which costs were determined.--Vehicle: 95 hp; cost $1956; estimated life, 4,000 hours; estimated trade-in value, $400. Tires: 7.50-20, 10 ply; cost $78.91 each, including tubes and taxes; estimated life: on woods road 4,000 miles; on improved road, 7,500 miles. Mileage: woods road, 5 miles per gallon; improved road, 9 miles per gallon; gasoline at $0.25. Average speeds: woods road, 5.4 mph; improved road, 22.6 mph.

Table 13.--Equipment costs, Oliver Cletrac 38 horsepower tractor

Item	: Indirect costs : per year	: Direct costs : per hour
	- - Dollars - -	
Depreciation	1,667	
Interest on investment, taxes, etc.1/	333	
Total	2,000	
Maintenance and repairs (estimated)		1.00
Fuel (0.37 gallon per hour)		.052
Lubricants (0.13 gallon per hour)		.104
Total		1.156
		($0.0193 per minute)

1/ Ten percent of average annual investment.

    Data from which costs were estimated. Estimated life of tractor: 6,000 hours; cost (1949), with winch: $5,000; estimated cost of 3,000-hour overhaul: motor overhaul $500, rollers and track $300, pins and bushings $200, total $1,000; period of operation to date: 374 hours; fuel consumed: 140 gallons at $0.14; oil used (all weights): 50 gallons at $0.80.

Table 14.--Hourly direct cost rates per crew

Crew	: Labor	: Tools	: Tractor	: Total per hour
	- - - - - Dollars - - - - -			
3-man power chain saw crew (one man at 90 cents, two at 75 cents per hour)	2.40	0.58		2.98
3-man crosscut saw crew (all at 75 cents per hour)	2.25	.06		2.31
2-man crosscut saw crew or two-man bow saw crew (all at 75 cents per hour)	1.50	.06		1.56
1-man with double-bitted ax (75 cents per hour)	.75	.02		.77
Skidding crew (one man at $1.25, two at $0.75 per hour)	2.75	.10	1.16	4.01

MARKING GUIDES FOR OAKS AND YELLOW-POPLAR IN THE SOUTHERN UPLANDS

Joseph L. Burkle and Sam Guttenberg

SOUTHERN FOREST EXPERIMENT STATION
Harold L. Mitchell, Director
New Orleans, La.
FOREST SERVICE, U.S. DEPT OF AGRICULTURE

CONTENTS

MARKING GUIDES FOR OAKS AND YELLOW-POPLAR IN THE SOUTHERN UPLANDS

Joseph L. Burkle and Sam Guttenberg
Southern Forest Experiment Station

This paper offers guides that the forest manager in the southern uplands can use to judge when his trees should be cut--that is, when they are financially mature. The paper was written with special reference to the territory about Birmingham, Alabama (figure 1), but the portions that deal with lumber logs and cross ties should be widely applicable in other upland hardwood forests in the South. The discussion is restricted to red oaks, white oaks, and yellow-poplar, and to the chief products into which these trees are cut: standard factory lumber, cross ties, and--in the Birmingham territory--mine props. Face veneer logs and tight cooperage bolts are not considered; forest managers who expect to grow large quantities of these products will need to make special studies following the principles outlined here.

The guides are not particularly complex in principle, but the alternatives to which they must be applied are numerous. Their use is facilitated by an understanding of the way in which they were constructed. Accordingly, this paper first describes how the guides were determined and how they apply to the various alternatives encountered in actual management. The last part (pp.26-27) condenses the guides into a simple set of marking rules for use in the woods.

The data on which the guides are based are the best available at present. Ideally, each forest manager would accumulate his own figures and work out his own guides following the principles described in this publication. This would, however, require data not always readily available; an alternative course is to adapt the guides offered here.

Figure 1.--Birmingham territory.

How the Guides Were Determined

The guides are based on the cutting of financially mature timber. The idea of financial maturity is this:

From time to time the forest manager visits each acre of his holdings and marks for cutting those trees whose quality or growth rate is below par. Bearing in mind how the typical tree develops--low value at first, but rapid increase in value; then a gradual slackening in the rate of increase until the stage is finally reached where the tree ceases to pay its way--the manager appraises his trees with the object of putting the ax to those that are reaching this crucial point. This is the point of financial maturity. Just where it falls is determined by the rate at which the forest owner expects his trees to earn money, the so-called alternative rate of return.[1] If he expects a return of 4 percent, he will cut any tree that cannot produce at this rate.

The chief problem with which this paper is concerned, therefore, is to determine the rate of value increase that can be expected from a given tree over a given period of time. This rate will be estimated in the following steps:

1. Calculate the value of trees according to their volume, grade, and diameter.

2. Formulate reliable rules for judging the vigor of trees, as an index of their prospective diameter-growth rate.

3. Determine the prospective rate of increase in tree value.[2]

Standard factory lumber makes up about 70 percent of total hardwood consumption and represents the highest general use of hardwood timber. In any operation aimed at growing lumber logs, however, cross ties and mine props will be harvested as a by-product from trees or portions of trees that for one reason or another are best not utilized as lumber logs. Too, at time of harvest those portions of lumber-log trees that can be more profitably utilized as face veneer or tight cooperage may be put to such use. In this paper, therefore, and following the steps listed above, trees will be appraised first for lumber logs, then for cross ties, and finally for mine props.

---

[1]. The choice of an alternative rate is conditioned by risk, taxes, business debts, and the like. See Duerr, Wm. A., and Bond, W. E. Optimum stocking of a selection forest. Jour. Forestry 50: 12-16. 1952.

[2]. For more complete treatment of the method, see Guttenberg, Sam, and Putnam, J. A. Financial maturity of bottomland red oaks and sweetgum. South. Forest Expt. Sta., Occas. Paper 117, 24 pp. 1951.

Lumber Logs

Step one--tree volume, quality, and value

With standard factory lumber in view, tree volume is here expressed in board feet by the International $\frac{1}{4}$-inch kerf rule, using standard volume tables as the guide.[3] Tree quality is developed in terms of the log grades formulated by the U. S. Forest Products Laboratory[4]--for example, a two-log tree with a grade 1 butt and a grade 3 second log will be classed as grade 1-3. Since a description of these grades is readily available, they will not be discussed here. It must be recognized, however, that facility in the use of the grades is absolutely essential in making financial maturity decisions. Special training (which can be self-training) and experience in the field are required. Familiarity with the log defects upon which the grades are based is especially important.[5]

Tree value will be expressed in terms of conversion surplus.[6] Conversion surplus is the difference between the sales value of the end product (in this case, standard factory lumber) and all the direct (variable) costs of producing this lumber--mainly labor and materials used in felling the tree and in making, transporting, and sawing the logs. Conversion surplus thus represents that part of a tree's gross product value which can be made available to increase the profit or reduce the loss of the business.

Since conversion surplus will here be based upon lumber value-- not log value or stumpage value--the examples to be developed will apply primarily to the timber owner who operates a sawmill and sells his product in the form of lumber. However, they will also apply to the log or stumpage seller who can count on receiving for each log or tree a price in line with the value of lumber it will yield. Indeed, prices would be in line on a freely competitive market if lumber producers and timber growers used such an analysis as a guide.

Lumber sales value.--The first ingredient of conversion surplus, lumber sales value, may be estimated for any tree. Knowing the grade and volume of each log in the tree and the lumber grade yield[7] of each grade of log, one may determine the amount of lumber, by grades, that

---

3/. Mesavage, C., and Girard, J. W. Tables for estimating board-foot volume of timber. 94 pp. 1946. Govt. Printing Office.
4/. Hardwood log grades for standard lumber: proposals and results. U. S. Forest Products Laboratory D 1737. 1949.
5/. Lockard, C. R., Putnam, J. A., and Carpenter, R. D. Log defects in southern hardwoods. U. S. Dept. Agr., Agr. Handbook 4, 37 pp. 1950.
6/. Guttenberg, Sam, and Duerr, Wm. A. A guide to profitable tree utilization. South. Forest Expt. Sta., Occas. Paper 114, 18 pp. 1949.
7/. As given in reference cited in footnote 4.

the tree will yield. With current sales prices for lumber, he may then calculate the weighted average price per M of lumber represented in the tree, and from this the total lumber value in the tree.

The problem is how to allow for changes in the lumber market. An estimate of financial maturity involves looking ahead at least one cutting cycle—perhaps 5 years or more—and the lumber price level probably will not be the same then as now. To help solve this problem, advantage may be taken of the fact that with all the ups and downs of the lumber market, the price of each grade tends to maintain a fairly stable relationship to that of other grades.8/ Grade prices may then be expressed as percentages (index numbers) that will have substantial validity over the years, and the value of lumber in a tree may be calculated in index terms instead of dollars, following the same procedure as would be used with dollars.

In figure 2, value index numbers are plotted for each grade of red oak, white oak, and yellow-poplar lumber. These index numbers were derived from annual average lumber grade prices, expressed as relatives in percent of the price of No. 1 Common and Selects for the oaks and No. 1 Common for yellow-poplar. The chart covers the period 1925-1950, omitting the abnormal war years 1942 through 1945. These are representative price data, f.o.b. mills in the Birmingham territory, for air-dried stock 4/4-inch and thicker. The lower grades of lumber have commanded relatively high prices in recent years as these grades have come into wider demand. Since this condition appears likely to

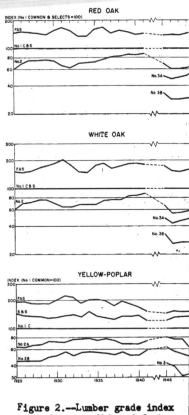

Figure 2.--Lumber grade index values per M board feet.

8/. Herrick, A. M. Grade yields and overrun from Indiana hardwood sawlogs. Purdue University Bul. 516, 59 pp. 1946.

continue, the index numbers used here were determined by averaging the
price relatives for the years 1940, 1941, and 1946 through 1950, allowing each year equal weight. The resulting indexes are as follows:

Lumber grade	Red Oak	White Oak	Yellow-poplar
Firsts and Seconds	144	168	154
Saps and Selects	...	...	132
No. 1 Common and Selects	100	100	...
No. 1 Common	...	...	100
No. 2 Common	71	68	...
No. 2A Common	...	...	71
No. 2B Common			54
No. 3 Common	...	...	35
No. 3A Common	52	49	...
No. 3B Common	30	28	

What these numbers mean--using red oak as an example--is that
in any marketing period, whatever price is being received for No. 1
Common and Selects, 144 percent of that price will probably prevail
for Firsts and Seconds. Seventy-one percent will prevail for No. 2
Common, 52 percent for No. 3A Common, and 30 percent for No. 3B Common.

·Direct costs.--The other ingredient of conversion surplus,
direct costs of lumber production from stump through mill, is estimated from whatever representative cost records are available. Direct
costs include only those additional outlays that arise because the
tree in question is being made into lumber. They specifically exclude
all fixed or overhead costs--all costs that are unaffected by whether
the particular tree is logged and.milled.

For Birmingham territory hardwoods, available logging and
milling study experience[2] was used to estimate direct costs on a
typical operation. Costs were estimated in dollars per M board feet
for trees of each species group, diameter class, and log height. Dollars were then converted to relatives in percent of the price of No. 1
Common and Selects for oaks and No. 1 Common for yellow-poplar, just
as was done for lumber value. This procedure is defensible on the
ground that prices and direct costs (primarily wages) tend to fluctuate together. It permits deriving conversion surplus as an index

2/. See publications cited on page 18 of the reference given in
footnote 6, and James, L. M. Logging and milling studies in the
southern Appalachian region. Southeastern Forest Expt. Sta. Tech.
Notes 62 and 63. 1946.

prices. The result of these calculations is given below under step
three.

Step two--vigor

Any woodsman can readily point out differences in tree vigor
Furthermore, there is no trick to spotting trees of very poor vigor
nor is there much question about cutting such trees for whatever va
they may contain. Here we are concerned with enhancing a skill already possessed by woodsmen, so that tree vigor may be classified a
used as an index of prospective growth. Three classes--high, mediu
and low (excluding decadent trees)--suffice for all practical purpo
and are clearly enough distinguished so that one can learn to ident
them with minimum use of an increment borer.

Tree vigor can be judged from external indicators. The sing
most reliable indicator is the bark. Bark features stand out so
strongly and are apparently so reliable that some others--site, age
root system--can largely be ignored. The crown, however, furnishes
additional evidence and should be considered.

Bark features for the various vigor classes are illustrated
in figures 3 through 5 (pages 9 - 11). The illustrations and the
specific bark features described below apply mainly to trees 12 inc
d.b.h. and over. The general indications of vigor furnished by the
bark and crown, however, apply to trees of all sizes.

Red oaks (fig. 3) of high vigor generally have smooth, thin
bark with wide, open fissures vividly colored at the bottom. As
vigor declines, the bark thickens and roughens, and the fissures
narrow, becoming discontinuous and obscure in low-vigor trees. Dec
dent red oaks, those below the class of low vigor, are usually deger
erate or dying. The bark is thick, dark, and very rough (fig. 3D)
else dull, bleached, and sickly. Excessive die-back of twigs often
makes the tree stagheaded.

White oaks (fig. 4) of high vigor generally have thick, dark
gray bark with distinct, long fissures; ridges are flat with few cr
breaks. As vigor declines, the bark thins and turns ash-gray, whil
fissures become obscure, ridges scaly, and cross breaks more numero
In white oaks the most conclusive indicator of decadence occurs whe
one or more faces have very thin bark.

Yellow-poplars (fig. 5) of high vigor generally have bark th
is shallowly ridged, somewhat corky, and light ash-grey. The fissu
display light-colored inner bark at the bottom. As vigor falls off
the ridges become more distinct and closer together, and the color

darkens. The inner bark tissue disappears as low vigor is reached.
Decadent yellow-poplars have very pronounced fissures, and the bark,
by contrast with healthy trees, is bleached and sickly.

Table 1 summarizes the features of the three vigor classes for
yellow-poplar (Liriodendron tulipifera) and for the better oak alternatives of the Birmingham territory--northern red oak (Quercus
borealis), southern red oak (Q. falcata), and white oak (Q. alba).
Vigor classes for scarlet oak (Q. coccinea) and black oak (Q. vel-
utina) follow closely those for northern and southern red oak. Chest-
nut oak (Q. montana) and post oak (Q. stellata), however, differ
markedly from white oak in the vigor indications furnished by the
bark, though not in those displayed by the crowns.

The growth rates to be expected from trees (including all oaks
mentioned above) in the three vigor classes are as follows:

Five-year average diameter growth (inches)

Vigor	Red oaks	White oaks	Yellow-poplar
High	1.50-2.00	1.00-1.50	1.75-2.25
Medium	1.00-1.50	.75-1.00	1.25-1.75
Low	.50-1.00	.50- .75	.75-1.25

Step three--value increase

With vigor classes defined, and with a method determined for
expressing tree value according to tree size and quality, we are
ready for the problem of growth in value--i.e., rate of increase in
tree value, the rate that is to be compared with the desired rate of
return to determine if a tree is financially mature. The problem will
be considered in two stages: first, rate of value increase when a tree
changes diameter but not log height or grade; second, rate of increase
when height or grade, as well as diameter, changes.

A 5-year cutting cycle will be assumed. That is, it will be
supposed--and this is realistic for the region--that the manager has
the choice of cutting a tree now or waiting at least 5 years. Consequently our interest centers on the rate of tree-value increase in
5 years.

No change in log height or grade.--Rates of value increase for
red oaks, white oaks, and yellow-poplar of each vigor class are worked
out in tables 2, 3, and 4. In each table, columns 2, 3, and 4 show,
in terms of index numbers as explained earlier, the lumber value, the
direct costs, and the difference between these two--the conversion
surplus--per M board feet as a weighted average for the tree. From
the number of board feet per tree (column 5), conversion surplus per

(Text continued on p. 18.)

Table 1.—Crown and bark characteristics of red oaks, white oaks, and yellow-po

High vigor	Medium vigor	

R E D   O A K S

Bark: Bark thickness, and color of the inner and outer bark, varies with the species. In general, the bark is healthy, fully normal in color, relatively thin and smooth, but with shallow fissures exposing fleshy, lighter-colored inner bark that contrasts markedly with dark outer bark. The bark is the most conclusive indicator.

Bark: Compared with high vigor, fissures less wide, inner bark duller and generally less conspicuous. Overall, bark is somewhat rougher and darker.

Bark: rowl; or : posed

Crown: 3/4 or more fully formed and without close competition. Full and thrifty. Profuse long, upward-reaching young branches and twigs, light-colored and lustrous. No dying leaders or dead stubs in upper crown. Foliage abundant and lustrous.

Crown: 1/2 or more well-formed, with abundant foliage, and without close competition. Some crowns may be entirely free of competition, but twigs will be thicker and fewer, and foliage scantier, than in high-vigor crowns.

Crown forme foli:

Crown quality and vigor more important than length or volume, except that a high ratio of crown length to stem length is significant indicator.

W H I T E   O A K S

Bark: Thick, dark-gray with distinct, long fissures. Ridges flat with few cross breaks running from fissure to fissure.

Bark: Thinner. Gray. Fissures shorter and less distinct, ridges somewhat scaly, cross breaks more common.

Bark: Fiss: scur(and dant.

Crown: 3/4 or more fully formed and without close competition. Full and thrifty, branches small, silvery, and ascending. No dying leaders or dead stubs in upper crown.

Crown: 1/2 or more fully formed and without close competition. Notably less full than those of high vigor trees. Branches darker, less ascending in aspect.

Crown form(foli:

Y E L L O W - P O P L A R

Bark: Corky but shallowly ridged, light ash-grey. Diamond-shaped fissures display light-colored inner bark.

Bark: Thicker. Ash-grey. Ridges more pronounced. Some inner bark visible in fissures.

Bark Fiss prom barl

Crown: 2/3 or more fully formed and without close competition. Full and healthy, composed preponderantly of small ascending leaders and twigs. Foliage abundant and lustrous. Sharp pointed crown an important indicator.

Crown: 1/2 or more fully formed and without close competition. Crowns notably less full and less pointed than those of high vigor trees. Crown limbs are larger and do not ascend as steeply as in high-vigor crowns.

Crow for: ate(tiv(and thai

- 8 -

Figure 3.--Southern red oak. <u>A</u>, High vigor. <u>B</u>, Medium vigor. <u>C</u>, Low vigor. <u>D</u>, Decadent.

Figure 4.--White oak. A, High vigor. B, Medium vigor. C, Low vigor. D, Decadent.

Figure 5.--Yellow-poplar. A, High vigor. B, Medium vigor. C, Low vigor.
D, Decadent.

Table 2.--Rate of value increase of red oaks, by log height, grade, diameter

Diam. breast high (inches) (1)	Per M board feet			Lumber per tree (5)	Conversion surplus per tree b				
	Gross lumber value (2)	Direct costs (3)	Con-version surplus (4)			High vigor		Medium	
					Now (6)	After 5 yrs. (7)	Annual increase (8)	After 5 yrs. (9)	A in
	- - - Index - - - - - -			Bd.ft.	- - Index - -		Percent	Index	

1-LOG-TREES, GRADE 1

18	92.2	61.4	30.8	112	3.45	5.40	9.4	4.85	
20	95.9	55.6	40.3	141	5.68	7.80	6.5	7.20	
22	97.6	51.0	46.6	174	8.11	10.55	5.4	9.85	
24	99.4	47.5	51.9	209	10.85	13.60	4.6	12.85	
26	101.3	44.7	56.6	248	14.04	17.05	4.0	16.20	
28	103.4	42.6	60.8	289	17.57	21.00	3.6	20.00	
30	105.6	41.2	64.4	334	21.51	25.28	3.3	24.20	
32	108.0	40.4	67.6	382	25.82	29.80	2.9	28.66	

GRADE 2

18	70.0	61.4	8.6	112	.96	2.15	17.5	1.	
20	71.9	55.6	16.3	141	2.30	3.70	10.0	3.	
22	73.8	51.0	22.8	174	3.97	5.60	7.1	5.80	
24	75.8	47.5	28.3	209	5.91	7.95	6.1	7.58	
26	78.0	44.7	33.3	248	8.26	10.55	5.0	9.90	
28	80.2	42.6	37.6	289	10.87	13.45	4.4	12.75	
30	82.6	41.2	41.4	334	13.83	16.80	4.0	15.95	
32	85.5	40.4	45.1	382	17.23	20.05	3.0	19.30	

GRADE 3

20	56.8	55.6	1.2	141	.17	1.00	55.1	.75	
22	57.5	51.0	6.5	174	1.13	2.10	13.2	1.85	
24	58.2	47.5	10.7	209	2.24	3.35	8.4	3.05	
26	59.1	44.7	14.4	248	3.57	4.80	6.1	4.40	
28	60.0	42.6	17.4	289	5.03	6.40	4.9	6.00	
30	61.2	41.2	20.0	334	6.68	8.20	4.2	7.80	
32	62.7	40.4	22.3	382	8.52	9.90	3.0	9.50	

2-LOG-TREES, GRADE 1-2

18	87.0	60.2	26.8	210	5.63	8.55	8.7	7.75	
20	88.3	54.5	33.8	266	8.99	12.60	7.0	11.55	
22	89.6	49.8	39.8	331	13.17	17.15	5.4	16.00	
24	91.0	46.2	44.8	397	17.79	22.40	4.7	21.05	
26	92.5	43.8	48.7	475	23.13	28.20	4.0	26.70	
28	94.1	41.9	52.2	554	28.92	34.80	3.8	33.05	
30	95.9	40.6	55.3	646	35.72	42.05	3.3	40.20	
32	97.9	40.0	57.9	743	43.02	49.50	2.8	47.65	

GRADE 1-3

18	74.0	60.2	13.8	210	2.90	5.35	13.0	4.60	
20	76.1	54.5	21.6	266	5.75	8.95	9.3	8.00	
22	78.2	49.8	28.4	331	9.40	13.00	6.7	12.00	
24	80.4	46.2	34.2	397	13.58	17.80	5.6	16.60	
26	82.7	43.8	38.9	475	18.48	23.20	4.6	21.90	
28	85.2	41.9	43.3	554	23.99	29.55	4.3	27.95	
30	87.7	40.6	47.1	646	30.43	36.50	3.7	34.75	
32	90.4	40.0	50.4	743	37.45	43.35	3.0	41.70	

Table 2 .—(Continued)

Diam. breast high (inches) (1)	Per M board feet			Lumber per tree (5)	Conversion surplus per tree by vigor class						
	Gross lumber value (2)	Direct costs (3)	Conversion surplus (4)		Now (6)	High vigor		Medium vigor		Low vigor	
						After 5 yrs. (7)	Annual increase (8)	After 5 yrs. (9)	Annual increase (10)	After 5 yrs. (11)	Annual increase (12)
	- - - Index - - - - - - -			Bd.ft.	- - Index - -		Percent	Index	Percent	Index	Percent
GRADE 2-3											
18	62.6	60.2	2.4	210	.50	2.25	33.6	1.80	29.2	1.25	16.5
20	63.9	54.5	9.4	266	2.50	4.80	13.9	4.15	10.6	3.50	6.9
22	65.4	49.8	15.6	331	5.16	7.75	8.5	7.00	6.3	6.20	3.7
24	66.8	46.2	20.6	397	8.18	11.20	6.5	10.35	4.8	9.50	3.0
26	68.5	43.8	24.7	475	11.73	15.15	5.2	14.20	3.9	13.20	2.4
28	70.3	41.9	28.4	554	15.73	20.00	4.9	18.80	3.6	17.55	2.2
30	72.5	40.6	31.9	646	20.61	25.55	4.4	24.10	3.2	22.75	2.0
32	75.4	40.0	35.4	743	26.30	30.60	3.1	29.40	2.2	28.20	1.4
34	77.2	40.1	37.1	841	31.20	34.80	2.2	33.85	1.6	32.80	1.0
GRADE 3-3											
22	53.0	49.8	3.2	331	1.06	2.60	19.6	2.15	15.2	1.75	10.5
24	53.3	46.2	7.1	397	2.82	4.42	9.4	4.00	7.2	3.55	4.7
26	53.7	43.8	9.9	475	4.70	6.55	6.8	6.00	5.0	5.45	3.0
28	54.2	41.9	12.3	554	6.81	8.95	5.6	8.35	4.2	7.85	2.9
30	54.9	40.6	14.3	646	9.24	11.43	4.3	10.80	3.2	10.15	1.9
32	55.9	40.0	15.9	743	11.81	13.80	3.2	13.20	2.2	12.15	1.4
34	56.7	40.1	16.6	841	13.96	15.55	2.2	15.15	1.6	14.70	1.0
3-LOG-TREES, GRADE 1-1-2											
22	91.5	49.5	42.0	475	19.95	26.60	5.9	24.60	4.3	22.65	2.6
24	93.1	45.5	47.6	580	27.61	.34.50	4.5	32.50	3.3	30.55	2.0
26	94.6	43.2	51.4	690	35.47	43.55	4.2	41.20	3.0	39.00	1.9
28	96.5	41.5	55.0	814	44.77	53.64	3.7	51.05	2.7	48.40	1.6
30	98.4	40.3	58.1	945	54.90	65.00	3.4	62.00	2.5	59.20	1.5
32	100.7	39.8	60.9	1,092	66.50	75.90	2.7	73.35	2.0	70.60	1.2
GRADE 1-1-3											
22	88.0	49.5	38.5	475	18.20	24.45	6.0	22.65	4.4	20.85	2.7
24	89.2	45.5	43.7	580	25.35	31.80	4.6	29.95	3.4	28.07	2.1
26	90.6	43.2	47.4	690	32.71	40.05	4.1	38.00	3.0	35.85	1.8
28	92.0	41.5	50.5	814	41.11	49.20	3.7	46:80	2.6	44.55	1.6
30	93.6	40.3	53.3	945	50.37	59.45	3.4	56.80	2.4	54.15	1.5
32	95.5	39.8	55.7	1,092	60.82	69.60	2.8	67.15	2.0	64.60	1.2
GRADE 1-2-3											
18	76.5	59.5	17.0	302	5.13	8.80	11.4	7.70	8.4	6.60	5.2
20	78.4	54.0	24.4	384	9.37	14.00	8.3	12.60	6.1	11.25	3.7
22	80.5	49.5	31.0	475	14.72	20.60	6.9	18.95	5.2	17.20	3.2
24	82.6	45.5	37.1	580	21.52	27.80	5.2	25.95	3.8	24.15	2.3
26	84.9	43.2	41.7	690	28.77	36.10	4.6	34.00	3.4	31.90	2.1
28	87.2	41.5	45.7	814	37.20	45.55	4.1	43.15	3.0	40.75	1.8
30	89.8	40.3	49.5	945	46.78	56.20	3.7	53.40	2.7	50.75	1.6
32	92.6	39.8	52.8	1,092	57.66	67.00	3.0	64.40	2.2	61.60	1.3

Diam. breast high (inches) (1)	Per M board feet Gross lumber value (2)	Direct costs (3)	Conversion surplus (4)	Lumber per tree (5)	Now (6)	High vigor After 5 yrs. (7)	High vigor Annual increase (8)	Medium vigor After 5 yrs. (9)	Medium vigor Annual increase (10)	Low vigor After 5 yrs. (11)	Low vigor Annual increase (12)
	- - - - Index - - - - -			Bd. ft. (5)	- - Index - -		Percent	Index	Percent	Index	Percent
1-LOG-TREES, GRADE 1											
18	95.5	58.7	36.8	112	4.12	5.59	6.3	5.19	4.7	4.86	3.4
20	98.5	53.5	45.0	141	6.34	8.09	5.0	7.57	3.6	7.25	2.7
22	101.5	49.0	52.5	174	9.14	11.24	4.2	10.52	2.8	10.10	2.0
24	104.4	44.9	59.5	209	12.44	14.81	3.5	14.07	2.5	13.62	1.8
26	107.2	41.7	65.5	248	16.24	18.82	3.0	18.02	2.1	17.52	1.5
28	109.8	39.3	70.5	289	20.37	23.19	2.6	22.33	1.8	21.75	1.3
30	112.2	37.7	74.5	334	24.88	27.75	2.2	26.79	1.5	26.24	1.1
32	114.6	37.2	77.4	382	29.57	32.50	1.9	31.58	1.3	31.03	1.0
GRADE 2											
16	67.2	64.1	3.1	86	.27	1.10	32.3	.81	24.6	.71	21.3
18	70.7	58.7	12.0	112	1.34	2.45	12.9	2.18	10.2	2.00	8.3
20	74.1	53.5	20.6	141	2.90	4.15	7.4	3.75	5.3	3.50	3.8
22	77.6	49.0	28.6	174	4.98	6.49	5.4	5.98	3.7	5.65	2.6
24	80.8	44.9	35.9	209	7.50	9.32	4.4	8.76	3.2	8.45	2.4
26	84.0	41.7	42.3	248	10.49	12.50	3.6	11.92	2.6	11.50	1.9
28	87.0	39.3	47.7	289	13.79	16.09	3.1	15.32	2.1	14.95	1.6
30	89.8	37.7	52.1	334	17.40	19.57	2.4	18.87	1.6	18.50	1.2
32	92.0	37.2	54.8	382	20.93	22.92	1.8	22.04	1.0	21.85	0.9
GRADE 3											
20	55.8	53.5	2.3	141	.32	.95	24.3	.73	17.9	.50	9.3
22	56.7	49.0	7.7	174	1.34	2.17	10.1	1.87	6.9	1.73	5.2
24	57.8	44.9	12.9	209	2.70	3.67	6.3	3.32	4.2	3.15	3.1
26	58.9	41.7	17.2	248	4.27	5.32	4.5	5.00	3.2	4.79	2.3
28	60.1	39.3	20.8	289	6.01	7.25	3.8	6.89	2.8	6.59	1.9
30	61.4	37.7	23.7	334	7.92	9.02	2.6	8.73	2.0	8.50	1.4
32	62.9	37.2	25.7	382	9.82	10.80	1.9	10.50	1.3	10.25	0.9
2-LOG-TREES, GRADE 1-2											
18	82.8	56.5	26.3	210	5.52	7.82	7.2	7.18	5.4	6.73	4.0
20	86.2	51.8	34.4	266	9.15	12.07	5.7	11.09	3.9	10.50	2.8
22	89.6	47.7	41.9	331	13.87	17.26	4.5	16.23	3.2	15.52	2.3
24	92.9	44.1	48.8	397	19.37	23.70	4.1	22.32	2.9	21.50	2.1
26	96.1	40.8	55.3	475	26.27	30.77	3.2	29.46	2.3	28.50	1.6
28	99.1	38.6	60.5	554	33.52	38.76	2.9	37.23	2.1	36.15	1.5
30	102.0	37.0	65.0	646	41.99	47.57	2.5	45.85	1.8	44.74	1.3
32	104.7	36.2	68.5	743	50.90	56.35	2.1	54.70	1.4	53.57	1.0
GRADE 1-3											
18	72.1	56.5	15.6	210	3.28	5.25	9.9	4.73	7.6	4.32	5.7
20	76.0	51.8	24.2	266	6.44	9.00	6.9	8.21	5.0	7.65	3.5
22	80.0	47.7	32.3	331	10.69	13.75	5.1	12.73	3.5	12.18	2.7
24	83.8	44.1	39.7	397	15.76	19.75	4.6	18.50	3.2	17.70	2.4

Table 3.—(Continued)

Diam. breast high (inches) (1)	Per M board feet			Lumber per tree (5)	Conversion surplus per tree by vigor class						
	Gross lumber value (2)	Direct costs (3)	Conversion surplus (4)		Now (6)	High vigor		Medium vigor		Low vigor	
						After 5 yrs. (7)	Annual increase (8)	After 5 yrs. (9)	Annual increase (10)	After 5 yrs. (11)	Annual increase (12)
	– – – – Index – – – – –			Bd. ft.	– – Index – –		Percent	Index	Percent	Index	Percent

GRADE 2-3

18	63.3	56.5	6.8	210	1.43	2.95	15.6	2.49	11.7	2.23	9.3
20	66.2	51.8	14.4	266	3.83	5.80	8.6	5.18	6.2	4.78	4.5
22	69.0	47.7	21.3	331	7.05	9.48	7.6	8.70	4.3	8.25	3.2
24	71.8	44.1	27.7	397	11.00	14.00	4.9	13.03	3.4	12.50	2.6
26	74.4	40.8	33.6	475	15.96	19.27	3.8	18.23	2.7	17.55	1.9
28	76.9	38.6	38.3	554	21.22	25.00	3.3	23.82	2.3	23.08	1.7
30	79.2	37.0	42.2	646	27.26	31.08	2.7	29.85	1.8	29.17	1.4
32	81.1	36.2	44.9	743	33.36	36.76	2.0	35.75	1.4	35.05	1.0

GRADE 3-3

20	53.6	51.8	1.8	266	.48	1.50	25.5	1.20	20.1	.97	15.1
22	54.5	47.7	6.8	331	2.25	3.57	9.7	3.22	7.4	2.85	4.8
24	55.4	44.1	11.3	397	4.49	6.25	6.8	5.71	4.9	5.28	3.3
26	56.4	40.8	15.6	475	7.41	9.28	4.6	8.73	3.3	8.32	2.3
28	57.6	38.6	19.0	554	10.53	12.72	3.8	12.04	2.7	11.55	1.9
30	58.8	37.0	21.8	646	14.08	16.42	3.1	15.65	2.1	15.20	1.5
32	60.1	36.2	23.9	743	17.76	20.03	2.4	19.41	1.8	18.92	1.3

3-LOG-TREES, GRADE 1-1-2

22	94.3	45.5	48.8	475	23.18	28.48	4.2	26.74	2.9	25.72	2.1
24	97.3	42.5	54.8	580	31.78	37.75	3.5	34.74	2.5	34.74	1.8
26	100.2	39.8	60.4	690	41.68	48.75	3.2	46.45	2.2	45.05	1.6
28	103.0	37.8	65.2	814	53.07	60.77	2.8	58.38	1.9	56.87	1.4
30	105.8	36.4	69.4	945	65.58	73.98	2.4	71.31	1.7	69.73	1.2
32	108.2	35.7	72.5	1,092	79.17	87.23	1.9	84.69	1.3	83.04	1.0

GRADE 1-1-3

22	87.4	45.5	41.9	475	19.90	24.98	4.7	23.30	3.2	22.35	2.4
24	90.8	42.5	48.3	580	28.01	33.76	3.8	31.95	2.7	30.78	1.9
26	94.0	39.8	54.2	690	37.40	44.19	3.4	42.00	2.3	40.73	1.7
28	97.1	37.8	59.3	814	48.27	55.49	2.8	53.23	2.0	51.75	1.4
30	100.0	36.4	63.6	945	60.10	68.23	2.6	65.54	1.7	63.98	1.3
32	102.8	35.7	67.1	1,092	73.27	80.74	2.0	78.30	1.3	76.82	0.9

GRADE 1-2-3

18	74.2	52.9	21.3	302	6.43	9.36	7.8	8.48	5.7	7.87	4.1
20	78.2	49.2	29.0	384	11.14	15.00	6.1	13.75	4.3	12.99	3.1
22	82.3	45.5	36.8	475	17.48	22.45	5.1	20.77	3.5	19.80	2.5
24	86.3	42.5	43.8	580	25.40	31.17	4.2	29.31	2.9	28.05	2.0
26	90.2	39.8	50.4	690	34.78	41.50	3.6	39.30	2.5	38.02	1.8
28	93.8	37.8	56.0	814	45.58	52.81	3.0	50.61	2.1	49.25	1.6
30	97.4	36.4	61.0	945	57.64	65.60	2.6	63.00	1.8	61.47	1.3
32	100.3	35.7	64.6	1,092	70.54	77.80	2.0	75.54	1.4	74.12	1.0

Table 4.—Rate of value increase of yellow-poplar, by log height, grade, diameter,

Diam. breast high (inches) (1)	Per M board feet			Lumber per tree (5)		Conversion surplus per tree by vigor				
	Gross lumber value (2)	Direct costs (3)	Con-version sur-plus (4)		Now (6)	High vigor		Medium vigor		
						After 5 yrs. (7)	Annual increase (8)	After 5 yrs. (9)	Annual increase (10)	
	- - - Index - - -			Bd. ft. - - Index - -			Percent	Index	Percent	

1-LOG-TREES, GRADE 1

18	99.2	51.5	47.7	112	5.34	7.68	7.5	7.05	5.7
20	100.7	46.2	54.5	141	7.68	10.39	6.2	9.65	4.7
22	102.0	42.3	59.7	174	10.39	13.38	5.2	12.55	3.8
24	103.4	39.4	64.0	209	13.38	16.72	4.6	15.85	3.4
26	104.6	37.2	67.4	248	16.72	20.26	3.9	19.40	3.0
28	105.7	35.6	70.1	289	20.26	24.01	3.5	23.10	2.7
30	106.6	34.7	71.9	334	24.01	27.89	3.0	26.95	2.3
32	107.5	34.5	73.0	382	27.89	31.60	2.5	30.65	1.9

GRADE 2

16	77.6	59.1	18.5	86	1.59	3.00	13.6	2.60	10.4
18	78.3	51.5	26.8	112	3.00	4.64	9.1	4.25	7.2
20	79.1	46.2	32.9	141	4.64	6.52	7.0	6.00	5.3
22	79.8	42.3	37.5	174	6.52	8.57	5.6	8.00	4.2
24	80.4	39.4	41.0	209	8.57	10.89	4.9	10.25	3.6
26	81.1	37.2	43.9	248	10.89	13.32	4.1	12.65	3.0
28	81.7	35.6	46.1	289	13.32	15.86	3.5	15.20	2.7
30	82.2	34.7	47.5	334	15.86	18.34	2.9	17.70	2.2
32	82.5	34.5	48.0	382	18.34	20.60	2.4	20.05	1.8

GRADE 3

16	68.0	59.1	8.9	86	.77	1.92	20.0	1.65	16.4
18	68.6	51.5	17.1	112	1.92	3.29	11.3	3.00	9.3
20	69.5	46.2	23.3	141	3.29	4.84	8.0	4.45	6.2
22	70.1	42.3	27.8	174	4.84	6.54	6.2	6.10	4.7
24	70.7	39.4	31.3	209	6.54	8.43	5.2	7.95	4.0
26	71.2	37.2	34.0	248	8.43	10.38	4.3	9.90	3.3
28	71.5	35.6	35.9	289	10.38	12.36	3.5	11.85	2.7
30	71.7	34.7	37.0	334	12.36	14.25	2.9	13.80	2.2
32	71.8	34.5	37.3	382	14.25	16.05	2.4	15.65	1.9

2-LOG-TREES, GRADE 1-2

18	90.7	50.0	40.7	210	8.55	12.37	7.7	11.35	5.8
20	91.7	45.2	46.5	266	12.37	16.91	6.4	15.75	5.0
22	92.6	41.5	51.1	331	16.91	21.80	5.2	20.60	4.0
24	93.6	38.7	54.9	397	21.80	27.50	4.8	26.10	3.7
26	94.4	36.5	57.9	475	27.50	33.30	3.9	31.80	2.9
28	95.1	35.0	60.1	554	33.30	39.86	3.6	38.20	2.8
30	95.7	34.0	61.7	646	39.86	46.21	3.0	44.60	2.3
32	96.1	33.9	62.2	743	46.21	52.40	2.5	50.85	1.9

GRADE 1-3

18	82.4	50.0	32.4	210	6.80	10.21	8.5	9.40	6.7
				266	10.21	14.27	6.0	13.20	5.1

Table 4.--(Continued)

Diam. breast high (inches) (1)	Per M board feet			Lumber per tree (5)	Conversion surplus per tree by vigor class						
	Gross lumber value (2)	Direct costs (3)	Conversion surplus (4)		Now (6)	High vigor		Medium vigor		Low vigor	
						After 5 yrs. (7)	Annual increase (8)	After 5 yrs. (9)	Annual increase (10)	After 5 yrs. (11)	Annual increase (12)
	- - - - Index - - - - -			Bd. ft.	- - Index - -		Percent	Index	Percent	Index	Percent

GRADE 2-3

16	72.3	56.7	15.6	162	2.53	4.87	14.0	4.30	11.1	3.75	8.2
18	73.2	50.0	23.2	210	4.87	7.71	9.6	7.00	7.5	6.30	5.3
20	74.2	45.2	29.0	266	7.71	11.09	7.5	10.10	5.5	9.40	4.0
22	75.0	41.5	33.5	331	11.09	14.73	5.8	13.80	4.5	12.85	3.0
24	75.8	38.7	37.1	397	14.73	18.95	5.2	17.90	4.0	16.80	2.7
26	76.4	36.5	39.9	475	18.95	23.16	4.1	22.00	3.0	21.00	2.1
28	76.8	35.0	41.8	554	23.16	27.91	3.8	26.75	2.9	25.55	2.0
30	77.2	34.0	43.2	646	27.91	32.32	3.0	31.20	2.2	30.10	1.5
32	77.4	33.9	43.5	743	32.32	36.70	2.6	35.65	2.0	34.60	1.4

GRADE 3-3

16	64.7	56.7	8.0	162	1.30	3.21	19.8	2.75	16.2	2.20	11.0
18	65.3	50.0	15.3	210	3.21	5.48	11.3	4.85	8.6	4.30	6.0
20	65.8	45.2	20.6	266	5.48	8.18	8.3	7.45	6.3	6.80	4.4
22	66.2	41.5	24.7	331	8.18	11.04	6.2	10.27	4.6	9.60	3.3
24	66.5	38.7	27.8	397	11.04	14.34	5.4	13.45	4.0	12.65	2.8
26	66.7	36.5	30.2	475	14.34	17.62	4.2	16.80	3.2	16.00	2.2
28	66.8	35.0	31.8	554	17.62	21.25	3.8	20.35	2.9	19.50	2.0
30	66.9	34.0	32.9	646	21.25	24.59	3.0	23.75	2.2	22.90	1.5
32	67.0	33.9	33.1	743	24.59	27.80	2.5	27.00	1.9	26.20	1.3

3-LOG-TREES, GRADE 1-1-2

22	96.8	41.0	55.8	475	26.50	34.45	5.4	32.40	4.1	30.40	2.8
24	97.7	38.3	59.4	580	34.35	42.99	4.5	40.80	3.4	38.75	2.4
26	98.5	36.2	62.3	690	42.99	52.42	4.0	50.00	3.1	47.75	2.1
28	99.2	34.8	64.4	814	52.42	62.75	3.7	60.05	2.8	57.50	1.9
30	99.9	33.5	66.4	945	62.75	73.05	3.1	70.35	2.3	67.80	1.6
32	100.3	33.4	66.9	1,092	73.05	81.82	2.3	79.70	1.8	77.43	1.2

GRADE 1-1-3

22	90.4	41.0	49.4	475	23.46	30.74	5.6	28.90	4.3	27.05	2.9
24	91.3	38.3	53.0	580	30.74	38.57	4.6	36.50	3.5	34.60	2.4
26	92.1	36.2	55.9	690	38.57	47.29	4.2	45.00	3.1	42.85	2.1
28	92.9	34.8	58.1	814	47.29	56.70	3.7	54.20	2.8	51.85	1.9
30	93.5	33.5	60.0	945	56.70	66.18	3.1	63.75	2.4	61.30	1.6
32	94.0	33.4	60.6	1,092	66.18	74.60	2.4	72.60	1.9	70.55	1.3

GRADE 1-2-3

18	83.5	49.0	34.5	302	10.42	15.44	8.2	14.20	6.4	13.00	4.5
20	84.6	44.4	40.2	384	15.44	21.23	6.6	19.80	5.1	18.30	3.5
22	85.7	41.0	44.7	475	21.23	28.07	5.7	26.35	4.4	24.60	3.0
24	86.7	38.3	48.4	580	28.07	35.47	4.8	33.60	3.7	31.75	2.5
26	87.6	36.2	51.4	690	35.47	43.63	4.2	41.60	3.2	39.60	2.2
28	88.4	34.8	53.6	814	43.63	52.54	3.8	50.20	2.8	48.00	1.9
30	89.1	33.5	55.6	945	52.54	61.37	3.1	59.05	2.4	56.85	1.6
32	89.6	33.4	56.2	1,092	61.37	69.00	2.4	67.15	1.8	65.25	1.2

tree (column 6) was determined, and, following this (columns 7
11), the prospective conversion surplus per tree after ·5 years
vigor class (growth rate), assuming no change in log height or
The ratio of value after 5 years to value now was computed, af
compound-interest tables were used to find the annual percenta
crease (columns 8, 10, and 12).

As an example of the use of tables 2, 3, and 4, take th
of a forest manager who is deciding whether to mark a 28-inch
poplar, grade 1-2, high vigor--or whether to let it grow for a
5 years. Large limbs above the second log permanently limit t
merchantable height, nor is the grade of either log likely to
soon. The rate of return that the manager follows as his guid
4 percent. Table 4 indicates that the yellow-poplar will earn
than this rate--only 3.6 percent--in the coming 5 years. It i
fore financially mature and should be marked for cutting.

Changes in height and grade.--Both log height and grade
change as a tree grows in diameter. Limbs die and fall off, a
wood covers the knots. Butt logs of small trees are subject t
cal improvement in grade once the limbs are dropped because of
increasing thickness of clear wood deposited over the heart ce
The same thing happens, though to a lesser degree, to upper lo
Furthermore, for reasons explained by the log grades already r
to, logs may improve in grade merely with increase in diameter
the upper bole, additional logs are created as girth is added
minimum size specifications and as the bole clears enough to m
quality specifications. Thus small, short hardwoods of low gr
often capable of developing into 2- and 3-log, high-grade tree
many exceptions are found, especially in upper logs, most of t
tential rapid improvement in grade has been realized by the ti
hardwood tree reaches 24 or 26 inches d.b.h. Large hardwoods
have their log height firmly established by the position and f
the crown. Imminent and potential changes in log height and g
with experience, be recognized by the forester or woodsman, an'
recognize them is essential in determining financial maturity.

Tables 2, 3, and 4 may be used for calculating (though
reading directly) the rate of value increase for trees expecte
grow in log height or to improve in grade. Suppose that our 2
grade 1-2 yellow-poplar is expected to add a third log in the
5 years, so as to become a 30-inch, grade 1-2-3 tree. Table 4
that the value now is index 33.30, and that the value in 5 yea'
(reading in the same column, opposite the expected grade and d
will be index 52.54. A calculation of the ratio of these inde
reference to compound interest tables discloses that the rate
increase is 9.5 percent. If a 4-percent return is satisfactor
tree is not yet mature.

in the same way. If the grade 1-2 yellow-poplar grows to grade 1-1-3 in 5 years, the rate of value increase will be even higher, 11.1 percent.

Table 5 (column 2), shows the probable course of grade development for a high-vigor and potentially high-grade red oak that grows from 18 to 28 inches d.b.h. As the tree goes from 18 to 22 inches, both logs will improve in grade. At 24 inches, the tree develops a third log and continues to improve in grade as it grows to 28 inches. Column 3 shows the rate of conversion surplus increase for a tree with this grade and height development. Column 4 shows the rates of value increase if there are no height and grade changes. The contrast between columns 3 and 4 indicates that any red oak tree near the borderline of financial maturity should be kept for additional growth if it promises to extend its log height or improve its grade; the same is true for white oaks and yellow-poplar.

Table 5.--Rate of value increase of high-grade, high-vigor red oaks, with and without log height and grade changes

Tree d.b.h. (inches)	Tree grade	Annual conversion-surplus increase	
		With height and grade changes	Without height and grade changes
(1)	(2)	(3)	(4)
		- - - - - Percent - - - - -	
18	2-3	54.7	33.6
20	1-3	14.5	9.3
22	1-2	8.1	5.4
24	1-2-3	6.8	5.2
26	1-1-3	5.0	4.1
28	1-1-2	...	3.7

At this point the reader may well ask, what about the competition between the tree in question and other trees? And how does the system provide for secondary products like cross ties and mine props? Before these two questions can be answered, it will be necessary to tabulate the rate of tree value increase in terms of ties and props.

Cross Ties

Currently, well over 90 percent of the standard grade-5 (measuring 7 inches by 9 inches by 8½ feet) hardwood cross ties produced in the Birmingham district are sawn; of these, about 80 percent are cut from red and white oak species. A typical mill for hardwood cross-tie production will, in a representative day's run, produce 150 to 250 grade-5

ties. In addition, it will cut
some 600 to 1,000 board feet of
side lumber per 100 ties, the
amount varying with the size and
quality of the timber. Since
side lumber is essential to a
profitable tie operation, tree
volume will be expressed in
board feet of ties and side lumber.

In a forest managed pri-
marily for lumber logs, tie cutting will be restricted to those
trees whose removal will expedite the production of lumber
logs. A tie, considered by itself, can be cut from any sound
timber. But since tie logs, on
the average, must afford an adequate cut of side lumber, the

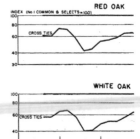

Figure 6.—Cross-tie
as a percent of
No. 1 Common and
per M.

following estimates of ties and side lumber are based on t
trees that the forest manager can be expected to channel i
duction.

In order to compare a tree's worth for standard fa
with its worth for cross ties, a common unit of value must
terms of the index values developed for lumber, it appears
the years ties (per M board feet of lumber equivalent) are
50 percent of the value of No. 1 Common and Selects (fig.
their index value is 50. Side lumber usually sells for tl
as 3A Common. In terms of index values, this is 52 for r
lumber, and 49 for that of white oak. For convenience, i
is used for both side lumber and cross ties. Table 6 sho
of ties and the amount of side lumber per tree, and the c
surpluses in terms of the lumber index. In addition, the
creases for 5-year periods and the annual rates of conver
increase are given for the three vigor classes.

Mine Props

Mine props are procured on a custom basis. That
ordered by specific sizes and quantities, to be delivered
mine head. Since the requirement of greenness precludes
stocking large quantities of the sizes ordinarily demande
production is a small-scale operation.

Table 6.--Cross-tie trees: rate of value increase, in terms of lumber index, by vigor class

Diam. breast high (inches)	Grade 5 ties	Side lumber	Ties and lumber	Gross pro-duct value	Direct cost	Now	Conversion surplus per tree by vigor class					
							High vigor		Medium vigor		Low vigor	
							After 5 yrs.	Annual increase	After 5 yrs.	Annual increase	After 5 yrs.	Annual increase
	No.	B.f. per tree		- Index per tree -			Index	Percent	Index	Percent	Index	Percent
						Red Oaks						
12	1	3	48	2.40	2.29	0.11	0.83	49.8	0.63	41.5	0.43	25.5
13	1 1/2	6	73	3.65	3.10	.55	1.18	16.4	1.00	12.7	.83	8.6
14	2	11	101	5.05	4.15	.90	1.50	10.7	1.35	8.4	1.18	5.6
15	2 1/2	27	139	6.95	5.70	1.25	1.71	6.4	1.62	5.3	1.50	3.7
16	2 3/4	33	157	7.85	6.28	1.57	1.87	3.6	1.80	2.8	1.71	1.7
17	3	50	185	9.25	7.50	1.75	1.99	2.6	1.94	2.1	1.87	1.3
18	3 1/4	63	209	10.45	8.55	1.90	2.07	1.7	2.04	1.4	1.99	.9
						White Oaks						
12	1	3	48	2.40	2.29	.11	.63	41.5	.47	33.7	.37	27.3
13	1 1/2	6	73	3.65	3.10	.55	1.00	12.7	.87	9.6	.78	7.2
14	2	11	101	5.05	4.15	.90	1.35	8.4	1.22	6.3	1.13	4.7
15	2 1/2	27	139	6.95	5.70	1.25	1.62	5.3	1.53	4.1	1.46	3.1
16	2 3/4	33	157	7.85	6.28	1.57	1.80	2.8	1.73	2.0	1.69	1.5
17	3	50	185	9.25	7.50	1.75	1.94	2.1	1.89	1.5	1.85	1.1
18	3 1/4	63	209	10.45	8.55	1.90	2.04	1.4	2.00	1.0	1.98	.8

Table 7.--Prop trees: rate of value increase, in terms of lumber index, by vigor class

Diam. breast high (inches)	5-foot props	Gross product value	Direct cost	Now	Conversion surplus per tree by vigor class					
					High vigor		Medium vigor		Low vigor	
					After 5 yrs.	Annual increase	After 5 yrs.	Annual increase	After 5 yrs.	Annual increase
	No.		- - Index per tree - -		Index	Percent	Index	Percent	Index	Percent
				Red Oaks						
6	2.0	0.33	0.28	0.05	0.21	33.1	0.17	27.5	0.13	20.8
7	3.0	.50	.34	.16	.28	11.9	.25	9.5	.22	6.6
8	3.8	.63	.39	.24	.32	5.9	.31	5.2	.28	3.1
9	4.4	.73	.44	.29	.35	3.8	.34	3.2	.32	2.0
10	5.4	.89	.56	.33	.38	2.9	.37	2.3	.35	1.2
11	6.2	1.02	.66	.36	.39	1.6	.38	1.1	.37	.6
				White Oaks						
6	2.0	.32	.27	.05	.16	26.1	.13	21.0	.10	14.9
7	3.0	.47	.33	.14	.24	11.3	.21	8.5	.19	6.2
8	3.8	.60	.37	.23	.29	4.7	.27	3.3	.26	2.5
9	4.4	.70	.42	.28	.33	3.3	.32	2.7	.31	2.1
10	5.4	.85	.54	.31	.35	2.4	.34	1.9	.33	1.3
11	6.2	.98	.63	.35	.37	1.1	.36	.6	.35	.0

Producers usually operate in trees ranging from 5 to 11 inches d.b.h. The most common size of prop is 5 feet long. Tree quality is relatively unimportant save for the straightness and soundness of the lengths cut.10/

While hardwood props have come into use in the Birmingham terri tory only in recent years, the current relation of prop trees to lumbe log and cross-tie trees may be assumed to hold for the near future. Sales values and direct costs of producing props have been worked out in dollars per tree and then converted to lumber index values per tree Thus comparisons can be made, tree by tree, between the alternatives of standard factory lumber, cross ties, and mine props. Table 7 shows the number of 5-foot props, conversion surpluses per tree, and the annual rates of conversion surplus increase for the three vigor classe

Effect of Other Trees

In judging financial maturity one cannot, of course, consider a tree solely on its own merits. Every forest tree is in more or less competition with other trees, and this competition affects the desira- bility of the tree as an investment. Some trees which, judged alone, are apparently financially mature, turn out not to be mature when one or more other trees are brought into consideration. Likewise trees may prove to be mature which considered by themselves were not so. The principle covering the effect of other trees upon financial ma- turity is this: the aim of the manager should be to maintain on each acre that volume of timber which, within the requirements of the sil- vicultural system and the program of regulation, has the greatest possible conversion surplus in trees not yet financially mature. For reasons made clear in the reference cited in footnote 1, following thi principle will result in securing, from each acre, the highest income that the desired rate of return will afford.

Therefore, where two competing trees both appear to be mature but where removal of one would raise the vigor of the other sufficient ly to make it no longer mature, a general rule is to retain that tree for growth which has the higher value now or prospectively--the greater size or higher grade, or the better chance of increasing its size or improving its grade. The other tree should be cut.

_____

10/. Osborn, R. M. Costs of producing mine props. South. For Expt. Sta., Occas. Paper 124, 22 pp., illus. 1951.

Griswold, N. B., and McKnight, J. S. Wood use by Alabama mine: South. Forest Expt. Sta., Occas. Paper 109, 12 pp., illus. 1947.

Take the case of two competing yellow-poplars. Both are of medium vigor, but if either is cut the vigor of the other will become high. Following are the pertinent data on these yellow-poplars, read from table 4:

	Tree A	Tree B
Diameter breast high (inches)	24	26
Grade	1-2	2-3
Conversion surplus	21.80	18.95
Medium vigor:		
Conversion surplus after 5 years	26.10	22.00
Rate of increase (percent)	3.7	3.0
High vigor:		
Conversion surplus after 5 years	27.50	23.16
Rate of increase (percent)	4.8	4.1

With a 4-percent desired rate of return, one of these trees is mature and should be cut, while the other should be left to grow one more cutting cycle. Since tree A, the 24-inch grade 1-2 yellow-poplar, has the higher value both now and prospectively, it should be left to grow.

A comparison of these alternative trees will show the reasoning behind the rule. For convenience, assume that conversion surplus equals dollars. Then the value added to the business over the next 5 years is:

	Cutting tree A	Cutting tree B
Value of tree cut, invested at 4 percent compound, earns	4.73	4.11
Growth in value of tree left:		
Tree A	...	5.70
Tree B	4.21	...
Total value added to the business	8.94	9.81

Since cutting tree B will increase the worth of the business by 87 cents more than the alternative of cutting A, tree A should be left to grow. The rule is designed to cover the vast majority of the numerous alternatives facing the forest manager. There will be exceptions, but the virtue of the rule is that it obviates making separate calculations for every alternative that arises.

The Maturity Dilemma

A glance at tables 2, 3, 6, and 7 will show that an oa
ipparently mature at different stages in its development, dep
vhether it is to be used for lumber, cross ties, or mine prop
:an this be? Should not there be only one point of maturity
given tree and a given rate of return? The answer is "yes."
orinciple to follow is that of electing maturity for the prod
vill afford the greatest amount of conversion surplus per uni
:ime and area for the desired rate of return.

This requires a judgment as to the number of prop tree
cross-tie trees that can be grown in the time and on the area
to bring a tree to maturity for lumber logs. Following is a
example of the kind of calculation necessary to resolve the d
of logs versus ties versus props.

Tables 2, 6, and 7 show that, where a 3-percent rate o
is desired, a medium-vigor red oak of potential grade 2-3 wil
for mine props at 9 inches d.b.h., at which time its index va
be 0.29. It will mature for cross ties at 15 inches d.b.h. (.
value 1.25) and for saw timber at 30 inches (value 20.61). I
time needed to grow the tree (at medium vigor) to lumber-log
about three crops of mine props can be raised, with perhaps 3
per crop. Similarly, about two crops of cross-tie trees coul
matured, with 2 trees in each crop. Therefore:

PROP TREE ALTERNATIVE
    Returns from first crop of props, invested
    at 3 percent compound interest   14

    Returns from second crop of prop trees, at
    3 percent compound interest   3

    Final crop of mine props . . . .   __
        Total conversion surplus   19

TIE TREE ALTERNATIVE
    Returns from first crop of ties, invested
    at 3 percent compound   14

    Final crop of cross ties . . . .   __
        Total conversion surplus   17

SAWLOG TREE ALTERNATIVE   20

In this example, the sawlog tree is clearly the best alternative, even without consideration of prop- or cross-tie trees that could have been raised in conjunction with it. If either the prospective grade or the growth rate (vigor class) of the sawlog tree had been lower, however, mine props or ties would have been the better alternative. A higher rate of interest than 3 percent would also have called for higher grade or faster growth if the tree was to be left for lumber-log maturity.

This example assumed that 3 prop trees or 2 tie trees could be grown in the space that the saw-timber tree would demand. Many different assumptions can and will be made, but the authors feel that for most situations any tree that is incapable of growing at medium or better vigor and, in addition, of producing a grade 1 or 2 butt log will mature for ties or props.

Similarly, where the choice falls between maturity for ties or props (because the trees have no prospect of achieving lumber-log quality), a reasonable generalization is to consider medium- or high-vigor red oaks and high-vigor white oaks mature for cross ties. The other vigor classes will mature for mine props.

The percentage rates of value increase and index values per tree are not directly comparable for trees of different species groups. Each species group has its own index base. The index values would be comparable only if No. 1 Common and Selects red oak sold for the same price as that of white oak, and of No. 1 Common yellow-poplar. Although differences in the base value do exist, they are not great and tend to maintain a fairly constant relationship to one another. White oak, red oak, and yellow-poplar base grades tend to average within 10 percent of each other. The highest price is usually paid for yellow-poplar, and the next highest for white oak; red oak sells for the lowest prices. Consequently, no great harm will result if direct comparisons are made, provided that where the decision is close a yellow-poplar is favored over either of the oak groups, and a white oak is preferred to a red.

Other considerations relating to the stand may affect the maturity decision. A tree may be otherwise financially mature and still not ready for harvest if it is needed as a seed source. Again, the need for holding the cut within a cutting budget, or of building the growing stock towards optimum diameter-class distributions, may well result in leaving for another cutting cycle some mature trees. In some areas there may be insufficient timber volume to support a commercial cutting operation despite a sprinkling of mature trees.

On the other hand, the owner may be forced to cut some trees in order to obtain a regular income or a regular supply of logs for a sawmill; in this case the desired rate of return has risen, at least temporarily.

Finally, there are two classes of management alternat:
have not been considered in this paper. The first of these :
possibility of managing for pine or for other hardwoods--chi
hickories, blackgum, and sweetgum. These hardwoods are prob
important alternatives to the species discussed in this pape
is. In most upland forests, pine forms such a valuable comp
the forest manager will wish to determine the grades and vig
of hardwood trees that should be cut to favor the pine. Som
thumb will serve this purpose. For example, the manager migh
that any pine showing promise of reaching sawlog size is a b
ternative than any hardwood incapable of growing at medium o
vigor and of producing at least a grade 2 lumber log.

The other important management alternative is the pos
replacing an existing tree with one or more superior pines o
grown from seed. This is a problem that merits the attentio
forest manager who is looking ahead more than the next cuttir
or two. The calculations are made in the same manner as the
between alternative existing trees. That is, against the pot
turns of the existing tree must be balanced the returns from
pected tree or trees that could replace it.

Simplified Marking Guides

For practical application in the woods, the forest mar
wish to prepare a set of readily usable marking instructions
preceding tables. To facilitate this task, the pertinent dat
lumber log trees are condensed in table 8. The table lists,
vigor class and two rates of return, the largest sizes of tr
should be left to grow for one more five-year period. The d
spread allows for differences in log height and grade.

In applying the simplified guides, the influence of o
should be recognized. When trees of low or medium vigor are
leased by cutting, they should be promoted one vigor class b
financial maturity is judged. Trees which show imminent pro
improvement in vigor class, grade, or stem length should be
to grow at least one more cutting cycle. Where one of two c
trees is to be cut, it is best to take the one of lesser siz
or vigor, or the one with poorer prospects for improvement.

Trees that mature for cross ties and mine props will
tinguishable from prospective lumber-log trees because they
fering with better trees, are unlikely to meet lumber-log st
now or in the future, or are otherwise of good quality but s
by fire, insects, or disease that they are poor risks. A wo
for such trees is to utilize those between 14 and 16 inches
for cross ties, and to regard those between 8 and 12 inches
financially mature for mine props.

Table 8.--<u>Simplified marking guides for lumber log trees</u>

Vigor class	Alternative rate of return	
	3 percent	4 percent
	- - - - D.b.h., inches - - - -	

RED OAKS

High	30-33	26-29
Medium	26-29	22-25
Low	21-24	18-20

WHITE OAKS

High	27-29	23-25
Medium	23-25	20-21
Low	21-22	18-19

YELLOW-POPLAR

High	30-31	26-27
Medium	26-27	22-24
Low	21-22	19-21

In trees below cross-tie size, where the choice is between maturity for ties or props, red oaks of medium and high vigor, and white oaks of high vigor, will mature for cross ties. The other vigor classes will mature for mine props.[11]

If these considerations are too complex to be used in the woods, the number of alternatives facing the timber marker may be still further reduced. For example, the spreads of diameters in table 8 could be replaced by single diameters for each species and vigor class. Further, by assuming that better quality and higher vigor trees can be grown from seed, all low-vigor trees and all grade-3 trees showing no promise of improvement can be considered mature for mine props, cross ties, or lumber logs, depending on tree size.

---

[11]. Some white oaks that mature for props or cross ties on the basis of this analysis may be worth keeping for tight cooperage.

SHORTCUTS FOR CRUISERS AND SCALERS

L. R. Grosenbaugh

SOUTHERN FOREST EXPERIMENT STATION
Harold L. Mitchell, Director
New Orleans, La
FOREST SERVICE, U.S DEPT OF AGRICULTURE

SHORTCUTS FOR CRUISERS AND SCALERS

L. R. Grosenbaugh
Southern Forest Experiment Station

This paper explains how to:

Cruise timber without plot boundaries or tree
diameters (page 2)

Simplify much mental arithmetic (page 9)

Compute individual tree volumes rapidly without
referring to volume tables (page 10)

Compute log volumes by any of four popular log
rules (page 11)

Compute cull percents for logs independently of
log rules (page 14)

Compute net log scale by machine accumulation
of recorded net lengths and diameters without
looking up individual log values (page 16)

Design sampling plans for cruising, scaling, or
tree measurement (page 20)

These methods also provide a basis for converting aggregate log
or tree volumes from one log rule to another, using only aggregate
length or aggregate product of length times diameter.

The last technique--random sampling design--has been thoroughly
covered by other authors, but it has been put into capsule form here.

PLOTLESS TIMBER CRUISING

The following pages describe a new and very convenient w
cruising timber. The method makes it unnecessary to measure or
tree diameter, plot radius, or strip width; and it greatly simp
office calculations. The cruiser merely stands at the center o
plot, looks through a simple hand-held angle-gauge, and counts
trees that appear larger than the angle laid out by the gauge.

Although based on a centuries-old concept, the use of an
gauge to accept or reject sample trees is new, the inspiration
European.[1] The first American exposition of the theory, publi
recently, adapted the method to American units of measure and d
veloped several new applications.[2] Readers should consult it
desire a more comprehensive account than can be given here.

For simplicity this paper assumes that the areas cruised
not slope more than 10 percent, but instrumental or computation
corrections can be made to adapt the method to steeper country.

Instrument

Before starting any field work, the cruiser will need t
an angle-gauge for optically defining an angle of 104.18 minute
its vertex at his eye.

His simplest course is to get a 33-inch stick and mount
peephole at one end with a line of sight bisecting a 1-inch met
crosspiece at the other end. The crosspiece, when viewed thro
the peephole, should exactly cover a 3-foot horizontal interce
a distance of 99 feet from the eye. The distance between cros
and peephole should be adjusted till this occurs. Hypsometer
Biltmore graduations may be put on the stick if desired.

A 4- to 7-power monocular with two vertical reticule li
30.3 mils apart constitutes a better instrument. Like the cro
device, it can be checked against a 3-foot intercept at 99 fee
cept that the vertex of the 104.18-minute angle will be 1 foca
in front of the objective lens instead of at the eye. Instrum
adjusting for slope are feasible, but will not be discussed he

---

[1] Bitterlich, W. Die Winkelzählprobe. Allgemeine Forst-
Holzwirtschaftliche Zeitung 59 (1/2): 4-5. 1948.
[2] Grosenbaugh, L. R. Plotless timber estimates—new, fas
easy. Journal of Forestry 50:32-37. 1952.

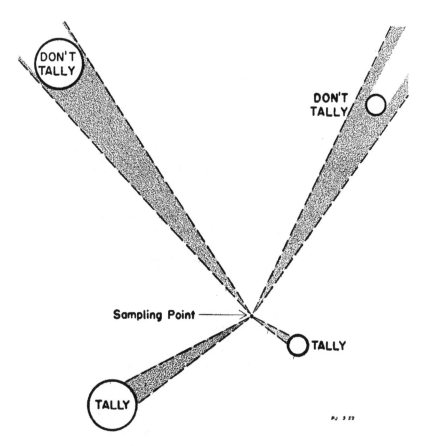

PJ 3 52

Figure 1.—Plotless timber cruising. Shaded areas represent 104.18-
minute angles optically established by the instrument described on
the opposite page. Circles represent cross-sections (at breast
height) of trees viewed from the sampling point. Cruiser merely
stands at the sampling point (analagous to a plot center), counts
every tree whose d.b.h. appears larger than the angle, and disre-
gards every tree whose d.b.h. appears smaller. All trees visible
from the sampling point must be counted or rejected. The count
of trees, multiplied by 10, gives an estimate of basal area per
acre. In the diagram, only two trees are counted, so the basal
area estimate is 20 square feet per acre. Reliable estimates
require more than one sample, of course.

Cubic feet can be divided by a locally appropriate conver
divisor to give rough stacked cords per acre (26.7 cords if loca
visor were 90 cubic feet per cord).

Miscellaneous Estimates

The number of trees tallied by the angle-gauge in any gi\
class tends to be directly proportional to the total basal area
acre in that class. This fact allows estimating basal area dist
tion by species, quality, vigor, diameter, length, or any other
sired criterion, by merely classifying tallied trees.

Basal area in each merchantable length class may be easi]
converted to various units of volume, because $\frac{\text{volume}}{\text{basal area}}$ ratios
merchantable length classes are relatively stable, regardless of
eter. For simplicity, it will be assumed that 40 trees of the s
species have been tallied at 5 sampling points and also classifi
to their merchantable length in terms of 16-foot sawlogs, thus:

        8 trees with zero logs
       10 trees with one log
       11 trees with two logs
        7 trees with three logs
        4 trees with four logs

Total: 40 trees tallied at 5 sampling points

It can be quickly estimated by earlier formulae that sta
basal area per acre is $(10)\left(\frac{40}{5}\right)$ = 80 square feet, of which 16 s
feet is in zero-log trees, 20 is in one-log trees, 22 in two-log
trees, 14 in three-log trees, and 8 in four-log trees.

Volume per acre in terms of International rule (1/4-inch kerf), Scribner rule, Doyle rule, or peeled cubic feet (without bark) could be estimated by multiplying tally in each merchantable length class by appropriate factor tabled below, summing the products, and proceeding as indicated.

Merch. length class	Volume factors				Tally	Example: Product of peeled cubic factor x tally
	Int. ¼-inch	Scribner	Doyle	Peeled cubic		
Zero-log	0	0	0	0	8	0
One-log	7	6	4	12	10	120
Two-log	13	11	8	20	11	220
Three-log	18	16	12	27	7	189
Four-log	23	20	15	34	4	136
Five-log	28	25	21	40	0	0

Total: 40 trees;   665 = sum of products

Estimated peeled cubic volume per acre

$$= (10) \left[\frac{\text{Sum of products}}{\text{Number of sampling points}} \right]$$

$$= (10) \left(\frac{665}{5} \right)$$

$$= 1,330 \text{ cubic feet}$$

Estimated board-foot volume per acre

$$= (100) \left[\frac{\text{Sum of appropriate products}}{\text{Number of sampling points}} \right]$$

$$= 8,620 \text{ bd. ft. Int. }\tfrac{1}{4}\text{-inch}$$

$$= 7,460 \text{ bd. ft. Scribner}$$

$$= 5,440 \text{ bd. ft. Doyle}$$

The ready-made volume factors given above will satisfy the accuracy needs of many cruisers, but it is possible to attain any desired precision by merely sampling local $\frac{\text{volume}}{\text{basal area}}$ ratios for each merchantable length class recognized (1/10 of that ratio has been used in case of board-foot factors).

Cubic feet can be divided by a locally appropriate conversion divisor to give rough stacked cords per acre (26.7 cords if local divisor were 90 cubic feet per cord).

Miscellaneous Estimates

The number of trees tallied by the angle-gauge in any given class tends to be directly proportional to the total basal area per acre in that class. This fact allows estimating basal area distribution by species, quality, vigor, diameter, length, or any other desired criterion, by merely classifying tallied trees.

Basal area in each merchantable length class may be easily converted to various units of volume, because $\frac{volume}{basal\ area}$ ratios within merchantable length classes are relatively stable, regardless of diameter. For simplicity, it will be assumed that 40 trees of the same species have been tallied at 5 sampling points and also classified as to their merchantable length in terms of 16-foot sawlogs, thus:

        8 trees with zero logs
       10 trees with one log
       11 trees with two logs
        7 trees with three logs
        4 trees with four logs

    Total: 40 trees tallied at 5 sampling points

It can be quickly estimated by earlier formulae that stand basal area per acre is $(10)\left(\frac{40}{5}\right) = 80$ square feet, of which 16 square feet is in zero-log trees, 20 is in one-log trees, 22 in two-log trees, 14 in three-log trees, and 8 in four-log trees.

Volume per acre in terms of International rule (1/4-inch kerf), Scribner rule, Doyle rule, or peeled cubic feet (without bark) could be estimated by multiplying tally in each merchantable length class by appropriate factor tabled below, summing the products, and proceeding as indicated.

Merch. length class	Volume factors				Tally	Example: Product of peeled cubic factor x tally
	Int. ¼-inch	Scribner	Doyle	Peeled cubic		
Zero-log	0	0	0	0	8	0
One-log	7	6	4	12	10	120
Two-log	13	11	8	20	11	220
Three-log	18	16	12	27	7	189
Four-log	23	20	15	34	4	136
Five-log	28	25	21	40	0	0

Total: 40 trees; 665 = sum of products

Estimated peeled cubic volume per acre $= (10) \left[\dfrac{\text{Sum of products}}{\text{Number of sampling points}} \right]$

$= (10) \left(\dfrac{665}{5} \right)$

$= 1,330$ cubic feet

Estimated board-foot volume per acre $= (100) \left[\dfrac{\text{Sum of appropriate products}}{\text{Number of sampling points}} \right]$

$= 8,620$ bd. ft. Int. ¼-inch

$= 7,460$ bd. ft. Scribner

$= 5,440$ bd. ft. Doyle

The ready-made volume factors given above will satisfy the accuracy needs of many cruisers, but it is possible to attain any desired precision by merely sampling local $\dfrac{\text{volume}}{\text{basal area}}$ ratios for each merchantable length class recognized (1/10 of that ratio has been used in case of board-foot factors).

Variable Plot Radius or Point-Sampling Concept

The 104.18-minute angle-gauge tallies all trees closer to the
sampling point than 33 times tree d.b.h. Whatever the size of the
tallied tree, the instrument guarantees that it is within a plot
radius determined by that particular tree's d.b.h.

Thus, a tree with a d.b.h. of 1/2 foot (and a basal area of
1/5 square foot) would be tallied inside a plot with a radius of
16-1/2 feet (and an area of 1/50 acre). The basal area of each 6-inch
tree must, then, be multiplied by 50 to place it on a per-acre basis.
This means that such a tree would contribute
$(50)\left(\frac{1}{5}\right)$ = 10 square feet to the estimate of basal area per acre. However, a tree with a d.b.h. of 1 foot (and a basal area of 4/5 square
foot) would be tallied inside a plot with a radius of 33 feet (and an
area of 4/50 acre). This means that each 12-inch tree would contribute
$\left(\frac{50}{4}\right)\left(\frac{4}{5}\right)$ = 10 square feet to the estimate of basal area per acre. Tallied trees of different sizes would have different basal areas, differ-
ent plot areas, and hence different blow-up factors, but the product
of each tallied tree basal area times its appropriate blow-up factor
would always be 10 square feet—the constant contribution of any tallied tree to the estimate of basal area per acre. This is why diameter
of a tallied tree and its distance from the sampling point are imma-
terial.

Some people can conceive the idea more clearly by visualizing
a sheet of paper with a known number of evenly spaced sample points
plotted on it. On the sheet are placed 2 transparent disks of different size, each exhibiting on its face a small concentric circle
with diameter 1/66 that of the disk, and with area $\left(\frac{1}{66}\right)^2$ or $\left(\frac{1}{4356}\right)$
that of the disk. Next, the number of disks sampled at each plotted
point is counted (it will be 0, 1, or 2 in this simple example) and
the counts at all points are summed. Those familiar with dot-counting
know that $\dfrac{\text{Sum of disk counts}}{\text{Number of sample points}}$ estimates $\dfrac{\text{Area of disks}}{\text{Area of sheet}}$, and
$\left(\dfrac{1}{4356}\right)\left(\dfrac{\text{Sum of disk counts}}{\text{Number of sample points}}\right)$ estimates $\dfrac{\text{Area of small circles}}{\text{Area of sheet}}$.
Multiplying this last by 43,560 square feet (one acre) estimates the
square feet of small circles per acre of sheet or
$(10)\left(\dfrac{\text{Sum of disk counts}}{\text{Number of sample points}}\right)$, which explains the reasoning behind
computations on page 4.

In cruising, the sheet of paper becomes the land to be cruised,
small circles become tree cross-sections, and disks become surrounding zones within which a sampling point must fall in order for the
104.18-minute angle-gauge to accept that particular tree for tally.

MENTAL ARITHMETIC

By memorizing the squares of numbers 1 through 25, a scaler or cruiser can greatly facilitate the ordinary volume calculations he performs mentally. The squares of numbers from 26 through 125 can be easily obtained from the first twenty-five, and multiplication can frequently be simplified.

To square numbers 26 through 75: subtract 25 from the number, suffix 2 zeros, and add the square of (50 minus the number).

Example: $(36)^2 = (36-25)00 + (50-36)^2 = 1,100 + (14)^2 = 1,296$
$(63)^2 = (63-25)00 + (50-63)^2 = 3,800 + (13)^2 = 3,969$

To square numbers 76 through 125: subtract 100 from twice the number, suffix 2 zeros, and add the square of (100 minus the number).

Example: $(83)^2 = (166-100)00 + (100-83)^2 = 6,600 + 289 = 6,889$
$(111)^2 = (222-100)00 + (100-111)^2 = 12,200 + 121 = 12,321$

To square numbers ending in $\frac{1}{2}$: square the whole portion, add the whole portion, and add $\frac{1}{4}$ (this last step can usually be ignored).

Example: $(25\frac{1}{2})^2 = 625 + 25 + \frac{1}{4} = 650\frac{1}{4}$

To multiply numbers: square the smaller, and add the smaller times the difference (useful where numbers differ by no more than 10).

Example: $(27)(36) = 729 + 243 = 972$

Alternatively, take $\frac{1}{4}$ the difference between the square of the sum and the square of the difference; or else take the difference between the square of the half-sum and the square of the half-difference (plus the smallest of the original numbers if fractions are dropped prior to squaring).

Example: $(27)(36) = \dfrac{(63)^2-(9)^2}{4} = \dfrac{3,969-81}{4} = \dfrac{3,888}{4} = 972$

$(27)(36) = (31)^2-(4)^2+ 27 = 961-16+27 = 972$

Squares to be memorized

N	N^2	N	N^2	N	N^2
1	1	11	121	21	441
2	4	12	144	22	484
3	9	13	169	23	529
4	16	14	196	24	576
5	25	15	225	25	625
6	36	16	256		
7	49	17	289		
8	64	18	324		
9	81	19	361		
10	100	20	400		

USEFUL NEW RULES FOR TREE VOLUME

The author has devised 6 simple rules for tree volume which can be easily memorized and mentally calculated for individual trees; 3 of them can be used to convert aggregate International $\frac{1}{4}$-inch board-foot volumes to other scales without reworking cruise data.

In all six rules:

    D stands for d.b.h. (outside bark) in inches.
    N stands for number of 16-foot logs to a merchantable
         sawlog top.
    M stands for number of 10-foot sticks to a merchantable
         pulpwood top.

Saw timber

Rule 1: Board feet (Int. $\frac{1}{4}$") $= \dfrac{(D-N)^2(N)}{2}$ where D lies between 6 an[c]

Rule 2: Cubic feet (peeled) $= \dfrac{\text{Int.}\frac{1}{4}\text{"bd.ft.}+2ND}{9}$ where D lies betwe[e]
    6 and 50

Rule 3: Board feet Scribner $= \text{Int.}\frac{1}{4}\text{"bd.ft.}-\dfrac{ND}{2}$ where D lies betwee[n]
    6 and 50

Rule 4: Board feet Doyle $= \text{Int.}\frac{1}{4}\text{"bd.ft.}-32N$ where D lies betw[een]
    11 and 32

Pulpwood

Rule 5: Cubic feet (including bark) $= \dfrac{(D+2)^2(M)}{40}$ where M lies betw[een]
    0 and 8

Rule 6: Number of trees per rough stacked cord $= \dfrac{3600}{(D+2)^2(M)}$ where
    M lies between 0 and 8

Experience will indicate when 5, 10, or 15 percent adjustments are needed to allow for peculiarities of local utilization or taper d[if] fering from Girard Form Class 80 in the case of sawtimber or Girard F[orm] Point (91-D) in the case of pulpwood. In the last case, local **experi**[ence] ence **may** indicate factors other than 40 or 3600 which will be **more** accurate.

If the sum of ND and the sum of N are accumulated on tally she[et] (along with International $\frac{1}{4}$-inch board-foot volumes according to loca[l] valid volume tables), rules 2, 3, and 4 allow easy conversion to othe[r] units, without need for developing different local volume tables or working up each tree volume anew.

POPULAR RULES FOR LOG VOLUME

Legend for all rules:

> d = average diameter of log inside bark at small end(inches)
> D = average diameter of log inside bark at large end(inches)
> L = length of log (feet)
> K = constant (2 for paraboloid, 3 for conoid, 4 for sub-
> neiloid)

General solid of revolution:

$$\text{Volume (cu. ft.)} = .005454154 \ L \left[Dd + \frac{(D-d)^2}{K} \right]$$

Conoid tapering 1 inch per 8 lineal feet:

$$\text{Volume(cu.ft.)} = .005454154 \ Ld^2 + .0006817692L^2d + .00002840705L^3$$

$$\text{Surface (sq. ft.)} = .2618029Ld + .01636268L^2$$

International log rule with taper allowance of $\frac{1}{2}$ inch per 4 lineal feet (master formulae):

With 1/8-inch kerf		With 1/4-inch kerf (=.904762 value for 1/8"kerf)	
+ .05500000	Ld^2	+ .04976191	Ld^2
+ .006875000	L^2d	+ .006220239	L^2d
- .2050000	Ld	- .1854762	Ld
+ .00028645833	L^3	+ .0002591767	L^3
- .01281250	L^2	- .01159226	L^2
+ .04666667	L	+ .04222222	L
Volume (bd. ft.) = sum of 6 terms		Volume (bd. ft.) = sum of 6 terms	

Coefficients for the International 1/8-inch rule are accurate to an infinite number of digits (if repeating decimals be extended), since they have been obtained by use of numerical progression summation formulae. Coefficients for the International $\frac{1}{4}$-inch rule and the conoidal rules have 7 significant digits.

Traditionally, International rule formulae have been applied only to lengths which are multiples of 4 feet, with linear interpolation for intervening values. Interpolated values are trivially higher than exact formulae values.

Final International rule values have usually been rounded to nearest 5 board feet.

Formulae for given diameters and varying lengths may be easily
derived from the master formulae and will contain only 3 terms. Formulae for given lengths and varying diameters may also be obtained
with only 3 terms, as illustrated below. When used for deriving a
formula for a given length, each coefficient below should be multiplied
by the desired length to eliminate L from all terms, but the special
form given below will be needed when accumulating length and diameter.

Coefficients for volume of logs of average
length by log rules allowing taper

Average log length in feet	Coefficients for board feet (Int. $\frac{1}{4}$-inch)			Coefficients for cubic feet (conoid, 1 inch per 8 feet)		
	Ld^2	Ld	L	Ld^2	Ld	L
4	+.0498	-.161	.000	+.00545	+.0027	+.0005
8	"	-.136	-.034	"	+.0055	+.0018
12	"	-.111	-.060	"	+.0082	+.0041
16	"	-.086	-.077	"	+.0109	+.0073
20	"	-.061	-.086	"	+.0136	+.0114
24	"	-.036	-.087	"	+.0164	+.0164
28	"	-.011	-.079	"	+.0191	+.0223
32	"	+.014	-.063	"	+.0218	+.0291
36	"	+.038	-.039	"	+.0245	+.0368
40	"	+.063	-.007	"	+.0273	+.0455
44	"	+.088	+.034	"	+.0300	+.0550
48	"	+.113	+.083	"	+.0327	+.0654

Coefficients for volume of logs of any
length by log rules ignoring taper

Average log length in feet	Coefficients for board feet Doyle			Coefficients for board feet Scribner (curved)		
	Ld^2	Ld	L	Ld^2	Ld	L
Any	+.0625	-.500	+1.000	+.0494	-.124	-.269

Volumes of 16-foot logs may be approximated adequately by 4 rules of thumb:

Int. $\frac{1}{4}$-inch bd. ft. per 16-foot log = $.8 (d - 1)^2$

Cu. ft. per 16-foot log $= \dfrac{(d + 1)^2}{A}$

where A is 11 if taper tends to be slightly more than
2 inches
where A is $11\frac{1}{2}$ if taper is 2 inches and linear
where A is 12 if taper tends to be slightly less than
2 inches

Scribner bd. ft. per 16-foot log = $.8 (d - 1)^2 - \dfrac{d}{2}$

Doyle bd. ft. per 16-foot log $= (d - 4)^2$

CULL PERCENT

Defect reduces the amount of lumber otherwise recoverable from a log. Most log rules are biased in their estimates of lumber yielded by a sound log, and the bias usually varies with log diameter. It would be logical to estimate the proportion of lumber lost through defect by a technique unaffected by the inconsistencies of badly biased log rules. Such a proportion could easily be translated into log-rule board feet by multiplying by the gross scale according to any desired rule, or it could be translated into equivalent cull length by multiplying by log length. This procedure is traditional where entire sections, wedge-like sectors, or sweepy lunes are lost, but interior defects have long been inconsistently boxed and computed in terms of mill tally. The author has devised the more universally consistent method described in (5) below. For convenience, formulae for cull proportion caused by four other common defects are also outlined. None of the five formulae depend on log scale. In all five, d is average diameter of log at small end in inches and L is length in feet.

(1) Proportion lost when defect affects entire section:

$$\left(\frac{\text{Length of defective section}}{L}\right)$$

(2) Proportion lost when defect affects wedge-shaped sector:

$$\left(\frac{\text{Length of defective section}}{L}\right)\left(\frac{\text{central angle of defect}}{360^\circ}\right)$$

(3) Proportion lost when log sweeps (or when its curved central axis departs more than 2 inches from an imaginary chord connecting the centers of its end-areas; ignore sweep less than 2 inches):

$$\left(\frac{\text{Maximum departure minus 2 inches}}{d}\right)$$

(4) Proportion lost when log crooks (or when a relatively short section deflects abruptly from straight axis of longer portion of log):

$$\left(\frac{\text{Length of deflecting section}}{L}\right)\left(\frac{\text{maximum deflection}}{d}\right)$$

(5) Proportion lost when average cross-section of interior defect is enclosable in ellipse (or circle) with major and minor diameters measurable in inches:

$$\left(\frac{(\text{Major})(\text{minor})}{(d-1)^2}\right)\left(\frac{\text{length of defect}}{L}\right)$$

In applying rule 5, defect in peripheral inch of log (slab collar) can be ignored, but the ellipse should enclose a band of sound wood at least 1/2-inch thick. Where it is necessary to use a rectangle instead of an ellipse to enclose the defect, the cull

percent will be 5/4 as much as for an ellipse with the same diameters as the rectangle. An obvious modification when a ring of
rot surrounds a sound heart with average diameter H (in inches)
is to estimate the proportion sound as $\frac{(H-1)^2}{(d-1)^2}$ and the proportion
defective as $1- \frac{(H-1)^2}{(d-1)^2}$.

In the rare case where cubic scale for other products than
sawlogs is being used, sweep ordinarily is not considered to cause
loss, and $(d+1)^2$ is used instead of $(d-1)^2$ as a divisor for
interior defect deduction.

When net log scale is desired at the time of scaling, as
is often the case, proportion defective should be multiplied by L,
and the equivalent cull length C thus obtained can be considered
as applicable to an equivalent cull section with the same scaling
diameter as the log. Direct multiplication of cull percent by
International gross volume will be preferable when scaling by that
rule, however, unless equivalent cull lengths are to be processed
as described in the following section.

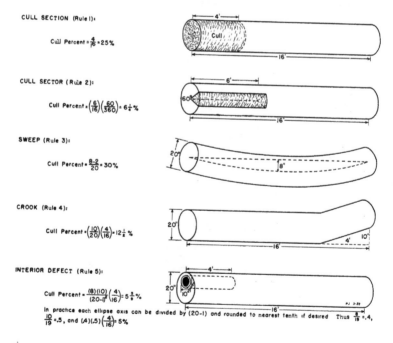

CULL SECTION (Rule 1):

Cull Percent $=\frac{4}{16} = 25\%$

CULL SECTOR (Rule 2):

Cull Percent $=\left(\frac{6}{16}\right)\left(\frac{60}{360}\right) = 6\frac{1}{4}\%$

SWEEP (Rule 3):

Cull Percent $=\frac{8-2}{20} = 30\%$

CROOK (Rule 4):

Cull Percent $=\left(\frac{10}{20}\right)\left(\frac{4}{16}\right) = 12\frac{1}{2}\%$

INTERIOR DEFECT (Rule 5):

Cull Percent $=\frac{(8)(10)}{(20-1)^2}\left(\frac{4}{16}\right) = 5\frac{1}{8}\%$

In practice each ellipse axis can be divided by (20-1) and rounded to nearest tenth if desired. Thus $\frac{8}{19} = .4$, $\frac{10}{19} = .5$, and $(.4)(.5)\left(\frac{4}{16}\right) = 5\%$

Figure 2.--Illustration of cull percent for a log 16 feet long and
20 inches in diameter.

· Where logs are scaled by a single rule, the usual
procedure and the cull deduction discussed above are qui
Where logs are to be scaled by several rules, however, c
desired to maximize field output and minimize field erro
technique is available.

Log diameter (d), log length (L), and equivalent
(or proportion cull x log length) are the only measureme
Logs longer than 16 feet will ordinarily be broken down
shorter logs, for each of which estimated d, L, C should
By the process explained below, one or two sets of 3 sum
accumulated in the office on a calculator (any fully aut
such as Marchant ACR-8M, ACT-10M; Monroe CAA-10, CSA-10;
models).

Steps in triple accumulation on the Marchant are

1. With the calculator conveniently tabbed, set up the
   starting on the left bank of the keyboard and simult
   set up its square (d^2) ending on the right bank of t

2. With the multiplying keys, multiply the above setup
   by the appropriate L.

3. Return the carriage to its original tabbed position,
   the keyboard but not the upper or middle dials.

4. Repeat the operation, using the next d, d^2, and appr
   until every d and L has been entered.

5. The upper dial will now read ΣL; the middle dial wi
   on the left and ΣLd^2 on the right.

6. After the totals are recorded and the machine cleare
   previous steps are repeated for d, d^2, and C.

The only special precaution needed is to place a
of any d, L, or C which is less than 10, so the accumula
correctly lined up. On the Monroe, L is set up before c
but otherwise the process is similar.

Following are examples that show how gross scale:
from ΣLd^2, ΣLd, ΣL; how cull scales are computed from
ΣC; or alternatively, how net scales are computed from
$\Sigma(L-C)d$, $\Sigma(L-C)$.

no.	d	L	c	Description of defect
	Inches	Feet	Feet	
1	12	16	4	Log with 5" of total sweep.
{2	14	08	3}	18' log with center rot in 8" x 10"
{3	15	10		ellipse 6' deep at small end.
4	15	12		
5	08	12		
6	24	14	3	Log with 90° cull sector 14' long.
{7	13	12	}	24' log with 2' butt section
{8	15	12	2}	entirely cull.
9	27	14		
10	16	16		

N=10: ΣL = 126   12 = ΣC
       ΣLd = 2,036   192 = ΣCd
       ΣLd^2 = 36,684   3,342 = ΣCd^2

Length of average entry = $\frac{\Sigma L}{N}$ = 12.6 ft.

     Coefficients for Doyle or Scribner are independent of average length and may be obtained directly from the table on page 12. Coefficients for International 1/4-inch or cubic feet must be interpolated from the first table on page 12, for average length 12.6 feet. As long as variation in L is not large (and it is kept low by breaking down long logs into several entries), interpolated coefficients will be quite accurate. Gross scales, then, by the various rules are (using 12.6-ft. average length):

Doyle   .0625 ΣLd^2 - .500 ΣLd + 1.000 ΣL = 1,401 bd. ft. gross

Scribner .0494 ΣLd^2 - .124 ΣLd - .269 ΣL = 1,526 bd. ft. gross

Int. 1/4" .0498 ΣLd^2 - .107 ΣLd - .063 ΣL = 1,601 bd. ft. gross

Cubic   .00545 ΣLd^2 + .0086 ΣLd + .0046 ΣL = 218.0 cu. ft. gross

     Exactly the same coefficients are used on ΣCd^2, ΣCd, ΣC, and give the following cull for the 3 board-foot rules (cubic feet of sawlog cull could be similarly calculated if desired):

Doyle   .0625 ΣCd^2 - .500 ΣCd + 1.000 ΣC = 125 bd. ft. cull

Scribner .0494 ΣCd^2 - .124 ΣCd - .269 ΣC = 138 bd. ft. cull

Int. 1/4" .0498 ΣCd^2 - .107 ΣCd - .063 ΣC = 145 bd. ft. cull

Where only net scale is desired, the cull deductio[n]
applied to L at the time of scaling, so that d, (L - C),
recorded. Before obtaining the length of average entry,
be added to Σ (L - C). Net scale thus directly obtained
same as subtracting the above cull from the previously co[mputed]
scales.

Entry no.	d	L - C	C
	Inches	Feet	Feet
1	12	12	4
{2	14	05	3}
{3	15	10	}
4	15	12	
5	08	12	
6	24	11	3
{7	13	12	}
{8	15	10	2}
9	27	14	
10	16	16	

N = 10; Σ (L - C)   114 | 12 = Σ C
          Σ (L - C)d   1,844
          Σ (L - C)d^2 = 33,342 | Av. gross length = $\dfrac{\Sigma(L-}{}$

Coefficients for these net accumulations will be e
same as for the gross or cull accumulations, since the 12
age length governs in both cases where taper affects the
Net board foot scales from the net accumulations, then, a

Doyle:

    $.0625 \; \Sigma(L-C)d^2 - .500 \; \Sigma(L-C)d + 1.000 \; \Sigma(L-C) = 1,276$

Scribner:

    $.0494 \; \Sigma(L-C)d^2 - .124 \; \Sigma(L-C)d - .269 \; \Sigma(L-C) = 1,388$

Int. $\frac{1}{4}$-inch:

    $.0498 \; \Sigma(L-C)d^2 - .107 \; \Sigma(L-C)d - .063 \; \Sigma(L-C) = 1,456$

Note that the sum of net plus cull equals previou[s]
might be expected. This is the method recommended for o
where only net scale is needed.

Those who desire to develop coefficients from local mill tallies instead of using those appropriate to a given log rule will find the subject is well covered in Schumacher and Jones 3/.

Had the individual gross scales corresponding to each entry been laboriously recorded (with cull by International ¼-inch rule), the results would have differed only trivially in the third digit, if at all. Thus,

Entry no.	d	L	Gross Doyle	Gross Scrib. Dec. C	Gross Int.¼-in.	Cull (Int.¼-in.)	Gross cubic[1]
	Inches	Feet	- - - - - Board feet - - - - -				Cu.ft.
1	12	16	64	80	95	25	14.8
2	14	8	50	60	65	25	9.2
3	15	10	75	90	95		13.3
4	15	12	91	110	115		16.2
5	08	12	12	20	25		5.0
6	24	14	350	350	370	90	47.3
7	13	12	61	70	85		12.4
8	15	12	91	110	115	20	16.2
9	27	14	463	480	470		59.4
10	16	16	144	160	180		25.2
			1,401	1,530	1,615	160	219.0
					less 160		
					net 1,455		

1/ Cubic volumes are for conoid with diameter increasing 1 inch per 8 lineal feet.

3/ Schumacher, F.X., and Jones. W. C., Jr. Empirical log rules and the allocation of sawing time to log size. Journal of Forestry 38: 889-896. 1940.

RANDOM SAMPLING DESIGN

Since systematic square grids of plots or mechanically spaced lines of plots tend to give more precise estimates than the same number of random plots or random lines of plots, the random error formulae given below should be used with caution and will provide only a rough upper bound for estimates of error where systematic sampling is employed. Those wishing to estimate the error of a systematic sample more precisely are referred to DeLury [4]. It should be remembered that the limits of error discussed below will be exceeded by 1 out of 3 sample means from a normal random population. Doubling the given coefficient of variation will result in more intensive sampling with only about 1 chance in 20 that the sample mean will not lie within the desired limits of error.

Individual sampling units chosen for volume measurement may comprise a population of point-counts, plot volumes, volumes of clusters of plots, volumes of lines of plots (as in conventional cruising), strip volumes, log volumes, or tree volumes. Variation in sample volumes (X) between (N) randomly selected individual units is conveniently estimated in terms of the mean volume and expressed as:

(A) Coefficient of variation in percent =

$$(100)\left(\frac{\text{standard deviation}}{\text{mean volume}}\right) = 100 \sqrt{\left(\frac{N}{N-1}\right)\left(\frac{N\Sigma X^2}{(\Sigma X)^2} - 1\right)}$$

Thus, if 5 sample-plot volumes (expressed in thousands of board feet per acre) were 2, 6, 10, 10, 12, with a mean of 8, the coefficient of variation would be:

$$100 \sqrt{\left(\frac{5}{4}\right)\left(\frac{1920}{1600} - 1\right)} = 100 \sqrt{.25} = 50 \text{ pct.}$$

The standard error of the mean of (N) random volumes from an infinite normal population may be estimated in terms of the mean volume and expressed in terms of the coefficient of variation:

(B) Standard error as percent of mean (infinite population) =

$$\frac{\text{Coefficient of variation in pct.}}{\sqrt{N}} = 100 \sqrt{\left(\frac{1}{N-1}\right)\left(\frac{N\Sigma X^2}{(\Sigma X)^2} - 1\right)}$$

Thus, the standard error (in percent) above would be:

$$\frac{50}{\sqrt{5}} = 22.4 \text{ percent}$$

---

[4] DeLury, D. B. Values and integrals of the orthogonal polynomials up to $n = 26$. Univ. Toronto Press, Toronto, Ont., 33 pp. 1950.

Where the population is finite, and an appreciable proportion has been sampled, the standard error in percent is further reduced, as indicated below (in an infinite population, of course, the unsampled proportion is always 1):

(C) Standard error as percent of mean (finite population) =

$$(\text{coefficient of variation in pct.}) \sqrt{\frac{\text{unsampled proportion}}{N}}$$

Thus, if the previous example, composed of 5 samples, had come from a population of only 25 possible sampling units (instead of one with infinite possible sampling units) the standard error (in percent) would have been:

$$50 \sqrt{\frac{.80}{5}} = (50)(.4) = 20 \text{ percent}$$

Where it is desired to design a random sample of an infinite population so that its standard error is in the neighborhood of a specified ± limit of error in percent, the following formula is useful; it is the basis for the table at the end of this section:

(D) Number of samples to be taken from infinite population = N_∞

$$N_\infty = \left(\frac{\text{coefficient of variation in percent}}{\text{specified limit of error in percent}}\right)^2$$

Thus, if an infinite population of plot volumes had a coefficient of variation of 50 percent, and it were desired to sample it randomly with a design whose standard error would be in the neighborhood of ± 10 percent, the number of random samples to select would be: $\left(\frac{50}{10}\right)^2 = 25$. This same result could have been read from the table at the end of this section.

Where the population is finite (composed of M individuals), N_∞ calculated by the previous formula must be replaced by N, as follows:

(E) $$N = \frac{1}{\frac{1}{N_\infty} + \frac{1}{M}} = \frac{1}{\left(\frac{\text{specified limit of error in percent}}{\text{coefficient of variation in percent}}\right)^2 + \frac{1}{M}}$$

Thus, if there had been only 25 possible individual plot volumes (instead of an infinite number) in the previous example, it would not have been necessary to sample all 25 to approach a ± 10 percent standard error. Instead, the number of samples to be taken would have been:

$$\frac{1}{\frac{1}{25} + \frac{1}{25}} = 12\text{-}1/2 \ (\text{or } 13)$$

The above formula is easily converted to the so-called "s
fraction" (or proportion of the finite population to be sampled)
dividing by the number in the population (M):

(F) Proportion of finite population to be sampled = $\frac{N}{M}$

$$\frac{N}{M} = \frac{1}{\frac{M}{N_\infty} + 1} = \frac{N_\infty}{N_\infty + M}$$

Thus, in the previous example, the proportion of the
tion to be sampled (where coefficient of variation is
percent, specified limit of error is ± 10 percent, ar
finite population contains 25 individuals) would be:
$\frac{1}{\frac{25}{25} + 1} = \frac{1}{2}$. This means that 1 out of every 2 individ

in the population must be sampled.

It is frequently advantageous to stratify a finite popula
into groups or classes of homogeneous individuals according to s
criterion other than actually sampled volume. Where reasonably
similar reduced coefficients of variation prevail for several gr
it is convenient and not too inappropriate to consider the large
of the reduced values as applying to the whole population thus s
fied. With optimum allocation of sampling to strata, formula (F
would apply to the population as a whole, although the proportic
would vary according to volume in each stratum. However, it is
convenient to adopt a constant or uniform sampling fraction, and
dispense with the correction for finite population as a margin c
safety, thus:

(G) Approximate proportion of finite stratified population
sampled = $\frac{N_\infty}{M}$.

(where coefficients of variation of strata tend to be s

Thus, if approximately 10,000 pine trees were to be c
and classified into 2-inch groups having within-stra·
coefficients of variation of 30 percent, and it was c
to sample individual tree volumes with a standard err
approaching 1-1/2 percent, the proportion to be samp
would be $\frac{400}{10,000} = \frac{1}{25}$ using the table below. Samplin;
tree for volume out of every 25 counted should be ac
plished in some unbiased manner.

As a rough guide to approximate coefficients of variation, the figures given below may be useful in the South in the absence of valid local information:

Individual sample	Population	Random coefficient of variation Percent
1/5-acre plot volume	Finite unstratified plots	100
Point-count (104.18' angle)	Infinite unstratified points	50
Tree volume	Finite pine trees, stratified in 2-inch groups	30
Tree volume	Finite pine trees, stratified in 4-inch groups	35
Tree volume	Finite pine trees, stratified in 6-inch groups	40
Tree volume	Finite hardwood trees, stratified in 2-inch groups	35
Tree volume	Finite hardwood trees, stratified in 4-inch groups	40
Tree volume	Finite hardwood trees, stratified in 6-inch groups	50

The following table gives N_∞ (the number of samples needed from an infinite population with a given coefficient of variation) so that the standard error of the samples will tend to be in the neighborhood of a given limit of error.

Number of samples to be taken from infinite pop

Coefficient of variation (percent)	Specified percent limit for s		
	± 1½ percent	± 5 percent	± 10 percent
		N_∞	
5	12	1	1
10	45	4	1
15	100	9	3
20	178	16	4
25	278	25	7
30	400	36	9
35	545	49	13
40	712	64	16
45	900	81	21
50	1,112	100	25
55	1,345	121	31
60	1,600	144	36
65	1,878	169	43
70	2,178	196	49
75	2,500	225	57
80	2,845	256	64
85	3,212	289	73
90	3,600	324	81
95	4,012	361	91
100	4,445	400	100
125	6,945	625	157
150	10,000	900	225
175	13,612	1,225	307
200	17,778	1,600	400

USDA

Southern Forest Experiment
Station Occasional Paper
No. 127+128 is out of print.
Please bind this in
its place.

کتابخانه ۴

CUTTING FINANCIALLY MATURE LOBLOLLY AND SHORTLEAF PINE

⌒ Sam Guttenberg and R. R. Reynolds

SOUTHERN FOREST EXPERIMENT STATION

Harold L. Mitchell, Director

OCCASIONAL PAPER 129   May 1953

CUTTING FINANCIALLY MATURE LOBLOLLY AND SHORTLEAF PINE

Sam Guttenberg and R. R. Reynolds
Southern Forest Experiment Station

This paper offers guides that the forester working with loblolly and shortleaf pine can use to judge when his trees become financially mature. The discussion is restricted to the production of large, high-quality saw timber. Pulpwood, poles, and piling are not considered; forest owners who expect to grow large quantities of these products will need to make special studies following the principles outlined here. The paper deals specifically with the conditions prevailing in the territory around Crossett, Arkansas (fig. 1), but the findings should be applicable on good sites (loblolly site index 80 or better) throughout the range of loblolly pine. (Only shortleaf pines that occur on loblolly sites are considered.)

The guides are not particularly complex in principle, but the alternatives to which they may be applied are numerous. Since their use is enhanced by an understanding of the way in which they were constructed, this paper first describes how the guides were determined and how they apply to the various alternatives encountered in actual management. The last part (pp. 15-16) condenses the guides into a readily usable set of marking rules for use in the woods.

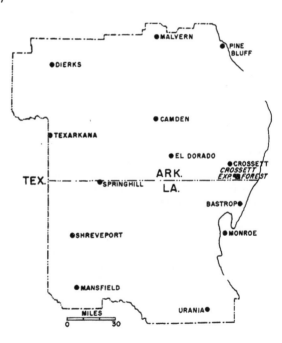

Figure 1. --Crossett territory.

These guides are no universal panacea. They are merely an
tempt to attach the dollar sign to the sum of the most important multi
factors that the silviculturist attempts to take into account when he ma
a tree for removal. They should, however, help the lumberman to u
derstand tree values. The data on which the guides are based are the
best available at present but are lacking in some respects. In the fir
place, there is not yet any very exact information as to the number a
grades of logs that managed trees will contain as they grow from sma
to large saw-timber diameter classes. Fully objective pine log grad
are still in the process of development. In addition, logging costs fo
trees of various size classes and for varying logging sites and condit
are far from complete. Ideally, each forester would accumulate his
own figures and work out his own guides following the principles desc
in this publication. This would, however, require data not always re
available; an alternative is to adapt the guides offered here, even tho
they fall short of the ideal, and to use them until more complete data
available.

Estimating Maturity

The guides are based on the cutting of financially mature timb
The idea of financial maturity is this:

The forester visits each acre of his holdings at regular interv
and marks for cutting those trees whose quality or growth rate is bel
par. Bearing in mind how the typical tree develops--low value at fir
but rapid increase in value, then a gradual slackening in the rate of i
crease until the stage is finally reached where the tree ceases to pay
way--the forester appraises his trees with the object of putting the a:
to those that are reaching this crucial point. This is the point of fina
cial maturity. Just where it falls is determined by the rate at which
forest owner expects his trees to earn money, the so-called alternat
rate of return. 1/ If he expects a return of 4 percent, he will cut an
tree that is not producing or will not produce at this rate.

The chief problem with which this paper is concerned, theref
is to determine the rate of value increase that can be expected from
given pine over a given period of time. This rate will be estimated :
two steps:

1/ The choice of an alternative rate is conditioned by risk, taxes, b
ness debts, and the like. See: Duerr, Wm. A., and Bond, W. E. O
mum stocking of a selection forest. Jour. Forestry 50: 12-16. 1952

1. Calculate the value of trees according to their volume, grade, and diameter.

2. Determine for typical growth rates the prospective rate of increase in tree value.

The proportions of the various sizes and grades of kiln-dried, dressed lumber (dimension and boards) typically produced by the large (Group I) mills of the Southern Pine Association are here used as the basis for appraisal. Many of these mills own and are progressive managers of forest land. Though their chief object is to grow high-grade saw timber, pulpwood and other products will inevitably be harvested from trees or portions of trees that for one reason or another are best not utilized for lumber. Too, appraising trees for lumber does not commit them irrevocably, for at the time of harvest a prudent business-man will put his trees to their most profitable use.

Step one--tree volume, quality, and value

With finished, graded lumber in view, tree volume is here ex-pressed in board feet by the International 1/4-inch kerf rule, using standard volume tables as the guide. [2/] Tree quality is developed in terms of the Crossett log grades (pp. 17 to 18). For example, a two-log tree with a grade-1 butt and a grade-3 second log will be classed as grade 1-3. Facility in the use of grades is absolutely essential in mak-ing financial maturity decisions. The Crossett log grades are very simple, but in applying them a considerable amount of judgment is re-quired. Such judgment is rapidly acquired with practice.

Tree value will be expressed in terms of conversion surplus. [3/] This is the difference between the sales value of the finished lumber and all the direct costs of producing this lumber--mainly labor and materials used in felling the tree and in making, transporting, and sawing the logs. Conversion surplus thus represents that part of a tree's gross product value which can be made available to increase the profit or reduce the loss of the business.

[2/] Mesavage, C., and Girard, J. W. Tables for estimating board-foot volume of timber. U.S. Dept. Agr., 94 pp. 1946. (For sale by the Supt. of Documents, U.S. Gov't Printing Office, Wash. 25, D.C., at 25 cents per copy.)

[3/] For a full discussion of conversion surplus, see: Guttenberg, Sam, and Duerr, Wm. A. A guide to profitable tree utilization. South. Forest Expt. Sta. Occas. Paper 114, 18 pp. 1949.

Since conversion surplus will here be based upon lumber value not upon log or stumpage value, the examples to be developed will app primarily to the timber owner who operates a sawmill and sells his product in the form of finished lumber. However, they will also appl to the log or stumpage seller who can count on receiving for each log or tree a price in line with the value of the lumber it will yield. Inde prices would be in line on a freely competitive market if lumber producers and timber growers used such an analysis as a guide.

Lumber sales value. --The first ingredient of conversion surplus, lumber sales value, may be estimated for any tree. Knowing the grade and volume of each log in the tree and the lumber grade yie of each grade of log, one may determine the amount of lumber, by grades, that the tree will yield. With current sales prices for lumbe he may then calculate the weighted average price per M of lumber rep resented in the tree, and from this the total lumber value of the tree.

The problem is how to allow for changes in the lumber market An estimate of financial maturity involves looking ahead at least one cutting cycle--perhaps 5 years or more--and the lumber price level probably will not be the same then as now. To help solve this probler advantage may be taken of the fact that with all the ups and downs of th lumber market, the price of each grade tends to maintain a fairly sta ble relationship to that of other grades (see fig. 2). Grade prices ma then be expressed as percentages (index numbers) that will have sub- stantial validity over the years, and the value of lumber in a tree may be calculated in index terms instead of dollars, following the same procedure as would be used with dollars.

In figure 2, values in index numbers are plotted for each of th main grades of southern pine lumber. These index numbers were derived from annual average lumber grade prices, expressed as relativ in percent of the price of No. 2 Common. The chart covers the peric 1925-51, omitting the abnormal war years 1941 through 1946. The prices used are f. o. b. mill averages received by large producers. Since the annual average price received by these companies for each grade of lumber cannot be far out of line with that obtained by other producers, the prices are assumed to be a close reflection of the enti southern pine market.

4/ As given in Reynolds, R. R., Bond, W. E., and Kirkland, B. P. Financial aspects of selective cutting in the management of second-growth pine-hardwood forests west of the Mississippi River. U. S. Dept. Agr. Tech. Bul. 861, 118 pp. 1944.

- 4 -

From figure 2A it can be seen that though the relative lumber grade prices vary from year to year, they generally fluctuate rather narrowly around constant values. In figure 2B is plotted the total volume of southern pine lumber produced in the United States and in figure 2C the average price received by the companies for their lumber. These three charts show that the relationship among the grade prices is more stable than, and relatively unaffected by, either the volume of production or the price level.

Postwar experience indicates a slight departure from the prewar record. In particular, both B and Better and No. 3 Common lumber grades are now closer in value to No. 2 Common than they were in prewar years. Accordingly, for present purposes, the postwar years are being used as a basis for estimating the price relations likely to prevail into the near future. The estimated index numbers are:

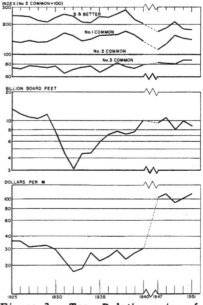

Figure 2. --Top, Relative price of the main grades of southern pine lumber. Middle, Total U.S. production of southern pine lumber. Bottom, Average price of southern pine lumber.

B and Better	185
No. 1 Common (including C finish)	140
No. 2 Common	100
No. 3 Common	85

Direct costs. --The other ingredient of conversion surplus, direct costs of lumber production from stump through mill, is estimated from whatever representative cost records are available. Direct costs include only those additional outlays that arise because the tree in question is being made into lumber. They specifically exclude all fixed or overhead costs--all costs that are unaffected by whether the particular tree is logged and milled.

The cost analyses of the Southern Pine Association indicate that over the years about 60 percent of the total cost of producing southern pine lumber is direct cost. The annual cost summaries[5] for Group I producers were used as a guide to estimate direct costs on a typical operation. [6] Costs were estimated in dollars per M board feet for trees of each diameter class and log height, using available logging and milling study experience. [7] Dollars were then converted to index numbers (percentages of the price of No. 2 Common lumber), just as was done for lumber value. This procedure is defensible on the ground that prices and direct costs (primarily wages) tend to fluctuate together. It permits deriving conversion surplus as an index number, and thus avoiding the problem of fluctuating costs and market prices. The result of these calculations is given in table 1 (pp. 8-10). In the table, columns 2, 3, and 4 list the lumber value, the direct costs, and the difference between these two--the conversion surplus-- per M board feet as a weighted average for the tree.

Step two--value increase

Studies on the Crossett Experimental Forest show that desirable growing stock trees will fall into one of the following average growth classes: 1-inch (low vigor), 1.5-inch (medium vigor), or 2-inch (high vigor) average diameter growth per 5 years. Taking these growth classes as given, and with the method described for expressing tree value according to tree size and quality, we are ready for the problem of growth in value. This rate of increase in tree value is the rate that is to be compared with the desired rate of return to determine if a tree is financially mature. The problem will be considered in two stages: first, rate of value increase when a tree changes diameter but not log height or grade; second, rate of increase when height or grade, as well as diameter, changes.

A 5-year cutting cycle will be assumed. That is, it will be supposed that the forester has the choice of cutting a tree now or waiting at least 5 years. Consequently, our interest centers on the rate of tree-value increase in 5 years.

---

[5] Southern Pine Association. Southern Pine Costs. New Orleans, La. Published annually.

[6] Those interested in allocating total costs will derive final answers not markedly different from the ones given herein.

[7] See publications cited in footnotes 3 and 4.

No change in log height or grade. --Rates of value increase for the given growth classes are worked out in table 1. In the table, columns 2, 3, and 4 show, in terms of index numbers as explained earlier, the lumber value, the direct costs, and the difference between these two--the conversion surplus--per M board feet as a weighted average for the tree. From the number of board feet per tree (column 5), conversion surplus per tree (column 6) was determined. Following this (columns 7, 9, and 11), the prospective conversion surplus per tree after 5 years for each growth class was calculated, assuming no change in log height or grade. The ratio of value after 5 years to value now was then computed, after which compound-interest tables were used to find the annual percentage increase (columns 8, 10, and 12).

As an example of the use of table 1, take the problem of a forester who is deciding whether to cut a 17-inch pine, grade 1-2-3, low vigor. Neither the vigor, merchantable height, nor the grade of the logs is likely to improve soon. The rate of return that the manager follows as his guide is 3 percent. Table 1 indicates that the tree will earn less than this rate--only 2.9 percent--in the coming five years. It is therefore financially mature and should be marked for cutting.

Change in height and grade. --Both log height and grade normally change as a tree grows in diameter. Limbs die and fall off, and clear wood covers the knots. Butt logs of small trees are especially likely to improve radically in grade once the limbs are dropped. Small, short pines of low grade thus are often capable of developing into high-grade trees with 3 or more logs. Imminent and potential changes in log height and grade can be recognized by the experienced forester or woodsman, and to recognize such changes is essential in determining financial maturity. As a general rule, the potential rapid improvement in grade has been realized by the time a pine reaches 22 to 24 inches d.b.h., and the potential merchantable length is usually established by the time the tree is 15 to 17 inches in diameter.

Table 1 may be used for calculating (though not reading directly) the rate of value-increase for trees expected to grow in log height or improve in grade. Suppose that our 17-inch, grade 1-2-3 pine is expected to add a fourth log in the coming 5 years, so as to become an 18-inch, grade 1-2-3-3 tree. Table 1 shows that the value now is index 20.8, and that the value in 5 years (reading in the same column, opposite the expected grade and diameter) will be index 30.1. A calculation of the ratio of these indexes and reference to compound interest tables discloses that the rate of increase is 7.7 percent. If a 3- or 4-percent return is satisfactory, the tree is not yet mature.

Table 1.--Rate of value increase of loblolly or shortleaf pine, by log height, grade,

D.b.h. (inches)	Per M board feet			Lumber per tree	Conversion surplus per t			
	Gross lumber value	Direct costs	Conversion surplus		High vigor			Med
					Now	After five years	Annual increase	After five years
(1)	(2)	(3)	(4)	(5)	(6)	(7)	(8)	(9)
	--- Index ---			Bd. ft.	-- Index --		Percent	Index

Tree grade 2-3

12	129.0	66.5	62.5	92	5.7	8.2	7.5	7.5
13	127.2	65.2	62.0	112	6.9	9.6	6.8	8.9
14	125.9	64.0	61.9	132	8.2	11.1	6.2	10.3
15	124.5	62.7	61.8	156	9.6	12.7	5.8	11.9
16	123.3	61.5	61.8	180	11.1	14.5	5.5	13.6
17	122.1	60.3	61.8	206	12.7	16.4	5.2	15.4
18	121.3	59.2	62.1	233	14.5	18.5	5.0	17.4
19	120.2	58.0	62.2	264	16.4	20.7	4.8	19.6
20	119.3	56.9	62.4	296	18.5	22.9	4.4	21.8
21	118.1	55.8	62.3	332	20.7	25.2	4.0	24.0
22	117.2	54.9	62.3	368	22.9	27.5	3.7	26.3
23	116.4	54.1	62.3	404	25.2	29.9	3.5	28.7
24	115.8	53.5	62.3	441	27.5	32.4	3.3	31.1
25	114.9	53.1	61.8	484	29.9	34.7	3.0	33.5

Tree grade 1-3

14	129.4	64.0	65.4	132	8.6	12.2	7.2	11.3
15	129.4	62.7	66.7	156	10.4	14.2	6.4	13.2
16	129.4	61.5	67.9	180	12.2	16.4	6.1	15.3
17	129.4	60.3	69.1	206	14.2	18.8	5.8	17.6
18	129.4	59.2	70.2	233	16.4	21.5	5.6	20.1
19	129.4	58.0	71.4	264	18.8	24.4	5.3	23.0
20	129.4	56.9	72.5	296	21.5	27.6	5.1	26.0
21	129.4	55.8	73.6	332	24.4	30.8	4.7	29.2
22	130.0	54.9	75.1	368	27.6	34.2	4.4	32.5
23	130.4	54.1	76.3	404	30.8	37.8	4.2	36.0
24	131.0	53.5	77.5	441	34.2	41.6	4.0	39.7
25	131.2	53.1	78.1	484	37.8	45.2	3.6	43.4
26	131.6	52.9	78.7	528	41.6	48.8	3.2	47.0
27	132.1	53.0	79.1	572	45.2	52.5	3.0	50.6

Tree grade 1-3-3

14	128.5	63.4	65.1	186	12.1	17.0	7.0	15.7
15	127.9	62.2	65.7	221	14.5	19.8	6.4	18.4
16	127.3	61.0	66.3	256	17.0	22.8	6.0	21.3
17	126.8	59.8	67.0	296	19.8	26.1	5.9	24.4
18	126.5	58.7	67.8	336	22.8	29.6	5.4	27.8
19	126.0	57.6	68.4	382	26.1	33.4	5.1	31.5
20	125.8	56.5	69.3	427	29.6	37.5	4.8	35.4
21	125.4	55.5	69.9	478	33.4	41.8	4.6	39.6
22	125.4	54.4	71.0	528	37.5	46.3	4.3	44.0
23	125.0	53.6	71.4	586	41.8	50.9	4.0	48.6
24	124.9	53.0	71.9	644	46.3	55.4	3.7	53.
25	124.8	52.7	72.1	706	50.9	60.1	3.4	57.
26	124.7	52.5	72.2	767	55.4	64.2	3.0	62.
27	124.6	52.7	71.9	836	60.1	69.0	2.8	66.

Table 1. --(Continued)

	Per M board feet			Lumber	Conversion surplus per tree by vigor class					
						High vigor		Medium vigor		Low vigor	
D.b.h. (inches)	Gross lumber value	Direct costs	Conversion surplus	per tree	Now	After five years	Annual increase	After five years	Annual increase	After five years	Annual increase
(1)	(2)	(3)	(4)	(5)	(6)	(7)	(8)	(9)	(10)	(11)	(12)
	– – – – Index – – – –			Bd. ft.	– – Index – –		Percent	Index	Percent	Index	Percent

Tree grade 1-2-3

14	132.7	63.4	69.3	186	12.9	17.9	6.8	16.6	5.2	15.3	3.5
15	131.6	62.2	69.4	221	15.3	20.8	6.3	19.3	4.8	17.9	3.2
16	130.8	61.0	69.8	256	17.9	24.0	6.0	22.4	4.6	20.8	3.1
17	130.2	59.8	70.4	296	20.8	27.5	5.7	25.7	4.3	24.0	2.9
18	130.0	58.7	71.3	336	24.0	31.2	5.4	29.3	4.1	27.5	2.8
19	129.5	57.6	71.9	382	27.5	35.3	5.1	33.2	3.8	31.2	2.6
20	129.5	56.5	73.0	427	31.2	39.7	4.9	37.5	3.7	35.3	2.5
21	129.4	55.5	73.9	478	35.3	44.5	4.7	42.1	3.6	39.7	2.4
22	129.5	54.4	75.1	528	39.7	49.5	4.5	47.0	3.4	44.5	2.3
23	129.6	53.6	76.0	586	44.5	54.7	4.2	52.1	3.2	49.5	2.1
24	129.9	53.0	76.9	644	49.5	59.9	3.9	57.3	3.0	54.7	2.0
25	130.2	52.7	77.5	706	54.7	65.3	3.6	62.6	2.8	59.9	1.8
26	130.6	52.5	78.1	767	59.9	70.8	3.4	68.0	2.6	65.3	1.7
27	130.8	52.7	78.1	836	65.3	76.4	3.2	73.6	2.4	70.8	1.6
28	131.4	52.9	78.3	904	70.8	82.1	3.0	79.3	2.3	76.4	1.5

Tree grade 1-1-3

16	132.0	61.0	71.0	256	18.2	25.0	6.6	23.2	5.0	21.5	3.4
17	132.4	59.8	72.6	296	21.5	28.9	6.1	26.9	4.6	25.0	3.1
18	133.0	58.7	74.3	336	25.0	33.1	5.8	31.0	4.4	28.9	2.9
19	133.2	57.6	75.6	382	28.9	37.7	5.5	35.4	4.1	33.1	2.8
20	134.0	56.5	77.5	427	33.1	42.7	5.2	40.2	4.0	37.7	2.7
21	134.3	55.5	78.8	478	37.7	48.1	5.0	45.4	3.8	42.7	2.5
22	135.2	54.4	80.8	528	42.7	53.6	4.7	50.8	3.5	48.1	2.4
23	135.6	53.6	82.0	586	48.1	59.5	4.4	56.5	3.3	53.6	2.2
24	136.3	53.0	83.3	644	53.6	65.4	4.1	62.4	3.1	59.5	2.1
25	137.0	52.7	84.3	706	59.5	71.6	3.8	68.5	2.9	65.4	1.9
26	137.8	52.5	85.3	767	65.4	78.0	3.6	74.8	2.7	71.6	1.8
27	138.4	52.7	85.7	836	71.6	84.5	3.4	81.2	2.6	78.0	1.7
28	139.2	52.9	86.3	904	78.0	91.0	3.1	87.7	2.4	84.5	1.6
29	139.9	53.4	86.5	977	84.5	97.5	2.9	94.2	2.2	91.0	1.5

Tree grade 1-2-3-3

16	129.2	60.9	68.3	322	22.0	30.1	6.5	28.0	4.9	26.0	3.4
17	128.7	59.5	69.2	376	26.0	34.5	5.8	32.3	4.4	30.1	3.0
18	128.3	58.2	70.1	430	30.1	39.2	5.4	36.8	4.1	34.5	2.8
19	127.9	56.9	71.0	486	34.5	44.3	5.1	41.7	3.9	39.2	2.6
20	127.7	55.6	72.1	543	39.2	49.8	4.9	47.0	3.7	44.3	2.5
21	127.4	54.5	72.9	608	44.3	55.7	4.7	52.7	3.5	49.8	2.4
22	127.3	53.4	73.9	674	49.8	61.7	4.4	58.7	3.3	55.7	2.3
23	127.0	52.6	74.4	748	55.7	68.2	4.1	64.9	3.1	61.7	2.1
24	127.0	51.9	75.1	822	61.7	74.6	3.9	71.4	3.0	68.2	2.0
25	127.0	51.6	75.4	904	68.2	81.3	3.6	77.9	2.7	74.6	1.8
26	127.0	51.4	75.6	987	74.6	87.8	3.3	84.5	2.5	81.3	1.7
27	127.0	51.5	75.5	1077	81.3	94.1	3.0	90.9	2.3	87.8	1.5

Table 1.—(Continued)

D.b.h. (inches)	Per M board feet			Lumber per tree	Conversion surplus per tree by vigor class					
	Gross lumber value	Direct costs	Conversion surplus		Now	High vigor		Medium vigor		Low v
						After five years	Annual increase	After five years	Annual increase	After five years
(1)	(2)	(3)	(4)	(5)	(6)	(7)	(8)	(9)	(10)	(11)
	– – – – Index – – – –			Bd. ft.	– – Index – –		Percent	Index	Percent	Index

Tree grade 1-1-3-3

16	131.2	60.9	70.3	322	22.6	31.5	6.8	29.2	5.3	27.0
17	131.2	59.5	71.7	376	27.0	36.3	6.1	33.9	4.7	31.5
18	131.4	58.2	73.2	430	31.5	41.4	5.6	38.8	4.3	36.3
19	131.6	56.9	74.7	486	36.3	47.1	5.3	44.2	4.0	41.4
20	131.8	55.6	76.2	543	41.4	53.2	5.1	50.1	3.9	47.1
21	132.0	54.5	77.5	608	47.1	60.0	5.0	56.6	3.7	53.2
22	132.4	53.4	79.0	674	53.2	66.8	4.7	63.4	3.6	60.0
23	132.8	52.6	80.2	748	60.0	74.1	4.3	70.4	3.2	66.8
24	133.2	51.9	81.3	822	66.8	81.6	4.1	77.8	3.1	74.1
25	133.6	51.6	82.0	904	74.1	89.4	3.8	85.5	2.9	81.6
26	134.1	51.4	82.7	987	81.6	97.1	3.5	93.2	2.7	89.4
27	134.5	51.5	83.0	1077	89.4	104.7	3.2	100.9	2.5	97.1
28	135.0	51.8	83.2	1167	97.1	112.2	2.9	108.4	2.2	104.7

Tree grade 1-1-2-3

16	132.8	60.9	71.9	322	23.2	32.5	7.0	30.1	5.3	27.7
17	133.2	59.5	73.7	376	27.7	37.6	6.3	35.0	4.8	32.5
18	133.7	58.2	75.5	430	32.5	43.1	5.8	40.3	4.4	37.6
19	134.2	56.9	77.3	486	37.6	49.2	5.5	46.1	4.2	43.1
20	134.9	55.6	79.3	543	43.1	55.8	5.3	52.5	4.0	49.2
21	135.4	54.5	80.9	608	49.2	63.0	5.1	59.4	3.8	55.8
22	136.2	53.4	82.8	674	55.8	70.5	4.8	66.7	3.6	63.0
23	136.8	52.6	84.2	748	63.0	78.5	4.5	74.5	3.4	70.5
24	137.7	51.9	85.8	822	70.5	86.7	4.2	82.6	3.2	78.5
25	138.4	51.6	86.8	904	78.5	95.3	4.0	91.0	3.0	86.7
26	139.2	51.4	87.8	987	86.7	103.9	3.7	99.6	2.8	95.3
27	140.0	51.5	88.5	1077	95.3	112.5	3.4	108.2	2.6	103.9
28	140.8	51.8	89.0	1167	103.9	121.1	3.1	116.8	2.4	112.5
29	141.7	52.3	89.4	1258	112.5	129.7	2.9	125.4	2.2	121.1

Tree grade 1-1-1-3

16	134.1	60.9	73.2	322	23.6	33.2	7.1	30.7	5.4	28.3
17	134.7	59.5	75.2	376	28.3	38.5	6.3	35.8	4.8	33.2
18	135.4	58.2	77.2	430	33.2	44.3	5.9	41.4	4.5	38.5
19	136.2	56.9	79.3	486	38.5	50.6	5.6	47.4	4.3	44.3
20	137.1	55.6	81.5	543	44.3	57.5	5.3	54.0	4.0	50.6
21	137.7	54.5	83.2	608	50.6	65.1	5.2	61.3	3.9	57.5
22	138.7	53.4	85.3	674	57.5	72.8	4.8	68.9	3.7	65.1
23	139.6	52.6	87.0	748	65.1	81.3	4.5	77.0	3.4	72.8
24	140.5	51.9	88.6	822	72.8	90.0	4.3	85.6	3.3	81.3
25	141.5	51.6	89.9	904	81.3	99.2	4.1	94.6	3.1	90.0
26	142.6	51.4	91.2	987	90.0	108.3	3.8	103.7	2.9	99.2
27	143.6	51.5	92.1	1077	99.2	117.5	3.4	112.9	2.6	108.3
28	144.6	51.8	92.8	1167	108.3	126.6	3.2	122.0	2.4	117.5
29	145.7	52.3	93.4	1258	117.5	136.3	3.0	131.4	2.3	126.6

Prospective increases in grade and vigor, as well as in height, may be evaluated in the same way. If the grade 1-2-3 low-vigor pine changes to a grade 1-1-3-3 at medium vigor growth in 5 years, the rate of value increase will be even higher, 11 percent.

Maturity Under Woods Conditions

Existing stands generally are comprised of trees of different grades and sizes. The greatest dollar returns will be realized when trees of the highest grade and vigor class are carried to maturity, because for any given desired rate of return, high-grade trees will afford more income than low-grade trees.

Cutting the low-grade or financially mature trees from a given stand does not take the timber manager from low to high grade timber in one jump. In stands composed of low-quality trees, the continuous cutting of financially mature trees may not increase the average grade to any great extent for many years. The use of the financial maturity concept does, however, provide a means by which the low-yielding trees can be recognized and the grade of the remaining trees gradually improved.

Financially mature timber has been harvested from the all-diametered second-growth pine-hardwood stands on the Crossett Experimental Forest for the past 15 years. In 1938, when cutting began, the stands had a volume of 4,800 board feet (Int. 1/4-inch rule) per acre in pines larger than 11.5 inches d.b.h. and a basal area of 48 square feet per acre in pines larger than 3.5 inches d.b.h. Grade of the stands in 1952 is shown in table 2. For stands similar to those on the Crossett Experimental Forest, the proportions in table 2 may approximately forecast the grade improvement to be realized through cutting financially mature timber over a period of 15 or 20 years.

Table 2. --Log grade of managed loblolly and shortleaf pines by tree diameter class, Crossett Experimental Forest, 1952

D.b.h. (inches)	Grade 1 logs	Grade 2 logs	Grade 3 logs
	Percent of cubic volume		
14	19	23	58
16	26	23	51
18	33	23	44
20	41	21	38
22	48	21	31
24	55	20	25
26	63	19	18

Part of the upward trend in log grade ma
trees is due to normal improvement in grade wi
Nevertheless, much of it is due to a conscious p
cially mature timber of all sizes.

Effect of other trees. --In judging financ
of course, appraise a tree solely on its own me
sidered as a part of the whole stand, because th
more trees changes the stand and has its effect
forest tree is in more or less competition with
ent or prospective, and this competition affects
tree as an investment. Some trees that seem fi
judged by themselves will improve in vigor whe
ing trees are cut. Other trees, despite high ea
cially mature because of their influence on the f
trees. The principle covering the effect of othe
maturity is this: The aim of the forester shoulc
acre that volume of timber which, within the re
cultural system and the program of regulation,
ble conversion surplus in trees not yet financial
this principle will result in securing, from eacl
come that the desired rate of return will afford.
in footnote 1 for additional explanation.)

Therefore, where two competing trees b
but where removal of one would raise the vigor
to make it no longer mature, a general rule is
growth which has the higher value now or pros
size or higher grade, or the better chance of i
improving its grade. The other tree should be

As an example, assume that two pines,
are competing, and that if either is cut the vigo
come high. Following are the pertinent data on
table 1:

	Tr
Diameter breast high (inches)	i
Grade	1-!
Conversion surplus now	4
Medium vigor:	
Conversion surplus after 5 years	!
Rate of increase (percent)	
High vigor:	
Conversion surplus after 5 years	!
Rate of increase (percent)	

If a 4-percent rate of return is desired, one of these trees is mature and should be cut. The other should be left to grow at least one more cutting cycle. Since tree A has the higher value now and prospectively, it should be left to grow.

A comparison of these alternative trees will show the reasoning behind the rule. If the market value of No. 2 Common is $85 per M, tree A is at present worth 41.4 index units; this equals 41.4 percent of $85 or $35.19. Computing the dollar values of these alternative trees shows that the value added to the business over the next 5 years is:

	Cutting Tree A	Cutting Tree B
Value of tree cut, invested at 4 percent compound, earns	$4.28	$3.57
Growth in value of tree left:		
Tree A	...	10.03
Tree B	8.33	...
Total value added to the business	$12.61	$13.60

Since cutting tree B will increase the worth of the business by 99 cents more than the alternative of cutting A, tree A should be left to grow. The rule is designed to cover the majority of the numerous alternatives facing the forester. There will be exceptions, but the virtue of the rule is that it obviates making separate calculations for every alternative that arises.

Two additional points need to be made: The first concerns the conflict between rate of increase in tree value and the ability to increase income. The second is that some decisions require looking ahead two or more cutting cycles.

In the example above, the trees increased in value at equal rates. A somewhat more typical example is that of a low-grade tree of high vigor competing with a high-grade tree of low vigor. The data for two such trees, taken from table 1, follow:

	Tree A (high vigor)	Tree B (low vigor)
Diameter breast high (inches)	18	16
Grade	2-3	1-1-3
Conversion surplus now	14.5	18.2
Conversion surplus after 5 years	18.5	21.5
Rate of increase (percent)	5.0	3.4

If the high-vigor tree (A) is cut, the other can be counted on to improve in growth to that of medium vigor. If 4 percent is the rate of return desired, however, it does not automatically follow that tree B should be cut. By the same calculations as before, and assuming that No. 2 Common is worth $85 per M, the present value of tree A is $12.32 and of tree B $15.47. Cutting either of these trees and computing the value added to the business over the next 5 years has the following results:

	Cutting Tree A	Cutting Tree B
Value of tree cut, invested at 4 percent compound, earns	$2.67	$3.36
Growth in value of tree left:		
Tree A	...	3.40
Tree B	4.25	...
Total value added to the business	$6.92	$6.76

Since cutting tree A will increase the worth of the business by 16 cents more than the alternative of cutting B, tree B should be left to grow. Now the conflict between the percent of tree-value increase and the ability to increase income can be resolved. The desired rate of return is simply a break-even point below which the forest owner refuses to carry investments. The fact that low-value, fast-growing trees will often increase in value at a faster rate than high-value trees whose growth and rate of increase they inhibit does not always determine the choice of alternatives. Where a choice must be made among 2 or more trees, the one to leave is usually the one that is expected to make the greatest contribution to income between now and the time it becomes financially mature.

While many doubtful cases can be resolved by considering the estimated growth rates over the next 5 years, situations will arise that require looking ahead two or more cutting cycles. This will be particularly true when sawlog-size trees of the lower grades are compared with potentially high-grade trees, including seedlings, that have not yet reached sawlog size. Such problems require looking ahead more than the next 5 years. The calculations of the break-even points follow the same pattern as for those of alternative sawlog-size trees. That is, the potential returns from the smaller trees or seedlings must be balanced against the more immediate returns from the sawlog-size tree.

Other considerations relating to the stand may affect the maturity decision. A tree may be otherwise financially mature and still not ready for harvest if it is needed as a seed source. Again, the need for holding the cut within a cutting budget, or of building the growing stock towards optimum diameter-class distributions, may well result in leaving for another cutting cycle some financially mature trees. In some areas there may be insufficient timber volume to support a commercial cutting operation despite the presence of financially mature trees; that is, in cutting such trees direct costs would run up far above those in table 1.

On the other hand, the owner may be forced to cut some trees in order to obtain a regular income or a regular supply of logs for a sawmill. In this case the desired rate of return has risen, at least temporarily.

Marking Guides

Those who will apply the foregoing principles in the woods must constantly keep in mind that a large percentage of the trees below 25 to 28 inches in diameter are continuously increasing in log height or in quality as they grow. Table 3 (column 2) shows the probable course of log-length and log-grade development of a typical high-grade loblolly or shortleaf pine, growing 3 inches d. b. h. per 10 years, as it increases from 12 to 24 inches in diameter. As will be noted, a third log usually develops at about 14 inches and a fourth at about 18 inches.

Table 3. --Rate of value increase of typical high-grade, medium-vigor loblolly and shortleaf pines, with and without log height and grade changes

D. b. h. (inches) (1)	Tree grade (2)	Annual conversion surplus increase	
		With height and grade changes (3)	Without height and grade changes (4)
- - - - Percent - - -			
12	2-3	11. 2	5. 6
14	2-2-3	7. 8	5. 1
16	1-2-3	7. 1	4. 6
18	1-2-3-3	5. 2	4. 1
20	1-1-3-3	4. 9	3. 9
22	1-1-2-3	4. 3	3. 6
24	1-1-1-3	...	3. 3

Column 3 shows the rate of increase in
a tree that develops in height and grade as indi‹
umn 4 shows the rate of value increase if there
changes from one diameter class to the next. '
columns 3 and 4 again emphasizes the fact that
ise of improvement in grade or log length shoul
at least one more cutting cycle, even though it
threshold of financial maturity. Furthermore,
trees on vigor should be recognized. When tre
vigor are to be released by cutting, they shoulc
class before financial maturity is judged.

Among trees that show no
promise of development in height
and grade, the decision to cut or
to leave will depend almost wholly
upon the rate of growth and the
rate of return desired. For such
trees, the data in table 1 can be
condensed into a readily usable
set of marking guides. Table 4
represents one way of setting up
such guides. The diameters refer to the largest size of tree that
should be left to grow one more
5-year period; the spread in diameter allows fc
height and grade.

Table 4. --M

Vigor class	]

High
Medium
Low

If it is desired to reduce the alternative
marker still further, the diameter spreads ma'
diameters for each vigor class. Or, by assum
and higher vigor trees can be grown from seed
above 18 inches, as well as any tree of perhap‹
does not have a grade-1 butt log, might be con‹

APPENDIX

Timber Appraisal

In the purchase or sale of timber on the stump, column 2 of table 1 can be used to judge the gross lumber value per M of the different sizes and grades of trees listed. These values can be applied to the estimated tree volumes to appraise the index value of the timber. Multiplying the index value by the prevailing price of No. 2 Common lumber will give the gross lumber value of the standing trees. If volumes per tree and costs as listed are acceptable, column 6 will give conversion index values of trees directly for appraisal purposes. Stumpage growers and producers can both use the values listed in column 6 as a basis for arriving at stumpage prices. It must be remembered, however, that these figures represent conversion surplus and are not direct stumpage values.

Crossett Log Grades For Pine

Pine Log Grade No. 1

Surface-clear logs 10.0 inches or larger in diameter inside bark at the small end, and logs over 16.0 inches in diameter at the small end containing not more than three 2- to 4-inch knots, or the equivalent (usually a maximum of about 6) in small knots. Length 10 feet or longer. Logs of this grade having 15 percent or more of volume lost because of sweep, crook, or other external defects are reduced one grade. A loss of over 40 percent reduces the log to grade 3. A loss of over 50 percent of the volume culls the log. Logs of this grade are expected to produce 25 percent or more of B and B grade lumber or 60 percent or more of B and B and No. 1 C combined.

Pine Log Grade No. 2

Surface-clear logs 8.0 to 9.9 inches d. i. b. at the small end, logs over 8.0 inches containing numerous small knots, or logs over 14.0 inches d. i. b. at the small end containing numerous small knots or up to six 2- to 4-inch knots. Length 10 feet or longer. Logs of this grade having 20 percent or more of the volume lost because of sweep, crook, or other external defects are reduced one grade. A loss of 40 percent or more of the volume culls the log. Logs of this grade are expected to produce 10 percent or more of B and B lumber or 50 percent or more of B and B and No. 1 C grade combined.

- 17 -

Pine Log Grade No. 3

 Knotty or crooked merchantable logs 8.
small end that do not fall in grade 1 or grade 2
Length 10 feet or longer. Logs of this grade w
of the volume lost because of sweep, crook, or
are culled. Logs of this grade are not expecte
cent of B and B lumber or more than 40 percen
combined.

Conditions Applicable to All Grades

 Knots. --Small knots are defined as any
stubs of any size up to and including 1. 9 inches
knots are 2. 0 inches or larger in diameter.

 Knots that are bunched at one end of a l
are not as serious as a number of knots scatter
faces. Logs having such bunched knots are not
as called for by the above definitions unless cr
ent.

 Rot. --Logs showing unmistakable evide
automatically reduced one grade below that ind
surface characteristics. This reduction in gra
caused by sweep or other defect.

FIFTEEN YEARS OF MANAGEMENT ON THE CROSSETT FARM FORESTRY FORTIES

—R. R. Reynolds

SOUTHERN FOREST EXPERIMENT STATION

Harold L Mitchell, Director

OCCASIONAL PAPER 130   July 1953

Cover photos: Fourth, tenth, and fourteenth
annual harvests from the Crossett good forty.

CONTENTS

Figure 1. --The good forty at Crossett. Although 175,000 board feet of logs, 312 cords of pulpwood, 418 fence posts, and 228 cords of firewood and chemical wood have been removed in the last 15 years, the pine stocking is greater and of better quality than when cutting began.

FIFTEEN YEARS OF MANAGEMENT
ON THE CROSSETT FARM FORESTRY FORTIES

R. R. Reynolds
Southern Forest Experiment Station

As late as 1935 management of the woodlands on the average farm in the South was, for all practical purposes, non-existent. Yet, farm forestry seemed to offer great opportunities. The small land-owner usually had as much land in woodland as in row crops. He also often had a stand of trees--of sorts. All he seemed to need was proof that he could manage his forest and make money by doing so.

Shortly after the Crossett Experimental Forest was established in south Arkansas in 1937, it was decided to study the management possibilities of these small tracts of timber. The first objective was to attempt to change previous cutting practice and see if well-managed woodlands containing a mixture of loblolly and shortleaf pine could not be a very profitable part of the average farm. The second objective was to try to put the timber crop on the same annual-return basis as row crops such as corn or cotton. Back in 1937, these were radical propositions, for timber had never brought the farmers much income and the average farm woodland was badly depleted.

What should be done with run-down woodlands? How long would it take to build them up to where good returns could be obtained? Once they had been improved, how much income would they yield?

THE CROSSETT FORTIES

To answer such questions, two "farm forestry forties" were established in 1937. One of these, containing 34 acres, was the most lightly stocked tract on the entire Crossett Experimental Forest. On this, the "poor forty," it was hoped that the annual cuts of forest products would pay current costs of holding the land while the stands were being restocked to good trees.

The second tract was well stocked. It was intended as an example of the returns possible after the run-down stands were built up. It contains 40 acres and is known as the "good forty."

The stands when management began

In 1937 the good forty had, per acre, 1,794 cubic fe
bark) of pine in trees over 3.5 inches in diameter at brea
(d. b. h.) and 602 cubic feet of hardwood in trees over 4.5 ir
(table 1). Red and white oaks accounted for 42 percent of t
volume, other merchantable hardwoods 26 percent, and un
hardwood species the other 32 percent.

Table 1. --Stocking per acre on the Crossett forties when management began in :

D.b.h. (inches)	Pine						Hardwood			
	Good forty			Poor forty			Good forty			Trees 1/
	Trees	Basal area	Volume	Trees	Basal area	Volume	Trees 1/	Basal area 2/	Volume 2/	
	No.	Sq. ft.	Cu. ft.	No.	Sq. ft.	Cu. ft.	No.	Sq. ft.	Cu. ft.	No.
4	20.9	1.8	16.7	15.7	1.4	12.6	(3/)	(3/)	(3/)	(3/)
5	19.2	2.6	32.6	14.1	1.9	24.0	11.3	1.5	22.6	23.5
6	16.5	3.2	51.2	11.5	2.3	35.6	9.3	1.9	30.7	12.6
7	13.6	3.6	70.7	8.1	2.2	42.1	6.8	1.8	32.6	8.3
8	12.1	4.2	94.4	5.9	2.1	46.0	5.4	1.8	36.2	3.8
9	6.4	2.8	68.5	4.8	2.1	51.4	3.9	1.7	34.7	2.8
10	7.0	3.8	98.7	4.3	2.3	60.6	3.5	1.9	39.6	1.9
11	6.5	4.3	117.0	3.4	2.2	61.2	2.7	1.8	37.8	1.0
12	5.5	4.3	123.2	2.9	2.3	65.0	2.4	1.9	40.3	.8
13	4.6	4.2	124.7	3.0	2.8	81.3	2.3	2.1	45.3	.6
14	3.8	4.1	122.0	3.3	3.5	105.9	2.6	2.8	59.1	.4
15	3.6	4.4	134.6	2.5	3.1	93.5	1.3	1.6	33.4	.3
16	3.4	4.7	146.5	1.9	2.7	81.9	1.1	1.5	31.5	.3
17	3.0	4.7	147.3	1.9	3.0	93.3	.9	1.4	28.8	.2
18	2.0	3.5	111.0	.9	1.6	50.0	.6	1.1	21.2	.3
19	1.4	2.8	87.2	.6	1.2	37.4	.6	1.2	23.3	.1
20	1.2	2.6	83.6	.2	.4	13.9	.4	.8	17.0	.1
21	1.0	2.4	77.9	.2	.5	15.6	.4	.9	18.7	.0
22	.4	1.1	34.6	.1	.3	8.6	.2	.6	10.2	.0
23	.1	.3	9.5	.0	...	...	.2	.6	11.2	.0
24	.3	.9	31.0	.0	...	...	.2	.6	12.4	.0
25	.1	.3	11.1	.0	...	...	.1	.3	6.8	.0
26	.0	...	....	.0	...	...	.0	..	...	.0
27	.0	...	...	.0	...	...	.1	.4	8.2	.0
Total	132.6	66.6	1,794.0	85.3	37.9	979.9	56.3	30.2	601.6	57.0

1/ Forty-five percent of the hardwoods on the good forty and 8 percent of those on the poor fo:
chantable species.
2/ Merchantable plus unmerchantable trees.
3/ Four-inch class hardwoods not recorded.

The stand was typical of many unmanaged second-growth pine-hardwood forests. Most of the larger pines had developed under competition in the virgin stand. A large percentage of them had cleaned up to a good height before they had been released by the cutting of the original larger trees. Their quality was generally good, but some were crooked, had heart rot, or were mature. In addition to these virgin residuals, there were some strictly second-growth pines that had grown up with little competition. Most of these were of very low quality. Pine quality was likewise low on three small old-field patches on the compartment.

On the poor forty, hardwoods up to 8 inches in diameter were more numerous than pine. Total pine volume, however, averaged 980 cubic feet per acre as compared to a hardwood volume of 290 cubic feet. Less than half of the ground area of this compartment was being used by trees, and, of the area that was occupied, nearly half was in hardwood or pine of very low quality.

Except for an occasional red or white oak, few of the hardwoods above 12 inches in diameter on either forty contained logs that would produce factory-grade lumber. On both forties, too, the hardwoods that were below sawlog size were of such poor potential quality that they were usually not worth saving.

Type of silviculture employed

Since even trees of the same species do not all grow at the same rate, and since on most areas of 20 acres or more the stands were essentially all-aged or all-diametered (table 1), it was decided to adopt a modified single-tree selection system of silviculture in managing the farm forties. Under such a system, trees could be removed singly or in small groups, and each tree could be cut or left according to its individual merits at the time it was examined. Little difficulty was expected in getting reproduction under this system.

Because large, high-grade sawlogs had a ready market and would net a much greater return per cubic foot of volume than pulpwood, it was decided to produce as many such sawlogs as possible. Long, high-class piling may produce greater returns than sawlogs, but so few small owners know piling specifications or have a market that production of piling did not seem warranted--especially as sales of this material almost always remove the most desirable growing stock from a stand.

The sawlog objective, however, did not rule out the production of pulpwood, chemical wood, firewood, posts, and other subsidiary items, and some of these have been cut nearly every year since 1937. Such products come from those trees and parts of trees that need to be removed but that are not of saw-timber size or grade.

Where cull hardwoods of no economic value interfered with the growth of pines, it was recognized that such hardwoods would have to be removed at some expense. This is consistent with good management, and is a very important part of insuring full stocking.

HOW THE CUTS ARE MADE

Inventory system

In order to minimize error in the estimation of growth, number of trees, or changes in quality with the passage of time, a 100-percent continuous inventory system was adopted for this study. All pines larger than 3.5 inches in d.b.h. and all hardwoods larger than 4.5 inches in d.b.h. are tallied each time an inventory is made. In addition, the merchantable sawlogs in trees larger than 11.5 inches in d.b.h. are tallied and graded.

An annual inventory is not necessary to the success of the study, and so the inventories have been repeated at 5-year intervals only. The first was made in the early summer of 1937, prior to the initiation of the study, and the others in the autumns of 1941, 1946, and 1951.

Computation of allowable cut

The first annual cut was made on the good forty in the fall of 1938 and on the poor forty in the fall of 1939. These were salvage and improvement cuts to remove merchantable pines and hardwoods that were of very poor form or had rot in them. No allowable cut was determined on either forty this first year, but neither were all poor trees removed. The next 4 annual cuts were based on an estimate of the average yearly growth on each forty. Subsequent cuts were based on the estimated average yearly growth on each forty, as revealed by the 5-year inventories.

Since the good forty was reasonably well stocked, it was decided to remove each year a volume in cubic feet approximately equal to the total cubic-foot growth of pine larger than 3.5 inches in d.b.h. This plan has been followed for each of the last 14 years. On the poor forty

- 4 -

the annual cuts have removed only about half of the total yearly pine cubic-foot growth in trees larger than 3.5 inches in d.b.h. This plan has been followed in order to build up the growing stock to a desirable level.

At the end of each 5-year period, a change is made to bring the annual allowable cut for each forty in line with the growth determined from the inventory.

Not all of the area of each forty contributes to each cut. Every year the whole of each forty is inspected and the portion in greatest need of attention is selected for the annual harvest. Usually the cut is taken from about 10 acres, but sometimes 20 acres are covered, and sometimes only 5. This flexible system also makes it possible to cut annually, from any place on the forty, trees that have been damaged by lightning, insects, or redheart.

Type of trees removed in cutting

The principle of always removing the poorest and most mature trees and saving the best ones has been followed in all marking and cutting. However, not all crooked, rough, or non-perfect trees have yet been cut. Many trees with crooked tops or a sweep in the bole have two or more clear logs, are growing 3 to 4 inches in diameter in ten years, and have crowns so high from the ground that the understory can develop for many more years without serious interference.

Nor has perfect spacing of the growing-stock trees been attained in the first 15 years. The distribution of stems is constantly improving, but perfection probably will never be had. Good immature stems are never sacrificed to get good distribution, and immature trees of a given diameter class are not cut just because there may appear to be a temporary oversupply of this size.

All in all, the aim has been to improve quality, spacing, and diameter distribution gradually, without appreciably diminishing potential earnings, and to make the best use of the potential growth and possible earnings of each tree before cutting it. A first step in carrying out this aim was to deal with the hardwood problem.

THE HARDWOOD PROBLEM

As has been noted, the hardwoods on the good forty in 19
of poor quality. It was thought, however, that good stems could
veloped. Since this was a farm woodland, it seemed necessary
at least enough volume of the better hardwood species of good fo
provide the equivalent of an annual cut of firewood.

By 1943 it was apparent that the hardwoods on these upla
would grow very slowly even under management and could not be
compete with pine from the standpoint of dollar returns per unit
growing space. Beginning in 1944, therefore, the annual allowa
was based on pine growth only. The volume of any hardwoods th
could or should be removed was additional.

By 1947 the farmers who were likely to be interested in t
management on pine sites were no longer using wood for heating
cooking. Most of them had switched to fuel oil or gas. It was tl
fore decided to remove the remainder of the merchantable hardv
during the next four years. This was done, the final cut of hard
products on the good forty being made in the fall of 1950.

The poor forty had so many fire-scarred, rough, and lov
crowned hardwoods in 1937 that all those with any merchantable
ucts were removed as fast as the pine could take over the growii
that the hardwoods were occupying. The last merchantable harc
were cut in 1945.

Cutting the merchantable hardwoods from the forties, of
did not remove all of the hardwood trees. Stems of all sizes frc
20-inch cull hickories and elms down to 1-inch sprouts of many
were still occupying considerable space and suppressing pine re
tion. In order to reduce this competition, a general timber stai
provement job was undertaken in the spring of 1951. In this, al
maining hardwoods larger than 3.5 inches in diameter were gir
and smaller hardwood sprouts were cut wherever they were ove
any needed pine reproduction.

Cost of timber stand improvement

This job required a total of 2.6 man-hours per acre on t
forty and 2.4 man-hours for the poor forty.

A farmer would probably do such stand improvement for
If he did, his cash outlay for the work would not be over 2 cents

per year over the first 15-year period. If the work were done by hired labor at $0. 80 per hour, the labor cost would be $2. 08 per acre on the good forty and $1. 92 per acre on the poor forty. The cost per acre per year for the 15 years has, therefore, averaged $0. 14 for the good and $0. 13 for the poor forty. Over a period of years an allowance of 0. 2 man-hour, or about $0. 15 per acre per year, is probably ample for timber stand improvement on most farm woodlands where pine or reasonably good hardwood seedlings are present or can be obtained by natural reproduction.

This has been the first non-revenue cutting on the forties. It undoubtedly will not be the last. However, as will be shown later, pine seedlings now have enough growing room so that the next improvement cut is several years in the future (fig. 2).

HARVESTS AND RETURNS

Products removed

Fourteen annual harvest cuts have been made from the good forty during the 1938-51 period. These have totaled approximately 174, 500 board feet, Doyle scale (205, 600 board feet, International 1/4-inch scale), of logs. The cuts also yielded 312 cords of pulpwood, 228 cords of chemical wood and firewood, and 418 white oak and post oak fence posts (table 2).

The average yearly harvest has been 12, 470 board feet, Doyle scale (14, 690 board feet International), of sawlogs; 22. 3 cords of pulpwood; 16. 3 cords of firewood and chemical wood; and 29. 9 fence posts. Per acre, the various cuts have removed 4, 360 board feet, Doyle scale (5, 140 board feet International), of logs; 7. 8 cords of pulpwood; 5. 7 cords of firewood and chemical wood; and 10. 4 fence posts.

On the poor forty the 13 annual harvests have totaled approximately 57, 500 board feet, Doyle scale (67, 700 board feet International), of logs; 153 cords of pulpwood; 158 cords of firewood and chemical wood; and 121 fence posts (table 3).

The average cut per year for the 13 years has been 4, 420 board feet, Doyle scale (5, 210 board feet International), of logs; 11. 8 cords of pulpwood; 12. 2 cords of firewood and chemical wood; and 9. 3 fence posts. Per acre, these cuts have removed a total of 1, 690 board feet, Doyle scale (1, 990 board feet International), of logs; 4. 5 cords of pulpwood; 4. 7 cords of firewood and chemical wood; and 3. 6 posts.

Figure 2. --Good fire protection and the removal of low-quality hardwoods have enabled fast-growing pines to occupy practically all of the growing space on both forties.

Table 2.--Volume and value of products from the good farm forestry forty (Compartment 51--40 acres)

Year	Volume	Logs Stumpage value		Mill value		Volume	Pulpwood Value per cord		Total value		Volume	Chemical wood and firewood Value per cord		Total value		Value of all products		
	Bd. ft., Doyle	Per M Doyle	Total	Per M Doyle	Total	Standard cords	Stumpage	Mill	Stumpage	Mill	Standard cords	Stumpage	Mill	Stumpage	Mill	Stumpage 1/	Roadside	Mill 2/
		Dollars		Dollars			Dollars		Dollars			Dollars		Dollars		Dollars		
1937	5,122	8.00	40 98	15 75	80 67	3 58	1.00	6 00	3 58	21 48						3 58	12 53	21 48
1938	15,426	8.00	123 41	15 75	242 96	60 90	1.00	6 00	60 90	85 40						60 90	213 15	85 40
1939	19,519	8.00	156 15	15 75	97 42	18 04	1.00	6 00	18 04	108 24	62 60	0.50	8 00	31 30	500.80	92 42	394 22	86 01
1940						12 25	1.00	6 00	12 25	73 50	34 91	.50	8 00	17 46	279 28	154 62	377 43	60 24
1941						43 49	1.00	6 00	4 49	260 94	6 00	.50	8 00	3 00	48 00	204 29	412 80	621 31
1942	16,723	8.00	133 78	15 75	263 39	42 09	1.00	6 00	42 09	252 54	11 33	.50	8 00	5 66	90.64	183 28	97 55	611 82
1943	14,938	8.00	119 50	15 75	235 27	35 19	1.00	6 00	35 19	211 14	9 50	.50	8 00	4 75	76.00	41 19	344 43	527 66
1944	13,904	13.17	183 11	23 67	329 11	18 32	1.50	10 00	27 48	183 20	10 26	.75	8 78	7 70	90 06	2 0 04	413 83	607 62
1945	13,144	12.36	162 47	22 86	00 49	15 33	1.50	10 00	23 00	153 30	11 97	.75	12 00	8 98	143 64	196 55	80 14	63 73
1946	11,760	14.69	172 70	27 19	319 70	24 24	2.00	10 50	48 48	254 52	6 58	.75	15 00	4 94	98 70	227 92	83 12	88 32
1947	11,661	17.73	206.75	31 23	84 16	7 81	2.00	11 62	15 62	83 96	17 43	.75	12 00	13 07	209 16	236 67	48 85	661 03
1948	13,864	24.74	84 72	39 74	51 02	5 33	2.25	11 60	11 99	61 83	23 72	.75	12 00	17 79	374 39	374 39	88 27	902 14
1949	12,631	24.12	38 12	81 54	3 44	11 35	2.00	6 88	39 04	7 10	24 21	.50	11 75	320.75	87 23	89 70		
1950	12,099	32.58	94 14	47 58	95 63	4 04	2.25	12 25	9 09	49 49	19 94	.50	11 15	9 97	2 33	44 85	83 63	852 40
1951	13,741	39.08	57 06	55 08	26 92	17 75	3.00	13 50	13 50	239 62						90 31	93 43	996.54
Total 3/174,532		2,877.81		4,808 28	311 80		1.32	7.56	411.33		228 45	.58	9.67	131.72	2,210.22	3,441 76	6,440 61	9,439.40
Average	16.49	27.55																

1/ Includes stumpage value of $20.90 for 418 posts produced 1939-50. 2/ Includes market value of $62.70 for 418 posts produced 1939-50. 3/ Equal to 205,600 board feet, Int. 1/4-inch scale.

Table 3.--Volume and value of products from the poor farm forestry forty (Compartment 56--34 acres)

Year	Volume	Logs Stumpage value		Mill value		Volume	Pulpwood Value per cord		Total value		Volume	Chemical wood and firewood Value per cord		Total value		Value of all products		
	Bd. ft., Doyle	Per M Doyle	Total	Per M Doyle	Total	Standard cords	Stumpage	Mill	Stumpage	Mill	Standard cords	Stumpage	Mill	Stumpage	Mill	Stumpage	Roadside	Mill
		Dollars					Dollars		Dollars			Dollars		Dollars		Dollars		
1939	0					20 25	0.67	3.50	13.50	70 88	74 46	0.46	3.70	34.25	275.50	151 38	203.12	2/354 85
1940	2,996	6 10	18.27	10.60	31.75	10 22	.67	3.75	6 82	38 32	3 40	.48	4.12	1.63	14.01	26.72	55 40	84.08
1941	2,748	8 83	24.26	14.83	40.75	10 84	1.00	5.15	10 84	55 83	00					35 10	65.84	96.58
1942	5,314	9.39	49.90	16.39	87.10	4 45	1.00	6.00	4 45	26 70	9 77	.75	5.50	7.33	53.74	41 68	114 42	167 54
1943	5,134	12.02	61.71	21.02	107.92	21 48	1.00	7.75	21.48	166.47	31 95	.70	8.77	22.36	280.20	83.19	178.79	276.39
1944	0					60 00	1.50	10.00	8.00	600.00	38 84	.75	11.98	29.13	465.30	112 36	96 28	880.20
1945	0					36	1.50	10.00	.54	3 60	00					29 67	249.28	468.90
1946	3,790	16.37	62.05	28.86	109.38	2 67	2.00	10.50	5 34	28 04	00					67 39	102 40	137.42
1947	4,521	17.08	77.23	30.58	138.26	5 16	2.00	10.75	10 32	55 47	00					87 55	140.64	193.73
1948	7,533	24.14	181.88	39.14	294.87	6 73	2.25	11.60	15 14	78 07	00					197 02	284 98	372.94
1949	10,309	26.38	271.98	40.38	416.32	00					00					271 98	44 15	416.12
1950	6,502	42.58	276.84	57.58	374.37	6 02	2.25	12 25	13 54	73 74	00					290.38	89 24	448.11
1951	8,627	39.70	342.46	55.70	480.50	3 00	3.00	13.50	13 50	63 45	00					36 56	80 26	543.95
Total 3/57,474		1,366.58		2,081.22	152 88		1.35	8.25	206 07		98 42	.60	6.87	94.70	1,088.75	1,670.98	3,054 99	4,439.01
Average	23.78	36.21																

1/ Includes stumpage value of $3.63 for 121 posts produced in 1939. 2/ Includes market value of $8.47 for 121 posts produced in 1939. 3/ Equal to 67,700 board feet, Int. 1/4-inch scale.

Sawlogs have produced the greatest return for a given volume of wood. Consequently, every tree that would make lumber was cut into one or more logs. Hardwood sawlogs made up only 9 percent of the total cut on the good forty and only one percent of the cut on the poor forty (table 4). These figures help to substantiate the statement that the hardwood was very low grade on both compartments.

Table 4. --Proportions of products in total fifteen-year cut from the farm forties

Product	Good forty			Poor forty		
	Pine	Hardwood	Total	Pine	Hardwood	Total
	- - - - - - - - - - Percent - - - - - - - - - -					
Sawlogs	30	9	39	30	1	31
Pulpwood	30	8	38	36	(1/)	36
Firewood and chemical wood	...	23	23	...	33	33
Posts	...	(1/)	(1/)	...	(1/)	(1/)
Total	60	40	100	66	34	100

Basis: Total cubic volume, inside bark, of all products removed.
1/ Negligible.

Of the total volume of products, only pine sawlogs--which made up 30 percent of the total cut from each forty--came from what might be considered permanent growing stock.

Returns from annual harvests

Good forty. --At the stumpage prices that prevailed locally each year in which the harvesting was done, the 174,500 board feet (Doyle scale) of sawlogs yielded $2,878 (table 2). The 312 cords of pulpwood returned $411. The 228 cords of chemical wood and firewood were worth $132, and 418 posts $21. Thus, the total stumpage value produced to date has been $3,442, or $86 per acre--$6.15 per acre per year. Had present-day prices ($40 per thousand board feet, Doyle scale, for logs; $3.00 per standard cord for pulpwood; $0.50 per cord for chemical wood and firewood; and $0.10 each for posts) prevailed throughout the 14 years, total stumpage returns would have equaled $8,073, or $202 per acre--$14.42 per acre per year.

The value of the products delivered to the mill, market, or railroad siding totaled $9,439 for the 14 crops. This amounted to $236 per acre, or $16.86 per acre per year. At present-day values, the returns would have been $16,808, or $420 per acre--$30 per acre per year.

Some farmers do not have the equipment to haul logs and pulp-wood. They do, however, usually have the means to cut and bunch the products alongside a road. It is estimated that the roadside value of the products harvested to date has been $6,441. This is approximately $161 per acre, or $11.50 per acre per year. Had present-day prices and costs prevailed throughout the study, the roadside value of the products to date would have been approximately $12,441, or $311 per acre--$22 per acre per year.

Poor forty. --It has been noted that, in order to build up the growing stock, the cuts on the poor forty have been held to only half of the estimated growth each year. Returns, however, have been far above expectations. The 13 annual cuts have produced 57,474 board feet (Doyle scale) of logs, worth $1,367 at prices prevailing in the year the harvesting was done. The 153 cords of pulpwood were worth $206. Chemical wood and firewood equivalent to 158 standard cords was worth $95, and 121 fence posts had a stumpage value of $4 (table 3).

Thus, the total stumpage return to date has been $1,671, or $49.15 per acre--$3.78 per acre per year for the 13 cuts. Since out-of-pocket costs, including timber stand improvement and taxes, would have averaged less than $0.35 per acre per year, the average owner would have netted better than $2.90 per acre per year for the 15 years of management.

If present-day stumpage prices had prevailed over the life of the study, the total stumpage return would have been $2,849, or $83.79 per acre--$6.45 per acre per year.

At the prices that were current during the year when harvesting was done, the roadside value of the products from the poor forty to date has been approximately $3,055, or $90 per acre--$6.91 per acre per year.

At present-day stumpage prices and costs, it is estimated that roadside value would have been $5,151, and total value delivered to mill or market $7,452. Thus, the roadside value per acre per year would have been $11.65, and the market value $16.86 for the 13 cuts made to date.

The yearly total average mill or market value of the products from the good forty has been nearly $700. The yearly cuts from the 34-acre poor forty have averaged $341. Under present-day prices these values would have been $1,200 per year for the good forty and $573 for the poor forty. Such returns are equal to the total returns that many farmers in the upland areas make from row-crop farming.

Cost of producing the forest crop

Forest management also has its costs.
in south Arkansas and north Louisiana vary fror
acre per year. The average is about $0.15. W
ed with the State fire protection agencies--at a (
per year. The owner undoubtedly would also sp
watching for or extinguishing fires that threaten
rarely would this entail additional out-of-pocket

Marking trees for cutting, supervising tl
the products have required less than 2 man-day:
However, since the average owner would have l(
trained personnel used on the Crossett forties,
forty per year has been allowed. On this basis,
these jobs would average about $3 per forty per

The total out-of-pocket costs and the lab
per acre for either of the forties are estimated

Dol

Taxes   0.
Fire protection   .
Timber stand improvement
Timber marketing   .|
Scaling and supervision   .

Total

The total 15-year costs for the good for
600 man-hours of labor. The 15-year costs for
poor forty would be $137.70 plus 510 man-hour

Assuming that all products were sold as
tree, the net return for the good forty would be
or $3,279.76. This is equal to $5.47 for each
pended in producing the crop; it of course inclu
ment as well as a return for labor.

The net stumpage return for 15 years of
forty is $1,670.98 minus $137.70 or $1,533.28
$3.01 per man-hour of labor, again including ir

Hourly returns for harvesting the forest crop

The foregoing are stumpage returns. If the owner harvested his timber himself he would be paid for his labor as well as for his wood. Tables 5 and 6 record the man-hours expended in producing the annual harvests from both forties. Table 7 gives the average man-hour requirements for producing a unit of each product.

Table 5. --Labor requirements for harvesting the good farm forestry forty (Compartment 51--40 acres)

Year	Logs			Pulpwood			Chemical wood and firewood			Total labor requirements[1]	
	Volume produced	Labor		Volume produced	Labor		Volume produced	Labor			
		To roadside	To mill		To roadside	To mill		To roadside	To mill	To roadside	To mill
	Bd. ft., Doyle	- - Man-hours - -		Standard cords	- - Man-hours - -		Standard cords	- - Man-hours - -		- - Man-hours - -	
1937					3.58	21.48	36.16				21.48	36.16
1938				60.90	365.40	615.09				365.40	615.09
1939	5,122	24.38	30.73	18.04	108.24	182.20	62.60	582.18	964.04	721.10	1,185.37
1940	15,426	73.43	92.56	12.25	73.50	123.73	34.91	324.66	537.61	476.09	759.90
1941	19,519	92.91	117.11	43.49	260.94	439.25	6.00	55.80	92.40	414.60	655.36
1942	16,723	79.60	100.34	42.09	252.54	425.11	11.33	105.37	174.48	442.76	706.93
1943	14,938	71.10	89.63	35.19	211.14	355.42	9.50	88.35	146.30	375.84	598.35
1944	13,904	66.18	83.42	18.32	109.92	185.03	10.26	95.42	158.00	276.77	433.45
1945	13,144	62.57	78.86	15.33	91.98	154.83	11.97	111.32	184.34	272.17	426.43
1946	11,760	55.98	70.56	24.24	145.44	244.82	6.58	61.19	101.33	268.01	423.91
1947	11,661	55.51	69.97	7.81	46.86	78.88	17.43	162.10	268.42	268.22	422.27
1948	13,864	65.99	83.18	5.33	31.98	53.83	23.72	220.60	365.29	323.22	508.50
1949	12,631	60.12	75.79	3.44	20.64	34.74	14.21	132.15	218.83	219.06	337.56
1950	12,099	57.59	72.59	4.04	24.24	40.80	19.94	185.44	307.08	272.22	427.07
1951	13,741	65.41	82.45	17.75	106.50	179.28	.00	...	...	171.91	261.73
Total	174,532	830.77	1,047.19	311.80	1,870.80	3,149.17	228.45	2,124.58	3,518.12	4,888.85	7,798.08

[1] Totals include values for 418 posts produced in the 1939-50 period man-hours to roadside 62 70, to market 83.60.

Table 6. --Labor requirements for harvesting the poor farm forestry forty (Compartment 56--34 acres)

Year	Logs			Pulpwood			Chemical wood and firewood			Total labor requirements	
	Volume produced	Labor		Volume produced	Labor		Volume produced	Labor			
		To roadside	To mill		To roadside	To mill		To roadside	To mill	To roadside	To mill
	Bd. ft., Doyle	- - Man-hours - -		Standard cords	- - Man-hours - -		Standard cords	- - Man-hours - -		- - Man-hours - -	
1939				20.25	121.50	204.52	74.46	692 48	1,146 68	[1]632.13	[1]1,375 40
1940	2,996	14 26	17 98	10.22	61.32	103.22	3.40	31.62	52.36	107 20	173 56
1941	2,748	13.08	16 49	10.84	65.04	109.48	.00	...	...	78 12	125 97
1942	5,314	25 29	31.88	4.45	26.70	44.94	9.77	90.86	150.46	142.85	227.28
1943	5,134	24.44	30.80	21.48	128.88	216.95	.00	...	...	153.32	247.75
1944	0	...	...	60.00	360.00	606.00	31.95	297.14	492.03	657.14	1,098.03
1945	0	...	...	.36	2.16	3.64	38.84	361.21	598.14	363.37	601.78
1946	3,790	18.04	22.74	2.67	16.02	26.97	.00	...	...	34.06	49.71
1947	4,521	21.52	27.13	5.16	30.96	52.12	.00	...	...	52.48	79,25
1948	7,533	35.86	45.20	6.73	40.38	67.97	.00	...	...	76.24	113.17
1949	10,309	49.07	61.85	.00	...	...	00	...	...	49.07	61.85
1950	6,502	30.95	39.01	6.02	36.12	60 80	00	...	...	67.07	99.81
1951	8,627	41.06	51.76	4.70	28 20	47.47	00	...	.	69.26	99.23
Total	57,474	273.57	344.84	152.88	917 28	1,544 08	158 42	1,473 31	2,439.67	2,682.31	4,352.79

[1] Includes 121 posts produced in 1939 man-hours to roadside 18.15, to market 24.20.

Table 7. --Average labor requirements per product unit

Product	To roadside	To mill or market
	- Man-hours -	
Logs, per MBM	4.76	6.0
Pulpwood, per cord	6.00	10.1
Chemical wood and firewood, per cord	9.30	15.4
Posts, each	.15	.2

The 14 annual cuts on the good forty have produced nearly ¡ man-hours of labor opportunity for harvesting products and delive them to mill or market. This is equal to 557 man-hours, approxi 3-1/2 man-months, per year. The labor opportunity for delivery the products to roadside has been nearly 5,000 man-hours--an ave of 349 man-hours, about 2 man-months, per year.

When total out-of-pocket costs of producing the forest crop deducted from total stumpage returns for the 1937-51 period, the ¡ is $3,279.76. Deducting the same costs from mill values leaves $9,277.40. The difference between the two figures, $5,997.64, i¡ return for logging and for hauling the products to mill or market. divided by 7,798 man-hours gives a return equal to $0.77 per ma; for labor and use of equipment. The hourly return has varied fror $0.25 per hour in the early years of the study, when prices and w¿ were low, to $1.11 per hour during 1946-51.

The return per hour for delivery of forest products, f.o.b. roadside, has been similarly computed. The average for the 14-y period is $0.61 per hour of labor, and the return for 1946-51 $0. ¡ per hour.

Since 1939 the poor forty has produced approximately 2,70 hours of labor opportunity for delivery of the forest products to r¢ and 4,350 man-hours for delivery to market. This is equal to 42 ¡ days per year for delivering the products to market. The return ¡ delivering the products to roadside has averaged $0.52 and to the ¡ or market $0.64 per hour. Over the last 5-year period, the retu¡ has averaged $1.23 per hour for delivery of products to roadside $1.70 per hour for delivery to mill or market. Returns per hour labor have been greater on the poor than on the good forty for the ¡ 5 years because pine sawlogs, which require the least labor per u volume of any product, have made up a larger proportion of the t¢ production on the poor forty than on the good.

THE STANDS AFTER 15 YEARS OF MANAGEMENT

Changes in stand structure and volume

Good forty. --The 14 annual cuts on the good forty have removed 51. 5 pines per acre from the original stand of 132. 6 pines per acre that were larger than 3. 5 inches in d. b. h. (table 8). These cuts have also removed an average of 48. 7 per acre of the original 56. 3 hardwoods larger than 4. 5 inches in d. b. h. The remaining hardwoods have been girdled. On the face of it, it looks as if sooner or later the area would run out of pine trees. This subject needs some consideration.

First, it must be remembered that until management started in 1937 the stand had never had any attention other than the cut of virgin timber to a 12-inch diameter in 1915, and fire protection starting in 1933. As a result, many of the smaller pines were badly suppressed or very crooked. A great many of these were removed in thinnings and salvage cuttings in the early years. Two ice storms took some more, only part of which could be salvaged. All together, salvage and thinning accounted for 38. 3 out of the total of 51. 5 pine trees per acre that were removed during the 15-year period. The harvest cutting of the more valuable sawlog-size trees removed only 13. 2 trees per acre--less than one tree per acre per year! Of equal importance, the number of trees that were more than 12 inches in diameter not only increased from 30. 4 to 36. 1 per acre during the period but the average size of these trees increased from 15. 6 to 16. 5 inches in d. b. h. (fig. 3).

From this it is reasonable to conclude that the number of trees cut each year in the second 15-year period will be considerably less than the number cut per year so far. However, it is also apparent that far fewer trees will have to be cut each year to get a given volume, and certainly to get a given return per acre.

The good forty now has 82 pines per acre that are 4 inches in diameter and larger (tables 9 and 10). If most of these grow into the 12-inch and larger diameter classes, as they are expected to do, they will yield annual cuts for about 80 years in the future if one sawlog-size tree per acre per year is removed. If two trees per acre are cut each year, the present stand 4 inches in d. b. h. and larger will support about 40 annual cuts without any new trees growing into merchantable sizes. The area is not likely to run out of pines at any time soon.

Nevertheless, a few more trees in the 4- to 11-inch diameter classes would be welcome. Fortunately, this is not a critical short-coming, and it will not last much longer. It arose largely because the

Table 8. --Pine cut per acre by diameter classes, 1937-51 1/

D.b.h. (inches)	Good forty			Poor forty		
	Trees cut No.	Basal area Sq. ft.	Volume Cu. ft.	Trees cut No.	Basal area Sq. ft.	Volume Cu. ft.
4	2.3	0.2	1.8	1.8	0.2	1.4
5	8.6	1.2	14.6	3.9	.5	6.6
6	8.4	1.6	26.0	3.5	.7	10.8
7	6.7	1.8	34.8	2.6	.7	13.5
8	4.8	1.7	37.4	1.9	.7	14.8
9	3.2	1.4	34.2	2.2	.8	20.3
10	2.4	1.3	33.8	2.2	1.2	31.0
11	1.9	1.2	34.2	2.0	1.3	36.0
12	1.8	1.4	40.3	1.8	1.4	40.3
13	1.6	1.5	43.4	1.4	1.3	37.9
14	1.2	1.3	38.5	1.1	1.2	35.3
15	1.1	1.3	41.1	1.5	1.8	56.1
16	.8	1.1	34.5	1.6	2.2	69.0
17	1.2	1.9	58.9	.9	1.4	44.2
18	1.0	1.8	55.5	1.4	2.5	77.7
19	.8	1.6	49.8	1.0	2.0	62.3
20	.8	1.7	55.8	.6	1.3	41.8
21	.8	1.9	62.3	.6	1.4	46.7
22	.9	2.4	77.8	.4	1.1	34.6
23	.5	1.4	47.4	.1	.3	9.5
24	.3	.9	31.0	.0	.3	10.3
25	.2	.7	22.2	.0	...	...
26	.1	.4	11.9	.0	...	...
27	.1	.4	12.7	.0	...	...
Total	51.5	32.1	899.9	32.3	24.3	700.1

1/ All merchantable hardwoods have either been cut or girdled. See table 1 for approximate number and volume by diameter classes.

Table 9. --Pine stocking per acre on the farm forestry forties, 1951

D.b.h. (inches)	Good forty			Poor forty		
	Trees No.	Basal area Sq. ft.	Volume Cu. ft.	Trees No.	Basal area Sq. ft.	Volume Cu. ft.
4	9.8	0.9	7.8	30.9	2.7	24.7
5	8.7	1.2	14.8	22.2	3.0	37.7
6	5.5	1.1	17.0	14.2	2.9	46.2
7	4.4	1.2	22.9	11.9	3.2	61.9
8	4.4	1.5	34.3	6.7	2.3	52.3
9	4.6	2.0	49.2	6.1	2.7	65.3
10	4.4	2.4	62.0	6.5	3.5	91.6
11	4.6	3.0	82.8	6.0	4.0	108.0
12	4.7	3.7	105.3	5.6	4.4	125.4
13	5.1	4.7	138.2	4.0	3.7	108.4
14	4.4	4.7	141.2	2.8	3.0	89.9
15	3.4	4.2	127.2	2.3	2.8	86.0
16	3.5	4.9	150.8	2.1	2.9	90.5
17	3.0	4.7	147.3	2.2	3.5	108.0
18	2.8	4.9	155.4	1.1	1.9	61.0
19	3.1	6.1	193.1	1.7	3.3	105.9
20	1.8	3.9	125.5	.9	2.0	62.7
21	1.8	4.3	140.2	.8	1.9	62.3
22	1.0	2.6	86.4	.5	1.3	43.2
23	.7	2.0	66.4	.3	.9	28.5
24	.4	1.3	41.3	.1	.3	31.0
25	.2	.7	22.2	.1	.3	11.1
26	.1	.4	11.9	.1	.4	11.9
27	(1/)	(1/)	(1/)	.0	...	...
28	.1	.4	13.5			
Total	82.5	66.8	1,956.7	130.0	57.5	1,513.5

1/ Negligible.

Table 10. --Pine growing stock per acre, by inventory periods

Item	Good forty				Poor forty			
	1937	1941	1946	1951	1937	1941	1946	1951
Trees:	- - - - Number - - - -				- - - - Number - - - -			
3.5 to 11.5 inches	102	91	51	46	68	71	68	105
11.5 inches +	30	31	31	36	17	21	19	25
Total	132	122	82	82	85	92	87	130
Basal area:	- - - Square feet - - -				- - - Square feet - - -			
3.5 to 11.5 inches	26	25	16	13	17	19	18	24
11.5 inches +	40	43	44	54	21	27	26	33
Total	66	68	60	67	38	46	44	57
Total volume:	- - - Cubic feet - - - -				- - - Cubic feet - - - -			
3.5 to 11.5 inches	550	524	354	291	334	394	373	488
11.5 inches +	1,244	1,339	1,378	1,666	646	819	808	1,026
Total	1,794	1,863	1,732	1,957	980	1,213	1,181	1,514
Saw-timber volume (Doyle):	- - - Board feet - - -				- - - - Board feet - - - -			
11.5 to 19.5 inches	4,027	4,413	4,791	5,571	2,191	2,840	3,057	3,399
19.5 inches +	1,047	1,353	1,663	2,724	150	295	644	1,184
Total	5,074	5,766	6,454	8,295	2,341	3,135	3,701	4,583
Saw-timber volume (Int. 1/4):	- - - Board feet - - -				- - - - Board feet - - - -			
11.5 to 19.5 inches	4,744	5,199	5,644	6,563	2,581	3,346	3,602	4,004
19.5 inches +	1,234	1,593	1,959	3,209	176	347	758	1,395
Total	5,978	6,792	7,603	9,772	2,757	3,693	4,360	5,399

hardwoods that occupied much of the area over the first 10 years kept out considerable pine reproduction. Some pine, however, did come in under the hardwoods, and an abundance of seedlings has shown up in the last several years.

Where these young pines occur in any kind of opening they are growing rapidly. Some that are now just under 3.5 inches in d.b.h. will soon overtake in diameter some of the older and larger trees that have been present for some time. Over the next few years, therefore, there will be a big surge of reproduction into the 4- to 8-inch size classes. These young trees will soon fill the diameter gaps in the present stand and will supply plenty of trees to support the present rate of cutting. In fact, they promise a big increase in future pulp-wood production.

Poor forty. --The 13 cuts on the poor forty have removed 32 of the original 85 pine trees per acre (table 8). In spite of this relatively heavy cut, the stand in 1951 had 130 pine trees per acre (tables 9 and 10) that were 4 inches in d.b.h. or larger--half again as many as when

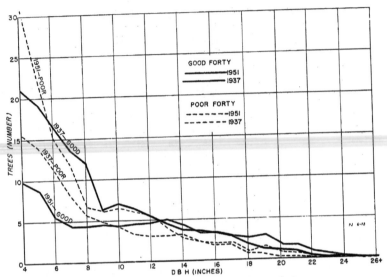

Figure 3—Number of pine trees per acre, good and poor forties

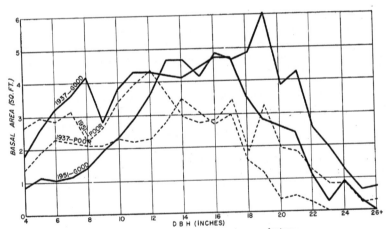

Figure 4—Pine basal area per acre, good and poor forties

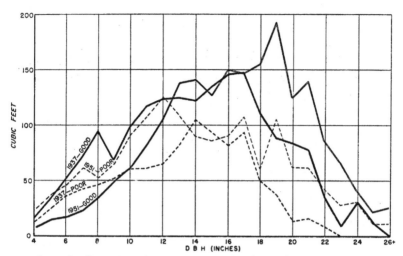

Figure 5.— Pine cubic volume per acre, good and poor forties

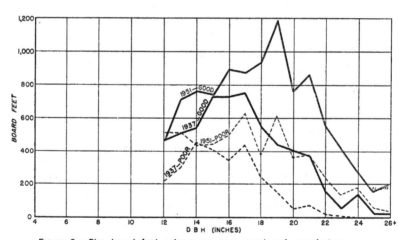

Figure 6 —Pine board-foot volume per acre, good and poor forties.

management started. The pine stems are still the same average size--
9 inches in d. b. h. --as in 1937. This means that it was not necessary
to remove too many of the largest trees merely to keep up the size of
the annual cuts. As figure 3 indicates, there has been a respectable
increase in number of pine trees in practically every diameter class.
Basal area of pine increased 19 square feet per acre--or 50 percent
(fig. 4). Over the 15 years the pine growing stock has increased 534
cubic feet per acre. Thus the stand had 54 percent more pine volume
in 1951 than in 1937 (fig. 5). Even though approximately 81 percent
of the original board-foot volume (International 1/4-inch scale) in
trees 12 inches in d. b. h. and above was removed in the 1937-51 period,
in 1951 the volume of pine was actually 2, 642 board feet--or 96 percent--
greater than in 1937 (table 10 and fig. 6).

        To date, neither forty has achieved an ideal stand distribution.
There perhaps are too many trees in some diameter classes and too
few in others. The basal areas of trees in the small, medium, and
large size classes are not in proper proportion, and not all diameter
classes are represented on each acre.

        However, since the production of large, high-grade sawlogs is
the goal of management, stems in the 12-inch class and below are
considered merely as a pool of reserve trees to be drawn on as needed to fill gaps which cutting leaves in the larger size classes. Present
indications are that greatest returns can be obtained by carrying less
than 20 percent of the total cubic volume of growing stock in the 10-inch
and smaller diameter classes. In any event, good large but immature
trees are never cut merely to get a superabundance of seedlings, to
obtain maximum height growth on saplings, or to give more nearly
perfect tree distribution among the various diameter classes.

Pine growth

        The four 100-percent inventories made at 5-year intervals gave
a good basis for estimating average annual growth for these three periods. The calculations have been made by determining differences in
growing stock volume since the last inventory, and adding to this figure
the volume removed in the particular 5-year period. The resulting total,
when divided by 5, gives average annual growth. All calculations are
made in cubic feet (inside bark), which are then converted to board feet
for an estimate of the volume growth in sawlog-size trees. Percentage
growth is determined by dividing the volume present at the beginning
of the period into the yearly growth and multiplying by 100. Thus these
values are net periodic annual growth percents (table 11); non-salvage-
able mortality has been excluded.

Period	Cubic volume		Board-foot volume		
	Cu. ft.	Percent	Doyle scale	Int. 1/4-inch scale	Percent
GOOD FORTY					
1937-41[1/]	81	4. 5	244	287	4. 8
1942-46	52	2. 8	372	439	6. 5
1947-51	98	5. 6	596	702	9. 2
POOR FORTY					
1937-41[1/]	86	8. 8	231	272	9. 8
1942-46	57	4. 7	251	295	8. 0
1947-51	118	10. 0	391	461	10. 6

1/. Because the 1937 inventory was made in the summer, annual growth for 1937-41 is computed by using a 4-year period.

The drop in cubic-foot and percentage growth during the 1942-46 period undoubtedly reflects the severe damage suffered in 1944, when a heavy load of ice broke or tipped over a large number of 4-to 9-inch trees. As much of this material as possible was salvaged, but a large volume was so badly split that it could not be used.

As a result of the early removal of the large volume of low-grade hardwoods, the big increase in the number of pine stems, and the large number of these pine stems growing into the merchantable diameter classes, the cubic-volume growth per acre is now greater on the poor than on the good forty. The pine stocking of the poor forty was approximately 1, 500 cubic feet per acre in 1951, and it is rapidly catching up to the 1, 957 cubic-foot stocking of the good forty (fig. 7). Because much of the growing stock volume is in rapidly growing small trees, the cubic volume percentage growth for the poor area is nearly double that of the more heavily stocked good area. The very satisfactory increase in board-foot growth on both forties would also make any land-owner happy. Much of it has been brought about by the removal of the low-grade hardwoods and of the mature and slow-growing pine. In-growth into the sawlog sizes has accounted for an unusual proportion of the board-foot growth on the poor forty. The number of fast-growing big trees is responsible for a considerable portion of the very good sawlog growth on the good forty.

Figure 7. --The poor forty after 15 years of management. Small pines have filled nearly all of the openings in the stand, and the forest is now very near to full stocking.

Pine stocking

     Since 1937 the trees on the two forties have produced one fair (1942) and two good crops (1939 and 1950) of pine seed. Much of the 1939 crop was lost because a great deal of the area was still occupied by low-grade hardwoods. Removal of many of the hardwood stems during the 1940's saved many of the seedlings of the 1942 crop. The timber stand improvement job in 1951, when all remaining hardwoods 4 inches in d. b. h. and larger were cut or girdled, released many of the suppressed seedlings of the earlier crops and allowed many seedlings from the 1950 crop to become well established. No fire, controlled or natural, has burned any part of either area since before 1933. What part of the total area is now occupied by pine?

     A survey of pine stocking was undertaken in the late summer of 1951. Temporary milacre (1/1000-acre) plots were taken at two-chain intervals on lines spaced two chains apart. Each plot was classified as to whether it was overtopped by a pine or hardwood crown (and hence did not need or could not use a pine seedling) or was free of overstory competition. Next, each plot was classified as to whether it contained a pine smaller than 3.5 inches in d. b. h., and whether this pine needed release.

     It was found that 91 percent of the milacre plots on the good forty and 93 percent of those on the poor forty were either stocked with small pines or overtopped by large pines (table 12). Such stocking must be considered excellent, for it is rarely exceeded even in the most successful pine plantations.

THE CROSSETT FORTIES AND THE SMALL LANDOWNER

     What can the average owner of a small tract of woodland expect from intensive management? There is plenty of room for difference of opinion on the subject, but 15 years of experience with the Crossett forties would seem to have the following significance to the farmer or small woodland owner:

     Most owners do not have the good stocking and large trees of the good forty, and consequently very few can expect, for some years to come, returns similar to those from this area. The good forty is primarily indicative of the eventual goal of management. The poor forty, however, does show how the more typical understocked and unmanaged farm woodland can be made to produce. As has been indi-

cated, it has been built up from a run-down to a highly p
est in 15 years. In 1951 the average stumpage return p
over $10.00.

Table 12. --Overtopping conditions, stocking, and need
pine, 1951

Stocking	Good for
	Percen
Overtopping conditions--trees larger than 3.5 inches in d. b. h.	
Overtopped by pine	59
Overtopped by hardwood	2
Free of overtopping tree	39
Available for pine reproduction	41
	100
Pine stocking	
Overtopped by pine	59
Available area stocked with pine 3.5 inches d. b. h. and smaller	32
Overtopped or stocked with pine	91
Need for release of pine	
Needing release	4
Requiring no release	87
Overtopped or stocked with pine	91

The owner of the average small woodland does n
knowledge of forestry to obtain returns nearly as good a
the Crossett forties--provided his stocking is similar.
knowledge of how trees grow and occasional help from a
that is required. A forester's help is most likely to be
owner has to determine inventory, growth, and allowab.
how to select trees for harvest; and to deal with unwant

Annual cuts have proved thoroughly feasible and
but are not required. The small landowner may choose
stumpage or make his cuts at longer intervals, althougl
than 5 years are not recommended. The shorter the pe
it is to salvage damaged trees and to anticipate mortali

The farmer with the necessary equipment can greatly increase his total return and have a profitable off-season job by doing his own logging.

Finally, it is believed that the timber-growing potentialities of the Crossett forties are not much different from those on most small forests in the same region. Since the typical small stand is considerably depleted, the owner will have to reduce his cut below growth for several years and otherwise approximate the measures used at Crossett. If he will do these things, he should be able to get fully 90 percent of the growth obtained on the Crossett forties.

SUMMARY

In 1937 a poorly stocked woodland of 34 acres and a well-stocked area of 40 acres were established on the Crossett Experimental Forest, in southern Arkansas, to determine the possible costs and returns from intensively managed small tracts of timber in the shortleaf-loblolly pine-hardwood type.

In order to create greatest possible interest in the management of farm woodlands, it was decided to determine if it was possible to make annual cuts of products.

Management has been on a modified single-tree selection system. The aim has been to produce a maximum of high-grade sawlogs, plus whatever pulpwood and other products would come from thinnings, tops of sawlog trees, and improvement cuttings.

The annual cuts on the good forty have been approximately equal to the yearly growth. Because of the need for building up the growing stock on the poor forty, annual cuts there have been limited to approximately one-half of the total pine cubic-foot growth each year.

To minimize errors in estimation of growth, a 100-percent inventory system was adopted. Complete reinventories are made every 5 years. The annual allowable cuts are determined from the differences between these inventories.

All cutting during the first 15 years has aimed at removing the poor, limby, defective trees first and saving the best for additional growth.

Since merchantable hardwoods on the poor forty w
grade, they were removed during the first seven years of
several years much of the hardwood on the good forty, ev
good quality, was saved to provide an annual cut of firewo
middle 1940's, however, it was apparent that other fuels
wood, and all merchantable hardwoods were cut and sold.

All non-merchantable hardwoods 3.5 inches and la
diameter were girdled in 1951. Cost of this operation wa
hours or $2.08 per acre on the good forty and 2.4 man-ho
per acre on the poor forty.

The fourteen annual harvests on the good forty hav
174,500 board feet (Doyle scale) of logs, 312 cords of pul
cords of firewood and chemical wood, and 418 white oak f
Thirteen annual cuts on the poor forty have yielded 57,50(
(Doyle scale) of logs, 153 cords of pulpwood, 158 cords o
chemical wood, and 121 fence posts.

At stumpage prices that prevailed during the year
cutting was done, the fourteen cuts on the good forty prod
stumpage return of $3,442, or $86 per acre--$6.15 per a
The poor forty yielded $1,671 of stumpage value from the
during the 15 years. This is equal to $49.15 per acre, o
acre per year.

Taxes, fire protection, timber stand improvement
marking, scaling, and supervision have averaged $0.27 p
year cash outlay plus one hour per acre annually of the ov

If all products had been sold as stumpage, the net
in growing and handling the crop would have been $5.47 p
the good forty and $3.01 on the poor forty.

The good forty produced nearly 8,000 man-hours
opportunity for harvesting products and hauling them to th
market. The poor forty produced 4,350 man-hours of lal
tunity. Thus, the good forty created an average of nearl
months of off-season labor per year and the poor forty ne
man-months.

With stumpage value and all out-of-pocket costs d
the mill value of the products, the return for labor and u
ment on the good forty over the 15 years has averaged $0
The return from the poor forty has averaged $0.64 per m

Following is a synopsis of the stocking per acre when management began, amount cut during the 15 years, growth for the same period, and stands present in 1951:

Table 13. --Changes in pine stocking per acre, farm forestry forties, 1937-1951

Item	Good forty				Poor forty			
	Stand in 1937	Cut 1937- 1951	Growth 1937- 1951	Stand in 1951	Stand in 1937	Cut 1937- 1951	Growth 1937- 1951	Stand in 1951
Trees over 3.5 in. d.b.h.								
Number	132	52	2	82	85	32	77	130
Basal area (sq. ft.)	66	32	33	67	38	24	43	57
Vol. (cu. ft., i.b.)	1,794	900	1,063	1,957	980	700	1,234	1,514
Trees over 11.5 in. d.b.h.								
Vol. (Doyle)	5,074	2,597	5,818	8,295	2,341	1,866	4,108	4,583
Vol. (Int. 1/4-inch)	5,978	3,059	6,853	9,772	2,757	2,199	4,841	5,399

Growth on the good forty has increased from 244 board feet (Doyle scale) per acre per year during the 1937-41 period to 596 board feet per acre per year during the 1947-51 period. The growth on the poor forty has increased from 231 to 391 board feet per acre annually.

Ninety-one percent of the ground area of the good forty is stocked with overstory pine or pine reproduction; four percent of the area in reproduction needs release from overtopping hardwoods. Ninety-three percent of the poor forty is stocked to pine. Three percent of the stocked area needs release from hardwoods.

EVAPO-TRANSPIRATION: EXCERPTS
FROM SELECTED REFERENCES

Compiled by
Howard W. Lull

VICKSBURG INFILTRATION PROJECT,
SOUTHERN FOREST EXPERIMENT STATION, U S FOREST SERVICE
In cooperation with
WATERWAYS EXPERIMENT STATION, CORPS OF ENGINEERS

OCCASIONAL PAPER 131   August 1953

Cover photo: An experimental site operated by the Vicksburg Infiltration Project. Soil moisture, solar radiation, and weather measurements are made at this site daily. The shelter on the left houses fiberglas soil-moisture unit terminals and recording potentiometers. In the center is an evaporation pan, anemometer, weather shelter and recording raingage. In the right background are total and net exchange radiometers and plots where soil-measure units are installed.

ACKNOWLEDGMENT

This publication was prepared in connection with the work of the Vicksburg Infiltration Project, Vicksburg, Mississippi. The Project is a cooperative effort of the Southern Forest Experiment Station, Forest Service, U. S. Department of Agriculture and the Waterways Experiment Station, Corps of Engineers, U. S. Army.

Mr. M. D. Hoover of the U. S. Forest Service and Dr. Earl L. Stone, Jr., of Cornell University reviewed the initial list of references and suggested several that were subsequently included. Mr. George B. Herring of the Vicksburg Infiltration Project was responsible for copying and proofreading the excerpts.

The compiler is grateful to the authors and publishers of the original references for permission to use copyrighted material.

CONTENTS

EVAPO-TRANSPIRATION: EXCERPTS FROM SELECTED REFERENCES

Compiled by Howard W. Lull
Southern Forest Experiment Station

Evapo-transpiration is the process by which water moves from the soil to the atmosphere. As the link in the hydrologic cycle influenced by vegetation, its importance matches its complexity. While recent research has provided a better understanding of the process, there is as yet no publication that summarizes the present state of knowledge in this field. The purpose of this compilation, which was originally intended only for use of the staff of the Vicksburg Infiltration Project, is to give the reader a birds-eye view of the subject as it is understood today. The best way of accomplishing this seemed to be to let some of the authorities speak for themselves.

The compilation may be useful in providing some source material, but by no means does it cover the entire subject. Those interested in reading further should consult the original references. Most of these will cite other works which bear study, ad infinitum, for there is a labyrinth of literature on the subject. Many studies will be found which deserve but are not given a place in this compilation. Space limitations and, unfortunately, oversight have prevented their inclusion.

The abundance of literature attests the complexity of evapo-transpiration. Largely, this complexity stems from the nature of the process: partly physical and partly physiological. Physically, evapo-transpiration involves the amount of energy received from the sun and the forces which hold water to the soil. The first varies principally with the time of day, season of the year, and latitude; the second, with the wetness and temperature of the soil, and the concentration of the soil solution. Physiologically, the plant is involved from root hair to uppermost stoma. To add to the complexity, evapo-transpiration can be studied in the laboratory or in the field, with portions of plants, entire plants, or plant communities--and with results that are sometimes as disparate as the methods.

With all these complexities, the effect of evapo-transpiration is relatively simple. Through the process the water stored on and in the soil is returned to the air. As the soil dries, water supplies for plant growth are decreased while storage space for the next rainfall is increased. Therein reside the practical implications of evapo-transpira-

tion for crop production, as a means of influencing
of water supplies, and for flood control. Knowled
piration rates is useful in scheduling delivery of in
The possibility of reducing transpiration by manipu
tion, and thus increasing streamflow, has received
Conversely, the maintenance of heavy growths of v
transpire maximum amounts of water and provide
of storage space has application in flood control.

These lines of application give importance
recent research on evapo-transpiration. Much of
fundamental, resulting in a better understanding of
opening new avenues of application. By contrast, 1
work was based on empirical studies of water requ
piration ratios.

From the more recent research, one concl
which is as important as it is simple: during the g
rates of evapo-transpiration are governed first by
of water supplies. When and where supplies are a
controlled by atmospheric factors, the nature of ve
acting together. Where supplies are limited, the 1
principally a function of the amount of available so
must be added the observation that, country-wide,
most part limiting.

These considerations render suspect determ
transpiration from vegetation growing in soil kept
content. Major reliance must be placed on studies
natural conditions of limited moisture supplies. 1
the forces by which vegetation can remove water a
which hold water in the soil. Root habits of vegeta
for it is by root growth that much of the soil moist
accessible to the plant. Soil structure and depth a
affecting root growth also become important.

Contents

The excerpts that follow cover several diff
evapo-transpiration. To delineate these and to pr
presentation of the references, six subject-matter
References which deal with one or more of these s
under the subject principally concerned.

A brief review of the material under each heading follows:

General discussion.-- The first sources consulted during the preparation of these excerpts were textbooks and papers which summarize present knowledge. These sources cover in a general way some of the more specialized references under the other five headings. Included in this section is material on such subjects as the means by which plant roots secure water (1, 4, 5); [1] the forces involved in the removal of water from the soil by vegetation (1, 2, 3, 4, 5, 6); the relation of root growth and distribution to water removal (4, 5, 7); the availability of water throughout the soil moisture range (3, 4, 5, 7); comparison of the evaporation and transpiration processes (3, 6); and the effect of soil moisture content (3, 6, 7), water table depth (1, 6), and climatic factors (1, 3, 7) on evapo-transpiration.

Evaporation.-- References concerning evaporation from bare soil deal principally with expositions of the process (10, 11, 14, 15); or the influence of various site factors upon it such as climate (8, 9, 10, 11, 12, soil temperature (13, 20), water table depth (11, 16), and, of considerable importance, soil-moisture content (8, 12, 20). Some consideration is given to evaporation under summer and winter conditions (9, 10, 11); to the moisture-conserving effect of dry surface soil (10, 11 19); to soil compaction (12) and puddling (15); and to the effect of rainfall on water table losses (17, 18). The last reference in this section (21) summarizes existing knowledge of evaporation from forest soils.

Transpiration.-- This section begins with an interesting reference (22) on the probable origin and essential wastefulness of the transpiration process. Certain physiological aspects are discussed in references 23, 26, 28, 29, 32, and 33. The effects of such weather factors as air temperature, humidity, wind, and light intensity are considered in several of the references (23, 24, 25, 26, 27, 28). Soil temperature is also discussed (25, 30). Transpiration during the winter is described (29, 30).

The influence of soil moisture on transpiration rate is noted in reference 31, dated 1905, and in several more recent works (25, 27, 31, 32, 33, 34, 35, 36). The distinction as to whether atmospheric factors or soil moisture supply govern transpiration rate is clearly brought out (34, 36), as well as the error in the concept of "transpiration ratio" (35).

---

[1] Numerals refer to numbers preceding reference titles.

Evapo-transpiration. -- A number (
45, 46, 47, 48, 49) deal with use of the ene
tion data in some form as a means of under
tion process or estimating its rates. All o
where water supplies are not limiting. Th⅃
estimating evapo-transpiration rates or sea
45, 49). Where water supplies are limited
affecting them are given for several differe
40, 41, 42, 43), together with methods for
tent for such areas (44, 67). How the remo
water losses is described for three areas (
(46, 47, 48) deal with the concept of potentɩ
one example (49) illustrates the use of this

Roots and soil moisture. -- Since rc
vegetation removes water from the soil, th
are of particular import to an understandin
Factors which affect the entry of water into
The extent of the root system and the relatɩ
various depths to rates of water removal aɪ
54, 55, 56). Other references show that r
making water available to the plant (57), bu
not be inhibited by dry soil (58, 61). Undeɪ
ized roots play an important part in water a
is dry, water vapor may furnish a source ɗ
move from the plant to the soil (61). The e
capacity of vegetation to remove water froɪ
64, 65, 66, 67).

Roots. -- An understanding of the iɱ
in evapo-transpiration gives a new significɑ
growth and habits, permitting inferences a
evapo-transpiration. For instance, the eff
of soil factors affecting root growth (68, 6
(70, 71, 72), and of stages of root growth iɪ
be easily inferred. Likewise the differenc
grown in different climates (74), and the eɟ
(75) and of different intensities of grazing (

GENERAL DISCUSSION

1. THE NATURE AND PROPERTIES OF SOILS

T. L. Lyon, H. O. Buckman, and N. C. Brady. 591 pp.
Macmillan, New York. 1952. (Reprinted by permission.)

[Pp. 210-213] "At any one time only a small proportion of
the soil water lies in the immediate neighborhood of the adsorptive
surfaces of plant root systems. Consequently, a query arises as to
how the immense amount of water necessary to offset transpiration
is so readily and steadily acquired by vigorously growing crops.
Two phenomena seem to make adequate and continuous contact possible
if the soil is in good condition for the plant growth: (1) the capillary
adjustment of the soil water and (2) the extension of the root system
of plants, especially when the elongation is rapid.

"Rate of capillary movement. When plant rootlets begin to
absorb water at any particular point or locality in a moist soil, the
thick water films at the capillary fronts are thinned and their curvatures increased. This intensifies the capillary pull in this direction
and water tends to move toward the points of plant absorption. The
rate of movement depends on the magnitude of the tension gradients
developed and the conductivity of the soil interstices.

"With some soils the above adjustment may be comparatively
rapid and the flow appreciable; in others, especially heavy and poorly
granulated clays, the movement will be sluggish and the amount of water
delivered meager. Thus, a root hair, by absorbing some of the moisture
with which it is in contact, automatically creates a tension gradient and
a flow of water is initiated toward its active interfaces.

"How effective the above flow may be under field conditions is
questionable. Many of the early investigators greatly overestimated the
distances through which capillarity may be effective in satisfactorily supplying plants with moisture. They did not realize that the rate of water
supply is the essential factor and that capillary delivery over appreciable
distances is very slow. Plants must have large amounts of water regularly
and rapidly delivered. Therefore, capillarity, although it may act through
a distance of a foot or two, if time be given, may actually be of importance
through only a few centimeters as far as the hour by hour needs of plants
are concerned.

"The above statement must not be taken to mean that the slower
and long-ranged capillary adjustments are in the aggregate not important.
They are important but in a broader and more seasonal way. Nor are

gravitational and vapor transfers to be ignored entirely even
hour-by-hour watering of plants.

"Rate of root extension. This limited water supplying
ity of capillarity directs our attention even more forcibly to t
of root extension and here early workers made an underestim
They failed to recognize the rapidity with which root systems
and the extent to which new contacts are constantly establishe
favorable growing periods, roots often elongate so rapidly tha
factory moisture contacts are maintained even with a lesseni
supply and without any great aid from capillarity. The mat o
rootlets, and root hairs in a meadow, between corn or potato
under oats or wheat is ample evidence of the minutiae of the ?
tions.

"The rate of root extension is surprising even to thos‹
in plant production, On the basis of the data available the e]
may be rapid enough to take care of practically all of the wat‹
a plant growing in a soil at optimum moisture. If this be the
plant is more or less independent of capillary adjustment for
water supply.

"As long as the force with which soil water is held is
low, that is, in the tension range centering around .5 of an at
vigorously growing plant should have little difficulty in obtain
rapidly enough to offset transpiration. Its roots are ever pus
moisture loosely held and absorption should be easy. In addi
conductivity should bring in some low-energy water from nei|

"But, if the root zone receives no new supply of water
condition soon develops. As the moisture of the soil is gradu
surface evaporation and plant absorption, the water remainin
ever-increasing tenacity. Absorption by the plant becomes n
difficult against the higher and higher tensions. At first the j
to adjust to the diminished intake. Soon, however, the tensic
the absorption by the plant will barely meet its transpiration
Obviously if water is not applied at this critical stage the pla
first temporarily, then permanently. And the phenomenon o‹
cause the plant is absorbing no water but because the intake,
negative tension of the soil moisture, is too slow to offset tra
The percentage of moisture within the zone of influence of the
when permanent wilting first occurs is called the wilting coef
critical moisture point."

The transpiration ratio (i.e., pounds of water transpired per pound of above-ground dry matter produced) ranges from 200 to 500 for crops in humid regions, and almost twice as much in arid climates.

[P. 222.] "Much of the variation observed in the [transpiration] ratios quoted arises from differences in climatic conditions. As a rule, the less the rainfall, the lower is the humidity and the greater is the relative transpiration. This accounts for the high figures obtained in arid and semi-arid regions. In general, temperature, sunshine, and wind vary together in their effect on transpiration. That is, the more intense the sunshine the higher is the temperature, the lower is the humidity, and the greater is likely to be the wind velocity. All this would tend to raise the transpiration ratio.

"The factors inherent in the soil itself are of special interest as regards transpiration, since they can be controlled to a certain extent under field conditions. In general, an increase in the moisture content of a soil above optimum results in an increased transpiration ratio. This has been established by a number of investigators.

"Moreover, the amount of available nutrients and their balanced condition are also concerned in the economic utilization of water. The data available show that the more productive the soil, the lower is the transpiration ratio provided the water supply is held at optimum."

Concerning factors affecting evaporation at the soil surface (pp. 225-226):

"Relative humidity. Any changes in the vapor-pressure gradient and hence in the rate of evaporation will be determined by fluctuations in the relative humidity of the atmosphere immediately above. The lower the relative humidity the more pronounced will be the vaporization tendency. Sometimes the relative humidity of the atmosphere approaches 100 per cent. Under this condition evaporation might not only cease but condensation could be induced. Since relative humidity fluctuates rather widely from time to time, it cannot but exert a variable yet important influence upon the loss of water from soil by evaporation."

"Temperature. In direct sunlight the soil and its water often have temperatures several degrees above that of the atmospheric air. This increases the vapor pressure and so markedly steepens the gradient that evaporation is greatly encouraged."

"In fact this temperature effect on vaporization is usually much more important than that resulting from a lowering of the relative

humidity of the air. For instance, raising the temperature of the wate
only 5 $^\circ$ C. above that of the atmospheric air is equivalent in its effect
on the vapor-pressure gradient to a lowering of the relative humidity
about 35 per cent. Hence, temperature difference is a major control
of the vapor-pressure gradient and, therefore, of the surface evaporation of water from soils.

"Wind. At the same time, if a dry wind is stirring, the accum
lated water vapor is continually swept away, the moist air being repla(
by that with a lower relative humidity. This tends to maintain the vap(
pressure gradient and evaporation is greatly encouraged. The drying
effect of even a gentle wind is noticeable even though the air in motion
may not be at a particularly low relative humidity. Hence, the capacit
of a heavy wind operating under a steep vapor-pressure gradient to
enhance evaporation both from soil and plants is tremendous.. Farmer
of the Great Plains dread the hot winds characteristic of that region."

"In respect to the depth to which soils may be depleted by this
evapo-capillary pumping, soil physicists are agreed that the distance
is far short of the 4, 5, or even more feet sometimes postulated. The
conditions that impede and interrupt the flow are such as to allow it to
deliver important amounts of water to the surface only through compa)
tively short distances. Some investigators think that the rate of evapo
tion is often so rapid as to cause the moisture column to break relativ
near the surface, and that this is one of the major inhibitions. But wha
ever the explanation, a 20- or 24-inch depth is probably a maximum
range."

2. SOIL CONDITIONS AND PLANT GROWTH

Sir E. John Russell (Recast and rewritten by E. Walter Russell).
635 pp. Longmans, Green and Co., London. 1950.
(Reprinted by permission.)

[Pp. 370-371.] "One consequence of the amount of water used by
crops depending primarily on the energy supply or the evaporation po\
of the air, is that all the water falling on the crop, whether as light dr)
or dew or whether as a definite rain storm, is as effective in contribu)
to the water requirements of the crop as is the water removed from tl
soil...every thousandth of an inch of rainfall, however long it takes to .
on being evaporated from the plant leaves or stems reduces the plants
demands on the soil water by this amount."

[Pp. 372-3.] "Plant roots can only extract water from a soil if they can apply a sufficiently great suction to move it out of the pore space. As the soil dries, so the suction needed to extract water rises and the rate of movement of water into a given length of root decreases, which may sometimes have the consequence that if the crop is growing in conditions conducive to high rates of transpiration, the actual maximum rate it can reach will decrease as the soil dries. The maximum suction roots can exert on the soil water does not appear to be a very definite quantity, but if the water is held at suctions higher than about 7 atm. --the first permanent wilting point of the soil--the roots appear to be unable to extract sufficient water to keep the whole plant of most farm crops turgid when placed in a saturated atmosphere, i.e., in an atmosphere when transpiration cannot take place; and if above about 20 to 30 atm. --the ultimate wilting point--to keep any leaves turgid. The water held between these two suctions is sufficiently available to the roots for the maintenance of life, but not for growth; that held at suction less than that corresponding to the first permanent wilting point may be, but is not necessarily, readily available for growth."

[P. 374.] "The soil water is only at a suction of between 10 to 15 atm. or over in a soil carrying a permanently wilted crop if the osmotic pressure of the soil solution is low. If it is appreciable, the plants wilt permanently when the soil water is at lower suctions than this. The ease with which plant roots extract water from the soil seems to depend, not on the suction of the water in the soil, but on its free energy; and these two are only equivalent when the water contains no dissolved substances. The free energy of a solution in a soil is approximately the sum of the free energy changes due to the dissolved salts and that due to the curved air-water menisci bounding the solution in the soil pores...Hence if the osmotic pressure of the solution is 3 atm. and it is under a suction of 3 atm. in the soil pores, then plant roots will have at least the same difficulty in using the water from this solution as they would if the soil contained pure water at a suction of 6 atm."

3. PLANT AND SOIL WATER RELATIONSHIPS

Paul J. Kramer. 347 p p. McGraw-Hill, New York. 1949.
(Reprinted by permission.)

[P.44.] "...Water flows under the influence of gravity, moves in capillary films, and diffuses as vapor, [always moving] along a gradient from regions of higher to regions of lower free energy."

[P. 57.] "It is generally agreed that transpiratior
exceed losses by evaporation where well-developed grass
forests occur. If evaporation removes water only from tl
foot of soil, the remainder of the soil moisture would ren
were it not for the roots of plants."

[P. 59.] "In general, it appears that, on an acre
ably more water is lost by transpiration if the supply is a
than if it is somewhat limited at times. This is partly be
shoots are produced when there is an abundance of water
cause the rate of transpiration in this case is less often r
wilting and stomatal closure."

[Pp. 65-66.] "From the standpoint of energy inv
ment of water from soil to plant, there can be little doubt
ture becomes less and less readily available as the moist
decreases from field capacity to the permanent-wilting pe
the moisture content of the soil decreases, there is inevi
crease in the amount of energy required to move a unit m
unit distance. In another sense, however, at least in ligl
moisture may be practically as readily available to the pl
contents just above the wilting percentage as at the field (
in the sense that under some conditions water may be abs
pired at the same rate in drier soils as in soils at the fie
This is because, as the moisture content of the soil and t
content of the plant decrease, the osmotic pressure and t
pressure deficit within the plant increase. While an inc
atmospheres in the diffusion-pressure deficit of the roots
the increased energy gradient necessary to maintain a hi
absorption, it does not appreciably reduce transpiration.
thoroughly permeated with roots, plants might reduce alr
soil mass nearly to the wilting percentage before transpi
or the plants exhibit symptoms of a deficit. In heavy soi
distribution is variable and sparse, as in the lemon orch
California and the pear orchards at Medford, Oregon, th
uniformly absorbed from the entire soil mass and one ca
is a definite moisture content above which water is availe
which it is unavailable."

[P. 67.] "For practical purposes, however, in r
water may be regarded as being equally available over m
from field capacity to permanent wilting. This is becaus
tension curve of most soils is hyperbolic and most of the
available water lies in the flat portion of the curve. Mos

available water is removed from light soils before the tension on the remainder exceeds 1 atmosphere, and only a small fraction is held with sufficient force to hinder absorption. This is not true, however, in heavy clay, where 50 per cent or more of the available water sometimes is held with tensions in excess of 1 atmosphere. In such soils water actually does become limiting to growth before the moisture content is reduced to the permanent-wilting percentage."

[P. 94.] "Since the field capacity is the amount of water held against gravity by a soil, it is obviously impossible to wet any soil mass to a moisture content less than its field capacity."

[P. 211.] "It seems certain...that when transpiration is rapid the active absorption mechanism responsible for root pressure is not only inadequate to supply the required amount of water but actually becomes inoperative, because of the increasing diffusion-pressure deficit in the root cells. Under these conditions, the intake of water is not a special function of the root cells but is a function of the entire plant, the root cells merely providing an absorbing surface, through which water is absorbed. To put it another way, in freely transpiring plants water is absorbed through the roots, rather than by the roots."

[P. 233.] "It is concluded that the principal cause of reduced intake of water by transpiring plants in cold soil is the physical effect of increased resistance to water movement across the living cells of the roots. This results from the combined effects of the decreased permeability and increased viscosity of the protoplasm of the living cells in the roots and the increased viscosity and decreased diffusion pressure of the water intake. Other factors, such as decreased root extension, water-supplying power of the soil, and metabolic activity of the roots, are of distinctly secondary importance."

4. SOIL MOISTURE IN RELATION TO PLANT GROWTH

F. J. Veihmeyer and A. H. Hendrickson.
Annual review of plant physiology 1:285-304. Annual Reviews, Inc.,
Stanford, Cal. 1950. (Reprinted by permission.)

"The upper and lower limits of water storage in the soil reservoir are fixed by two soil-moisture conditions, which might be called soil-moisture constants. In fact, the authors believe they are the only ones of any practical value for consideration in connection with plant growth. The first of these, the field capacity, is the amount of water

held in a soil after excess water has drained away and the ra
ward movement has materially decreased. This usually take
within two or three days after rain or irrigation in pervious
uniform structure and texture. Below the first foot and in th
of vegetation the field capacity persists for months without m

"The lower limit of the soil reservoir or the moistur
at which we may consider it to be emptied, since it no longer
sufficient water to maintain normal growth and vigor of plant
permanent wilting percentage."

"Briggs and Shantz' conclusion that all plants wilt at
moisture content when grown on the same soil has been quest
number of investigators... The authors have tested many pla
ing them in the same kind of soils in small containers and al
trials. The results substantiate the conclusion of Briggs and
which now seems to be accepted by most investigators."

"... The moisture-extraction curves from field samp
only a slight reduction after the permanent wilting percentag
This reduction is small, often not more than 1 per cent, ever
or four months. These extraction curves practically coincid
year, both as to slope and as to minimum moisture content r

"The authors [1945] believe that the wilting range is n
er for plants in the field than for those in containers. They
permanent wilting percentage not as a unique value but as a s
of soil-moisture contents within which permanent wilting take
This range need not exceed 1 per cent for fine textured soils
cent for coarse textured ones."

"Typical vapor pressure curves for soils show the sm
in vapor pressure from the field capacity to the permanent w
centage and the very rapid decrease thereafter. The position
permanent wilting percentages on these curves near the regi
tightness with which the water is held by the soil increases v
is significant."

The authors review literature on the relation of trans
photosynthesis, and plant growth to soil moisture content. T
conflicting evidence and make a strong case that these proc
affected as long as the moisture content is above the perman
percentage.

"Physical measurements show that the energy required to remove water from the soil changes materially as the moisture content decreases, but it does not follow that the availability of the water to plants also decreases. By far the greatest drop in energy in the soil-moisture plant system occurs at the surface of the leaf cell walls which surround the sub-stomatal surface as Gradmann [1928] and Edlefsen [1942] have pointed out. The latter, from Thut's [1939] data, has calculated the drop in free energy between the leaf tissues and the outside air [1] of -9.231×10^8 ergs per gm. The total free energy of the water surrounding the roots at the permanent wilting percentage may be taken to be about -0.16×10^8 ergs per gm. At 40 percent relative humidity, that of the air is -9.4×10^8 ergs per gm. or an overall drop of -9.24×10^8 ergs per gm. from the soil to the air. The increase in energy required when the soil moisture is reduced from the field capacity to the permanent wilting percentage is unimportant when the system as a whole is considered."

"The reason that plants wilt may be explained by the position of the permanent wilting percentage on the energy soil moisture curve in the region where a slight decrease in moisture content results in a great increase in resistance to removal of the water. Failure of the water supply to the plant, of course, may be due to the slowness of movement of water into the mass of soil dried by the roots."

"Another cause of water deficiency at soil-moisture contents near the permanent wilting percentage may be the failure of roots to elongate rapidly enough into regions where there is still water above the permanent wilting percentage."

"Whether water is readily available to plants or not in the final analysis must be decided by empirical experiments. While the results of growing plants in containers may indicate trends, they should not be taken as being conclusive unless confirmed by field trials."

'One difficulty with plants in the field is the sparse root-development of some plants. Sometimes, either due to soil conditions or inherent characteristics of the plants, roots will not thoroughly permeate the soil. Consequently, neither soil sampling nor measurements of soil properties which are related to soil-moisture contents, made with physical instruments inserted into the soil, will give reliable records of the actual moisture content of the soil in contact with the absorbing portion of the roots. Thus, erroneous conclusions may be drawn."

---

[1] From correspondence with F. J. Veihmeyer: "In references to the total free energy required at 40% relative humidity and 30 ° C, that of the air is -12.8×10^8 ergs per gram in place of -9.4×10^8. This will give an over-all drop of -12.6×10^8 ergs per gram."

"Much of the material not reviewed does not cont
data to permit an analysis because they were based on te
which obviously are faulty. For instance, those in which
ed to maintain a predetermined moisture in the soil in wl
growing were not given consideration."

"The results of investigations on the relation of p
soil moisture show that the plants grow well throughout a
soil moisture, but some investigators question whether t
equal facility throughout the entire range from field capa
nent wilting percentage. The permanent wilting percenta
important soil moisture constant. The accuracy of its d
highly important."

5. SOIL WATER AND PLANT GROWTH

L. A. Richards and C. H. Wadleigh. Soil physical (
and plant growth, pp. 73-251. Academic Press, In
New York. 1952. (Reprinted by permission.)

[Pp. 82-83.] "On the basis of experiments by L
it is inferred that in the plant-growth moisture range and
of temperature gradients, the movement of water in soil
primarily in the liquid phase through the pore channels (
sorbed film phase over the surface of the soil particles,
pared with these processes, the movement of water in tl
is negligible. This is reasonable, for the whole plant-g
range corresponds to a relative humidity range of less t
and therefore vapor-pressure gradients would always be
isothermal conditions.

"If there is an appreciable temperature gradient
vapor transfer of water through field soils may be more
film flow, particularly at moisture contents near the wi
However, little quantitative data on this point appear to
Field measurements in central California by Edlefsen a
indicate that an upward movement of water takes place (
apparently in response to the temperature gradient. Th
did not distinguish between vapor and film transfer, but
that vapor transfer was significant. Hilgeman [1948] m
moisture content of a bare soil in Arizona to a depth of
period of 22 months. The total loss of water was 9.8 ir

47 percent of the water available for plant growth. The most rapid
losses occurred in summer. In this case, also, it appears likely
that vapor transfer under the action of temperature gradients
played an important part in moisture loss from the subsurface soil."

[P. 84.] "The high tension in the soil moisture in the
vicinity of the root sets up a tension gradient and thereby a force
action in the soil-water system that tends to move water toward the
root. This tendency of water to move toward plant roots in response
to tension gradients is of considerable importance for perennial plants
with large developed root systems because a small distance of movement over a considerable combined length of root system would account
for an appreciable volume of water, even at the slow rates at which
water moves through dry soil. However, for young plants with a newly
developing root system this movement is so slow that sufficient water
for normal growth would not be supplied unless the plant roots are
able to extend themselves outward into a fresh soil-moisture supply.
Therefore, as has been described by Davis [1940], when a new corn
plant is developing, the available moisture is extracted in the vicinity
of the base of the plant and the soil approaches the wilting percentage,
whereas just a few inches farther away from the plant the soil may be
at or near field capacity. The roots of the newly developing plant must
extend themselves outward in order to maintain a continuous supply of
available water. When the roots of the plant have permeated the soil
region in which they can grow well, the soil-moisture content throughout the soil region occupied by roots will be reduced into the wilting
range. Unless additional moisture is supplied by rain or irrigation,
vegetative growth will cease and the plant will wilt."

[P. 85.] "In irrigated areas the pattern of moisture extraction
for mature perennial tree crops has been extensively studied by
Hendrickson and Veihmeyer [1929, 1934, 1942]. Starting with a deep
permeable soil when the whole profile was wet, they found that moisture
is continuously extracted at all depths down to 6 or 8 feet or deeper,
depending on the soil and the species of tree. The surface 2 or 3 feet
may approach the permanent wilting percentage at about the same rate.
Often, however, the rate of extraction is greater near the tree and near
the soil surface, so that available water is first depleted from the surface layers of soil.

"Veihmeyer and Hendrickson [1938] have used the pattern of
moisture extraction as an indication of the probable root distribution,
and state: 'The fact that soil samples taken at any place within the
experimental plots in mature peach, prune, and walnut orchards which

have had an even application of water, agree at compar
shows that these trees, under conditions existing at Da'
uniform distribution of roots.'

"Such observers assume that moisture depletior
region is evidence that active roots traverse that regioı
effective distance through which water in the available
toward the root is certainly of the order of inches and r
pattern of moisture extraction in soils is therefore larg
of the active root distribution. Root distribution, as re
by Kramer [1949] is mainly determined by the genetic c
plant but is modified by plant spacing as well as by soil
factors."

[P.103.] "Measurements and observations by \
Hendrickson [1927, 1938]; Aldrich, Work, and Lewis [1
have shown reasonably conclusively that moisture will ı
root-free soil at a moisture content below field capacit
quate to supply roots in adjacent soil at distances of the
number of centimeters away. Nevertheless, the fact tł
of soil 1 centimeter deep can be brought in 24 hours to
a membrane supporting a pressure difference of 15 atm
indicates that unsaturated permeability is not negligible
range above the wilting percentage. This fact further i
in a root zone where the maximum distance of soil fron
than 1 centimeter, soil-moisture tension gradients may
pear during the overnight period when transpiration is l

[Pp. 108-9.] "The specific characteristics of t
al plant must be taken into account in considering its re
moisture conditions. The nature of the root system is
tinent. Kramer [1949] has given well-merited emphasi
teristics of root systems in his recent text on water rel
plants. The extensive studies of Weaver [1926] and We.
[1927] show that the roots of various crop plants differ
inherent capacity to penetrate deeply into the soil. Asp
alfalfa roots grow to depths of 12 to 15 feet or more, if
soil is favorable. Under such crops, soil-moisture stu
the surface foot or even the surface 3 feet might well b
On the other hand, soil-moisture investigations under s
rooted crops as onions and potatoes would need to be la
with the surface 2 feet of soil, for few roots of these cı
beyond this depth. Differences among varieties of a gi
actually be involved in this consideration. Kiesselbach

noted that upon hybridization, the depth of penetration and combined length of all main roots of corn increased materially in the first generation. This observation is probably related to the superior yields obtained with hybrid corn.

"In addition to the extent of root penetration, consideration must also be given to root proliferation or the special density of root distribution. Owing to the slow rate of unsaturated flow in soils, the exhaustion of soil moisture by plants is very dependent on thorough permeation of the soil mass by fine rootlets. Onions and celery characteristically have root systems exhibiting poor proliferation and permeation. Successful celery culture is largely dependent upon supplemental irrigation, even in humid climates. On the other hand, many of the grasses thoroughly permeate the soil with their fine roots. In consequence, grasses not only efficiently remove available moisture from the fine interstices of the soil but also are especially effective in generating good structural characteristics in the soil. It appears that these two effects are to some extent related."

[Pp. 144-5] "From the irrigation and soil-moisture experiments mentioned in the foregoing sections it is apparent that there is considerable evidence that significant differences in growth rates occur along with varying degrees of moisture depletion within the so-called available soil-moisture range. In the interpretation of the statement that soil moisture is equally available until moisture is depleted about to the wilting percentage, several factors should be kept in mind. Both from the standpoint of supporting evidence and applications, use of the term 'available' in this connection should be restricted in its meaning to the rates of soil-moisture extraction and water use by plants. Various experimenters have found that during an irrigation cycle the rate at which an established root system removes water from a soil root zone is approximately uniform down to about the wilting percentage. In this statement the interpretation of 'about' and 'wilting percentage' is somewhat variable. If by 'about' is meant 2 or 3 percent of soil moisture then in many cases this moisture-content range covers the major part of the soil-moisture tension range over which plants can grow. In the rest of the moisture range above and below field capacity, soil-moisture tension changes only slowly with soil-moisture content because of the hyperbolic nature of the soil-moisture-release curve. Also, the wilting condition of plants, both temporary and permanent, corresponds to an appreciable range in soil moisture content. In the field, on successive days temporary wilting occurs during an increasing fraction of the diurnal cycle and merges indistinguishably into a range of permanent-wilting stages as has been pointed out by several investigators. Throughout the moisture-depletion process the soil-moisture stress increases continuously, and much experimental evidence

- 17 -

supports the hypothesis that the growth rate of various plants
creases markedly in the available soil-moisture range and tha
vegetative growth is completely inhibited by the time the soil
ture is depleted to the permanent-wilting range."

6. SOME PLANT-SOIL-WATER RELATIONS IN WATERSHEI MANAGEMENT

Leon Lassen, Howard W. Lull, and Bernard Frank. 64 pp
U. S. Dept. Agr. Cir. 910. 1952.

"The depth to which evaporation from a bare soil exter
depends on soil porosity and depth to water table. On fine-te:
upland soils where the water table does not influence evaporat
water losses are generally limited to the first foot of soil. C
textured soils and cracked soils possess more and larger ave
for escape of the vapor particles, and here evaporation may r
water from depths as great as 5 to 6 feet."

"Where the water table lies close to the surface so tha
capillary flow feeds water to the surface for evaporation, soi
again is the limiting factor, but this time in relation to the mo
of water in its liquid state rather than in its vapor state. Sin
pores conduct capillary water farther than large pores, evapo
from fine-textured soils affects water tables to greater depth:
coarser-textured soils. For a coarse sand the limiting depth
14 inches; for a clay, 3 to 4 feet [Penman, 1946]. Thus, in ar
evaporation opportunity is great because of high water tables,
to which evaporation is effective is greater in clay than in sar
in upland soils, where water tables are not a factor, the reve

"The amount of water evaporated varies not only with
and wind movement, but also with the supply of available wate
tion opportunity). This factor outweighs all others. Even du
temperatures of summer, evaporation losses will be no great
during the cooler winter months if the soil is not frequently w
rainfall. During winter, when the soil is usually at or near f
evaporation rates will be limited, not by the supply of water,
ditions due to low temperatures."

"...The rate at which roots extract moisture decreas
moisture content is decreased, or, in relation to energy, wh
force is required to overcome the attraction of moisture to tf

- 18 -

particle. When moisture becomes available at a certain depth, as at the surface following a light rain, the rate of extraction at this depth will be increased. During this time interval moisture extraction will continue at other depths, but at comparatively higher tensions and lower rates.

"The foregoing analysis indicates the operation of a governing relationship which tends to reduce the difference between the magnitudes of the forces involved in the extraction of moisture at the various depths. Thus at high moisture contents, soil moisture will be reduced more rapidly than at lower moisture contents, tending to bring the energy levels--and the moisture contents--together. Simply expressed: the higher the moisture content the faster the loss."

"These considerations point to some salient distinctions between transpiration and evaporation, particularly in connection with the manner in which they create available retention storage. One is that whereas transpiration removes water simultaneously throughout the entire depth occupied by roots, evaporation proceeds downward from the surface. Another is that by acting on a greater volume of soil, transpiration removes a greater amount of water in a unit of time than does evaporation.

"Given a soil in which evaporation and transpiration reach to equal depths, a comparison of water loss from a bare area (where water is removed only by evaporation) with that from a well-vegetated area (where transpiration is the principal agent) will show that water is removed more rapidly--that is, retention storage opportunity is created more rapidly--in the vegetated soil. But assuming no addition of moisture to either area, evaporation will eventually remove a greater amount of water because the evaporation process is not governed by the physiological factors which limit transpiration losses."

"It must be kept in mind, however, that comparisons between evaporation from a bare soil and transpiration serve merely to indicate the manner and magnitudes of water losses. Actually, large expanses of completely bare areas--where only evaporation operates--are uncommon. Where they do occur, as in arid regions, lack of water severely limits evaporation losses. Bare spots within sparsely vegetated areas have a hydrologic importance that is related more directly to the rate at which water can enter the soil surface than to evaporation losses."

7. PLANT AND SOIL WATER RELATIONS ON THE WATERSHI

Paul J. Kramer. Jour. Forestry 50 (2): 92-95. 1952.
(Reprinted by permission.)

"Since most of the water is lost from the leaves, and sin
there are large differences among species with respect to leaf a
and leaf area, differences in rate of water loss would also be ex

"It is often supposed that trees bearing thick, heavily cu
leaves have lower transpiration rates than trees bearing thin lea
but this is not necessarily true. Trees bearing thick leaves act
often have higher transpiration rates per unit of leaf surface tha
bearing thin leaves. According to data of Caughey [1945], Ilex gl.
and Gordonia lasianthus, which bear thick, leathery leaves trans
more per unit of surface than poplar, which has thin leaves. Hc
[1931] observed that bur oak and hickory transpire more per unit
area than red oak and linden, which have much thinner leaves.
[1943] found the transpiration rate of Ilex opaca to be higher tha
coleus and tobacco. He concluded that leaf structure is not a re
indicator of differences in transpiration among species."

"It is generally assumed that pines lose less water than l
leaved species. This is true in terms of water loss per unit of :
area, but is not always true in terms of water loss per tree fror
equal crown volume. During a careful comparison made in our
loblolly pine transpired much less rapidly than yellow poplar an
red oak on a leaf area basis. The total leaf area of the pines wa
three times that of the two hardwoods; hence the total loss per t
higher for the pines. The crown volumes of the various species
these studies were similar. Groom [1910] many years ago point
that conifers do not always have low transpiration rates and that
sometimes lose as much or more water per tree as hardwoods (
size."

"Large seasonal variations in transpiration occur. Few
ments of transpiration of trees have been made over an entire g
season, but it is obvious that water loss increases rapidly in th
as new leaves unfold. According to Weaver and Mogensen [1919
a gradual decline in transpiration during the early autumn befor
fall. In general, it is probable that transpiration reaches its m
about the time maximum leaf area is attained, then decreases s
increasing proportion of the leaves become senescent, and decr
rapidly as leaf fall begins."

"Studies by Weaver and Mogensen [1919] and by Kozlowski [1943] indicate that winter transpiration of conifers is almost as low as transpiration from bare branches of deciduous trees, at least under some climatic conditions. This probably is at least partly because cold soil hinders water absorption by conifers in the winter. It is probable that in the autumn and during periods of mild winter weather, transpiration of conifers may be much higher than transpiration of bare deciduous trees."

"Another factor affecting the amount of water removed by vegetation is the depth of root systems. Obviously the greater the volume of soil occupied by roots, the greater the volume of water removed. There are considerable differences among species in respect to depth of root systems, dogwood being a notably shallow-rooted species, while pines often send roots to a depth of many feet. In heavy soils where most of the roots are concentrated near the surface, species differences are probably of less significance than in well aerated soils where deep penetration is possible. In mature Piedmont forests the surface soil is completely occupied by roots regardless of the species, and age of stand is probably more important than species. In well aerated soils deep-rooted species often remove water to depths of many feet. Wiggans [1937], for example, found apple trees in Nebraska absorbing water from a depth of over 30 feet. Some herbaceous species, such as alfalfa, are also very deep rooted in certain soils and absorb water from considerable depths."

"An important question is the extent to which rate of transpiration decreases as the soil dries. Veihmeyer and Hendrickson [1950] claim that water is equally available from field capacity down almost to permanent wilting, but there is some evidence that this is not always true, especially in clay soils. The forces with which water is held increases as soil moisture decreases; hence less absorption would be expected from dry soil than from soil at field capacity. Schopmeyer [1939] found that transpiration of pine seedlings decreased as soil moisture decreased, long before permanent wilting was reached. Kozlowski [1949] obtained similar results with oak and pine seedlings. Transpiration of oak was decreased less than that of pine, probably because the former had more extensive root systems. Colman [1949] reported that at San Dimas evapotranspiration rates were higher during the rainy season than at any other time, at least partly because of the higher soil moisture content at that season.

"The highest water losses occur from vegetation growing at the margins of streams and bodies of water where the roots are always in contact with soil wetted to field capacity or higher. In general,

a forest stand; and it is quite certain that the total w
where the soil is kept wetted nearly to field capacity
occasionally dries down nearly to permanent wilting

"Soil temperature is a limiting factor on wate
during the winter. Experiments by Kozlowski [1943]
[1942] show that cooling the soil greatly reduces wate
pine seedlings. Furthermore, there are differences
absorption by northern species being reduced less th
southern species. It seems possible that in winter,
equal, a white pine stand might lose more water thar
stand of similar age and density of stocking."

EVAPORATION

8. SOIL PHYSICS

L. D. Baver. 398 pp., John Wiley and Sons, Inc.,
New York. Ed. 2, 1948. (Reprinted by permission.)

[P. 247.] "Evaporation of water from any source can occur
only when the atmosphere in contact with the water is not saturated
with water vapor, if both air and water have the same temperature
.... Any meteorological effect that tends to increase the vapor-
pressure gradient away from the soil will increase evaporation."

Baver cites temperature, relative humidity, and wind
velocity as climatic factors; soil moisture content, depth to water
table, texture, aggregation, color, exposure, and mulch as soil
factors.

"... An analysis of the evaporation data from 243 monthly
records from 29 meteorological stations throughout the United States
has indicated that evaporation from a free-water surface varies
approximately with the square of the mean monthly temperature in
Fahrenheit degrees   [Baver, 1937]."

[P. 248.] "The degree of saturation with moisture is the
most important soil factor affecting the amount of evaporation."

Fisher (1923) and Keen et al. (1926) show that rate is
practically constant at high moisture contents.

9. DRAINAGE AND EVAPORATION FROM FALLOW SOIL AT
   ROTHAMSTED

H. L. Penman and R. K. Schofield. Jour. Agr. Sci. 31: 74-109.
1941.

"In winter the soil does not dry at the surface. Winter evaporation is, therefore, much the same as would be obtained from a
water surface and extra rainfall does not affect it."

"In summer the surface remains moist only a short time after
rain has fallen; the air gradient is then much steeper than in winter.
For the rest of the time the surface is drier and there is also a vapour
pressure gradient in the soil. Hence (1) there is more rapid evapora-

tion while the surface is wet, (2) the total amount of evaporation is dependent upon both total rainfall and its distribution in time, (3) the later stages of evaporation are more dependent upon soil conditions than on air conditions, and (4) the total evaporation is much less than from open water."

10. LABORATORY EXPERIMENTS ON EVAPORATION FROM FALLOW SOIL

H. L. Penman. Jour. Agr. Sci. 31:454-465. 1941.

Penman determined water loss from a clay loam and sandy soil packed in 12-inch-deep cylinders under two treatments: (1) isothermal, when air and soil temperatures were equal; (2) non-isothermal, when soil surface was heated about 10 $^{\circ}$ C. above air temperature (typical June condition) by a 750-watt electric radiator suspended 2 feet above the soil surface for 8 hours a day.

In the isothermal treatment, soils lost moisture at a unit rate equal to that of open water until surface drying was apparent. With radiated soils, moisture loss was rapid for 2 days; "Thereafter the rate of loss is nearly constant and there is clear-cut evidence of conservation as compared with isothermal evaporation."

"Cumulative evidence suggests that some equilibrium is eventually attained whatever the nature of the initial behaviour, and that the differences between isothermal condition, intermittent radiation, and mulching lie in the rapidity with which this equilibrium rate is attained and the gross amount of water lost in attaining it."

"The result of an incomplete survey of the annual cycle of soil surface and air temperatures indicates that conditions may be regarded as isothermal when the mean air temperature is below 48 $^{\circ}$ F. and as non-isothermal above 48 $^{\circ}$ F...Thus the broad difference between winter and summer evaporation is that between isothermal and radiated condition."

From these experiments Penman assumed "that the action of radiation, or of very high temperature, is to dry out a shallow layer at the surface more quickly than it can be replenished by liquid flow from below "

"The liquid movement depends upon the capillary conductivity and the suction gradient, both being functions of moisture content; the vapour movement depends upon the relative humidity of the soil air and this is not nearly so dependent upon moisture content as the liquid variables are."

11. SOME ASPECTS OF EVAPORATION IN NATURE

H. L. Penman. Roy. Col. Sci. Jour. 16:117-129. 1946.

"... In the Rothamsted clay soil, there is little or no water movement from a water table lying at 3 or 4 feet below the surface, and for the coarse sand the limiting depth is about 14 inches. Water tables below these limiting depths (i.e., measured from the surface of bare soil or from the lower limits of plant roots in a cropped soil) will thus make no significant contribution to the soil as a source."

Concerning factors affecting evaporation: "In summer the most important factor is rainfall; the more often the soil is wetted the more evaporation there is. In winter, rainfall is unimportant because the soil is wet, and further rain cannot increase the wetness. The winter evaporation tends to remain constant...."

Plotting mean saturation deficit by months against mean evaporation per day by months "... shows that evaporation from bare soil is almost completely independent of air conditions in summer, but is greatly dependent on these conditions in winter.... In summer, the surface temperature of bare soil considerably exceeded air temperatures...."

"... Water movement in soil with even a slight moisture deficit is extremely slow, even with very great moisture gradients. As the deficit increases, the reluctance to move increases enormously. Drying conditions at the surface of bare soil, initially holding as much water as gravity will permit, tend to set up a liquid movement from below the surface. If the drying rate is small, the flow of water will be able to keep pace with it, and a steady drying rate will be maintained in which the soil surface behaves very nearly as if it were an open water surface... If the drying rate is rapid, the flow of soil water cannot keep pace with it, and the top layer of soil will dry, even although completely moist conditions exist only a few millimeters below... The result is that evaporation takes place, not at the soil-air surface, but a few millimeters

below, and the vapour has to diffuse through the dry soil
reaching the sink, thereby adding considerably to the tot
of molecular diffusion and reducing subsequent evaporat
to very small amounts."

"Surface hoeing, other than that necessary for ki
is a redundant operation as far as moisture conservatioi
ed; by the time the land is dry enough to be cultivated th
the job."

12. FACTORS AFFECTING THE EVAPORATION OF M(
    FROM THE SOIL

    F. S. Harris and J. S. Robinson. Jour. Agr. Res.
    7(10): 439-461.

    In laboratory experiments, evaporation increase
initial quantity of moisture in the soil. The increase wa
with higher percentages as with lower, and there seeme
number of critical points when the rate of loss changed i
the soils were saturated, evaporation was higher from f
particles than from coarser, but the differences were n(
rate of evaporation from a moist soil very rapidly decre
humidity increased. Air currents greatly increased eva
after a velocity of about 10 miles per hour was reached,
was slight. Reducing the intensity of sunshine greatly r
of evaporation, as did slight reductions in temperature.

    Compacting the soil (loam) in 2-inch layers gave
tion. Soils compacted in the first and second 2-inch lay
heavily than soils with packed layers farther from the s

13. THE EFFECT OF FREEZING ON SOIL MOISTURE
    EVAPORATION FROM A BARE SOIL

    Henry W. Anderson. Trans. Amer. Geophys. Uni
    27(6): 863-870. 1946. (Reprinted by permission.)

    "At Northfork, California, where nightly freezin
thawing of the bare soil frequently occurred, it was obs

- 26 -

during rainless periods of from one to three weeks the surface half-inch or so of the soil would become and remain very wet--near the liquid limit of plasticity. A study of the soil-moisture data and the associated soil freezing was made to determine what effect freezing had on the soil moisture in the profile and on evaporation from the soil."

"...The soil was an immature gravelly, sandy-clay loam of the Holland series....Soil-moisture samples were taken from under the brush and from a 14-foot square plot that was kept bare of vegetation and trenched annually to prevent entry of roots from the surrounding brush vegetation, which was 20 ft. distant from the bare area."

"By selecting early winter periods, before the soil profile had become wetted through, it was possible from successive soil-moisture samplings to study the changes in the moisture content for various depths. Since during these periods the deeper depths of the soil were warmer than the upper layers, movement of liquid moisture and vapor in the soil was upward, and no downward losses of moisture from the soil occurred. These were the only periods when evaporation losses could be determined with certainty by soil-moisture sampling.

"The results for a bare soil showed a rapid and relatively large upward movement of water and large evaporation losses during freezing periods.... The daily depth of freezing varied from 0.1 to 0.7 in., averaging 0.44 in. For the ten-day period, net water losses from the soil profile by foot depths were: from the zero- to 12-inch layer, 0.24 in.; from the 12- to 24-inch layer, 0.17 in.;from the 24- to 36-inch layer, 0.53 in.; and from the 36- to 48-inch layer, 0.08 in. Therefore, the net evaporation loss from the bare soil for the period was 1.02 in. of water, over one-half of which had been drawn upward from the 24- to 36-inch depth. The moisture content of the 30- to 36-inch layer was reduced to less than nine per cent, despite the fact that the moisture equivalent of this soil was 16 per cent (the permanent wilting percentage was six per cent). Of the 9.13 in. of water in the soil at the start of the period, 4.47 in. were required to satisfy the wilting percentage, leaving 4.66 in. available for plant growth. The ten-day evaporation of 1.02 in. amounted to a loss of 22 per cent of the total amount of water available for plant growth.

"In contrast to these results from a bare soil, no freezing occurred under the brush-covered area during the same period, and there the transpiration-evaporation loss of water from the soil profile amounted to only 0.25 in. Free water-surface evaporation from a standard

Weather Bureau pan was only one-twelfth of the loss from the bare soil--only 0.08 in. for the period. "

"The second method of study permits evaluation of the effect of freezing on the surface soil moisture during the whole winter period and offers a means of estimating evaporation losses affected by freezing.

"The method includes: (1) Determination of the relationship of soil moisture to various climatic factors, including freezing; (2) expression of the change of soil moisture as a function of time following a rain, when freezing was zero. This change equals the evaporation rate, with certain limitations as to the length of the period to which it can be applied; (3) expression of the evaporation rate as a function of the soil-moisture content. From this the evaporation rate and total evaporation during an interval is expressed as a function of the climatic factors, including freezing. The last is based on the assumption that as freezing affects the soil-moisture content, the evaporation rate in turn is affected; (4) comparison of the evaporation as determined by the equations with the actual measured evaporation. "

The author developed regression equations predicting total evaporation during freezing and non-freezing periods. The actual and calculated evaporation values were as follows:

Dates	Bare soil				Brush-covered soil, without freezing, actual values	Free water surface
	With freezing		Without freezing			
	Actual	Calcu-lated	1/ Actual	Calcu-lated		
	- - - - - - - -Inches per day - - - - - - - - -					
Jan. 9-19 1939	0.10	0.09	0.03	0.03	0.025	0.008
Dec. 20-31 1937	.09	.10	.03	.03	.025	.008

1/ Not for the same periods, but for similar 6- to 10-day periods.

In a discussion of the mechanism of water movement upward during soil freezing, the author points out that vapor movement due to vapor pressure and thermal gradients would account for only a small part of movement.

"Movement under tensions set up in the water films adjacent to the ice crystals seems most probably to be the principal mechanism. An indication of the magnitude of such tensions is obtained from the freezing point depression (3°C). According to an equation given by ANDERSON, FLETCHER, and EDLEFSEN [1942], if it is assumed that the ice freezes out at one atmosphere pressure, the adjacent water is under a tension of 36 atmospheres. Under such tension the water could be expected to flow through the connected water films which extend downward into the soil."

14. SOME FACTORS AFFECTING THE EVAPORATION OF WATER
     FROM SOIL

   C. A. Fisher. Jour. Agr. Sci. 13: 121-143. 1923.

     Fisher describes four stages of drying evident when rates of drying are plotted against moisture content. At high moisture contents, no change in rate for evaporation takes place from a free water surface. Then the rate breaks off from the horizontal and drops sharply, becoming proportional to water content. In the third stage, rates drop off because of the influence of capillarity, and because the rate of water movement from within the soil is less than the rate of surface evaporation. In the final stage, the curve bends to the origin, evaporation diminishing as adsorbed water is given off.

     Clay, silt, and loam all had drying curves of somewhat different shapes. "The moisture contents at which the rates of evaporation begin to fall off, i.e., at which the vapour pressure of the retained moisture begins to diminish, are characteristic for each soil. They are some function of the total surface of the soil grains which in turn is dependent on the average size of the grains. This water content may therefore be used as a means of characterizing soil."

THE MOVEMENT AND EVAPORATION OF SOIL WATER IN RELATION TO pF

C. M. Woodruff. Soil Sci. Soc. Amer. Proc. 6:120-125. 1941
(Reprinted by permission.)

"Small jelly tumblers were filled with soil, saturated with
r and set in the laboratory to dry. One set of samples were ex
1 to the still air of the laboratory, the remainder were placed
·e an electric fan. The relative humidity was near 50% and the
erature fluctuated between 24° and 28° C. Tumblers of water
intermingled with the tumblers of soil. All were weighed at
vals to observe the quantity of water lost from the soil in relat
e quantity lost from the free water surface. The loss from the
water surface was used as a functional measure of time. The
studied included white sand uniformly 0.08 mm. in diameter,
y loams, silt loams, and clay loams. The latter two were gra
and of good tilth."

"A typical curve expressing the quantity of water evaporated
ι the soil with respect to time...possesses two distinctly differ
϶onents. The rate of evaporation from the soil was constant an
ιtly less than the evaporation from the free water surface durin
nitial stage. This stage was followed abruptly by a much slowe
of evaporation that decreased regularly with time. The transi
ι the initial stage to the final stage occurred at a moisture cont
coincided with a moisture potential of pF3...The surface of th
was moist in appearance before the transition. It began to dry
e point of transition and as evaporation continued, an abrupt dι
ιt boundary moved deeper into the soil. Similar phenomena we
rved when the soil was exposed to stagnant air. However, the
of evaporation in still air was one-tenth of that in moving air ε
ransition between the two portions of the curve was less abrup

"Similar results were obtained for other soils. The outstaι
eatures of this study were the formation of a dry crust on the
ιce of the soil and the transition in the shape of the evaporatioι
e at pF3 for fine sandy loams, silt loams, and granulated claγ
ιs. Interpreted in terms of the previous results for sand of un
ιcle size, the results indicate that in the graded system of porε
ϩ, water will move as a liquid by capillarity below pF3 and as
r by diffusion above pF3."

"Six highly weathered soils...puddled upon wetting. These soils failed to show a well-defined boundary between the dry surface and the moist soil beneath. Moisture continued to move in the liquid phase from these soils at moisture potentials above pF3. Movement of water as a liquid at moisture potentials above pF3 may occur in soils dominated by pore sizes that are small enough for the adsorptive forces to stabilize the contents of the pores at high curvatures of the air water interface.

"The moisture content of the soil beneath the dry surface crust corresponded closely with that at pF3 in 16 of the 27 soils studied. Results comparable to these are to be expected for the majority of the medium-textured soils of good tilth, because the dominant group of pore sizes in these soils falls within the range where surface tension forces impose a condition of instability on the water within the pores.

"Movement of water through the soil in the liquid phase ceases when the surface of a short column of soil becomes dry. Water loss may then occur only by evaporation beneath the surface, and the rate of evaporation is related inversely to the square of the thickness of the dry layer of soil through which the vapor diffuses. Practically, this means that it will require four times as long to dry out the second inch of soil as it does to dry out the first inch."

16. EVAPORATION FROM SOILS AND TRANSPIRATION

F. J. Veihmeyer. Trans. Amer. Geophys. Union
19:612-619. 1938. (Reprinted by permission.)

"Since the motion of water at moisture-contents lower than field-capacity is slow, it would seem that the rate of movement to the surface and hence the rate of evaporation-loss would be small. Our experiments show that the loss of moisture by direct evaporation from the soil-surface [clay loam] is largely confined to a shallow surface layer of about eight inches. Furthermore, the evaporation-loss has been shown to be very small compared to the amount of water taken from the soil as transpired by plants."

"Some of the tanks with bare uncultivated soil were kept under observation for long periods but were covered during rains so that no water was applied to them after the beginning of the tests. One of the tanks lost only 57 pounds of water during four years of exposure to

evaporation and still there was moist soil belo
57 pounds loss, which is 18.9 pounds to the sq
exposed to evaporation, is equivalent to a dept
of water for a period of four years."

"When the soil is in contact with the fre
losses of water by surface-evaporation may be
indicated. Table 1 gives the average evaporat
soils at Davis, expressed in inches per day.]
is indicated in the first column. Depths were
water-level regulators. It is interesting to no
by evaporation from soils with a water-table a
face is about the same as the average use of w
plants. Also, it is interesting to note that witl
four feet from the surface, the loss by surface
mately one-hundredth of that from soils with a
or in other words, the transpiration-losses of
as much in one day as that which would occur l
from soils with a water-table at four feet in at

Table 1.--Average evaporation from wa
[Calif.]

Depth to water	1936	May 15 to Sept.1, 1937
feet	inch/day	inch/day
0.0	0.328	0.317
0.5	0.215	0.230
1.0	0.209	0.194
1.5	0.093	0.086
2.0	0.079	0.055
3.0	.....	.....
4.8	.....	.....

17. EXPERIMENTS TO DETERMINE RATE OF EVAPORATION
FROM SATURATED SOILS AND RIVER-BED SANDS

Ralph L. Parshall. Trans. Amer. Soc. Civ. Engin.
94:961-999. 1930. (Reprinted by permission.)

"The cooling effect of the rain on the soil increases the surface
tension of the capillary moisture drawn up from the water-table. Rain
water falling on the soil also dilutes the soil solution and if the solution
is alkaline it increases the rate of evaporation. It is evident that,
although adding moisture to the soil at the time, light showers may later
cause a more rapid depletion of the moisture already within the soil."

Soil tanks with water tables 12 inches below the surface were
sprinkled in quantities approximating a shower of rain. Evaporation loss
increased materially. For one tank the ratio of loss compared with that
from free water surface was 32 percent during the unsprinkled period
and 76 percent during the sprinkled period; for another tank comparable
values were 11 and 33 percent, respectively.

18. RAINFALL, RUNOFF AND SOIL MOISTURE UNDER DESERT
CONDITIONS

Forrest Shreve. Ann. Assoc. Amer. Geog. 24(3):131-156. 1934.
(Reprinted by permission.)

Soil-moisture determinations were made in Arizona in the
alluvial clay, or abode, of the Santa Cruz floodplains and on the adjacent
bajada soil.

"It is obvious...that the winter rains are far more effectual
than the summer ones in building up the moisture of the soil. This is
due in part to the high runoff that has been shown to characterize the
summer rains, and in part to the much more active evaporation from the
soil surface at that season. Of the six rainy seasons comprised in the
period of observation there was only one which influenced the moisture
below 60 cm., its effect extending to 150 cm. It is evident that much of
the water which enters the soil is destined to return by the same path and
never to reach the lower levels of higher moisture, where a long capillary
journey protects the moisture from possibility of evaporation. There is
abundant a priori evidence that the moisture of the lower levels is more
securely maintained by the existence of a dryer soil above. There is also

evidence in the present work that a profound wetting such as the soil received in the winter of 1931-32 may do more to reduce than to increas the moisture of the lower levels. It will be noted that the moisture at 180 cm. stood at 15-16% from September, 1930, until May, 1932, and that the same percentage was sometimes found at 150 cm. Following th heavy wetting of the winter, 1931-32, the moisture at 180 cm. fell to 13-14% and with various aberrancies has stood at that percentage ever sinc It is highly probable that this permanent fall was due to the establishmei of a better system of capillary films from 180 cm. to the surface than h: existed for some time, and that the lower levels of the soil lost water m rapidly through their existence than had been possible under the dry sur face conditions of the previous 20 months."

"The depth of the alluvial clay places a large store of water at 1 disposal of the mesquite, the roots of which sometimes penetrate to a de of 10 meters. The shallowness of the volcanic clay limits the extent of the good moisture supply available to the plants of that soil. The low ra of moistures in the bajada loam, together with the almost universal occu rence of caliche in the outwash of volcanic hills, renders the conditions severe for plants and permits only the open occurrence of the most drou resistant shrub of the region. The march of soil moisture conditions se therefore, to be one of the most important conditions differentiating the vegetation in a region in which securing an adequate water supply is the immediate problem of all plants."

19. STUDIES ON SOIL TEMPERATURES IN RELATION TO OTHER FACTORS CONTROLLING THE DISPOSAL OF SOLAR RADIATION

R. K. Dravid. Indian Jour. Agr. Sci. 10: 352-387. 1940.
(Reprinted by permission.)

"Weekly determinations of soil moisture [black cotton soil] at d of 0.2 in., 4 in., 6 in., 8 in., 12 in., and 18 in., were made from the midd of July, 1935, on the bare plot of the Central Agricultural Meteorologic Observatory. The data up to the end of November 1936, i.e., over a per of about sixteen months...illustrate the variation of moisture both with c and with time."

During rainy spells the soil had a high moisture content through the upper 18 inches."...During the frequent breaks in the monsoon rains at Poona, the moisture content fluctuates rapidly in the first six inches c the soil. After the withdrawal of the monsoon the surface layers of the are subjected to more or less unbroken desiccation during the long spel! dry weather extending from the first week of November 1935 to the begi! of June 1936."

- 34 -

After the initial desiccation the soil moisture content remained nearly constant during the dry season, 5 percent near the surface and 25 percent at a depth of about 1 foot. This illustrates the protecting influence of the dry surface soil on the layers below.

20. CHANGES IN SOIL MOISTURE IN THE TOP 8 FEET OF A BARE SOIL DURING 22 MONTHS AFTER WETTING

R. H. Hilgeman. Jour. Amer. Soc. Agron. 40: 919-925. 1948. (Reprinted by permission.)

Observations were made in the Salt River valley, Arizona. The soil was Cajon silt loam--a bare area, kept free of weeds. The distance above the water table was 65 to 70 feet.

"...Between May 1 and Nov. 1, 1944, when the soil moisture content was relatively high, 5.69 inches of water were lost from the upper 8 feet. The effect of moderate showers in May and July could not be detected from the samples obtained. Between Nov. 1, 1944, and Mar. 29, 1945, rainfall added to the amount of water in the upper 2 feet of soil. During this interval no significant changes occurred in the moisture content of the soil in the 24- to 96- inch zone. Beginning May 1 and extending to November 1, 1945, 2.59 inches of water were lost."

Decrease in moisture content during 22-month period

Depth (in.)	Loss (in.)	Available water lost (percentage)
0-12	1.35	67
12-24	1.51	75
24-36	2.42	75
36-48	1.72	78
48-60	1.39	51
60-72	0.66	20
72-84	0.54	18
84-96	0.24	10

"The rate of loss of water appeared to be related to temperature and the moisture content of the soil.... More water was lost during the first summer than during the second. Assuming the temperatures were equal, the rate of loss was a function of the moisture content of the soil."

1. FOREST INFLUENCES

Joseph Kittredge. 394 pp. McGraw-Hill, New York
(Reprinted by permission.)

From the author's summary of influences of fore
vaporation, p. 156:

"Evaporation as measured by losses from soil sι
sually less than from free water or wetted instrumenta
nd the differences become progressively greater as the
s drier.

"The depth to which evaporation is effective in lo
ree water level in initially saturated soils increases as
ɜxture becomes finer from abóut 14 in. for coarse sand
2 in. for heavy loam.

.   "Water rises more quickly by capillarity in coar
ine-textured soils, but the evaporation losses from any
xtremely slow after the surface 8 in. becomes dry.

"Evaporation from bare soil decreases as the an
noisture in the surface foot decreases and as the depth
able increases.

"Insofar as vegetation reduces wind velocity, it 1
vaporation also and in proportion to its density. "

"The evaporation from soil covered by forest flo
o 80 per cent of that from bare soil. The reduction vaɹ
f floor and increases as the floor becomes thicker at lɛ
hickness of 2 in.

"The evaporation from water or wetted surfaces
egetative cover progressively more by mesophytic and
erophytic and preclimax types; more by dense- than by
pecies, more by tolerant than by intolerant species; mɕ
tocked than by understocked stands; more by stands n
otation age than those older or younger; and more by d
·y evergreen species during the warm season. "

TRANSPIRATION

22. TRANSPIRATION AND TOTAL EVAPORATION *

Charles H. Lee. Physics of the Earth--IX. HYDROLOGY, pp. 259-330. Dover Publications, Inc. New York. 1942.

[Pp. 273-274.] "The relative advantages and disadvantages of transpiration have been the subject of much debate, the extreme positions being (1) that the process is an unavoidable evil and (2) that it is vitally important to the normal functioning of the plant organism. Among the arguments in support of the second position is that transpiration cools the leaves and prevents injury or death by high temperatures. Experimental work has shown that the degree of cooling rarely exceeds 2° to 5° C., which, in view of the present state of knowledge of protoplasm, could be of little aid in preventing injury from heat (Miller, 1938). The death of leaves in hot weather is apparently due to excessive loss of water from protoplasm rather than excessive increase in temperature.

"Another argument is that transpiration increases the rate of absorption from the soil of inorganic salts that are required for plant food. It has been proved conclusively, however, that there is no relation between rate of transpiration and absorption of mineral salts from the soil (Miller, 1938). It is also seriously questioned whether the transpiration stream aids materially in transporting absorbed salts from roots to leaves, leading authorities stating that transpiration can have little or no effect upon the movement of nutrients (Miller, 1938). Viewed in all its aspects, transpiration is unquestionably a very wasteful process, and its harmful effects upon plant life appear to exceed by far its beneficial effects. The benefits, if any, do not compensate for the elements of danger.

* Reprinted through permission of Dover Publications, Inc. and the National Research Council, from HYDROLOGY by O. E. Meinzer, Dover Publications, Inc., N. Y. ($ 4.95)

"It has been pointed out (Barnes, 1902) that the ex
of this anomalous condition may be found in the origin and
ment of the transpiration process. Probably in the primi
plants lived in water and derived the carbon dioxide **and** o
needed in the process of photosynthesis and respiration di
from this source. As development took place plants have
materially changed their method of obtaining these gases.
present organized, with foliage exposed to the atmosphere
of to water, wet cell walls are exposed to drying by evapo
well as absorption of carbon dioxide, and except under sp
conditions have developed no effective means of protection
loss of water. Transpiration, although a constant source
to plant life, is thus unavoidable and must be recognized i
study of the hydrologic balance."

23. INFLUENCES THAT AFFECT TRANSPIRATION FRC LEAVES

Burton E. Livingston. Sigma Xi Quart. 26: 88-101.
(Reprinted by permission.)

Livingston notes the distinction between two kinds
transpiration, cuticular and stomatal.

"Two kinds of foliar transpiration are to be distir
cuticular transpiration and stomatal transpiration. In cui
transpiration vaporization of water occurs just at the leaf
(from the cuticle and from trichomes of hairs, if present;
water vapor produced escapes to the surrounding air as i
In stomatal transpiration, on the other hand, vaporization
the peripheries of the substomatal chambers, into the int
phere of the leaf, and the water vapor thus produced diffu
quently through open stomata into the surrounding air.

"The fundamental driving pressure of cuticular transpiration
is just the vapor pressure at the exposed surface of the liquid water held
in the ultra-microscopic pores of the cuticle. Other conditions being
constant, it is of course greater as foliar temperature is higher, and
conversely. Because of the wax-like nature of cuticle and its low water
content even when saturated, cuticular vapor pressure never approaches
in magnitude the corresponding vapor pressure of a free surface of pure
water, or of aqueous solution like the solutions within the cells and
vessels. Cuticular transpiration from a leaf surface is therefore generally much slower than stomatal transpiration when open stomata are
frequent, but it is often relatively important when stomata are sparse or
predominantly closed."

"The driving force of stomatal transpiration is the vapor tension
of water vapor in the substomatal chambers. It is commonly greater
than the vapor tension of the outside air but not as great as the vapor
tension of water-saturated air at the same temperature. Its magnitude
depends on the rate of outward diffusion of vapor into the outside air and
on the vapor pressure of the imbibitionally wet cell-wall surfaces that
surround the chambers; in health these surfaces are regularly free of
superficial liquid."

"The internal influences that affect the transpiration rate
directly may be usefully summarized as: (a) the area of the leaf surface, (b) the water-vapor pressure of the cuticle, (c) the frequency of
stomata, (d) the average size (or diffusive capacity) of the stomatal
openings and (e) the vapor tension in the substomatal chambers. Both
the vapor pressure of the cuticle (its tendency to drive water vapor into
the air by cuticular transpiration) and that of the cell walls about the
substomatal chambers (their tendency to drive water vapor into the
chambers and outward through open stomata) are dependent on the nature
of the surfaces involved and upon their temperature, as well as on their
relative degree of wetness, and also upon the current rate at which heat
from absorbed radiant energy is being received at the vaporizing surfaces without corresponding rise in temperature. Wetness of walls or
cuticle naturally depends partly on the previous rate of water renewal
and partly on the previous rate of vaporization therefrom. The influences
of absorbed radiant energy, air temperature, air humidity and air movement are to be considered as external influences."

"The only external influence that affects transpiration directly
is the resistance offered by the adjacent layer of external air. But
air movement generally tends to reduce that resistance, by continually
replacing the adjacent air layer with drier ambient air, while absorbed

radiation and air temperature influence the vapor]
surfaces. Therefore air movement, the temperatι
of the ambient air and the drying effect of sunshine
sidered as the external influences that take part—al
the determination of transpiration rates. "

"The sweeping influence of wind, to acceleι
always of very great importance, but its mechanic
pressure bends and waves plant leaves.to and fro a
exposure) is not to be neglected in this connection.

"The pronounced and rapid changes of light
commonly occur in early morning and in late after
very great indirect influence on transpiration throι
responses of motile stomata. The stomata of man
morning because of increasing light intensity and c
because of decreasing light intensity. The comple
means of which these photeolic responses are brοʋ
be considered here further than to remark that the
guard cells increases with the increasing light inte
and decreases with the decreasing light intensity o
that the opening and closing movements are due to
decrease, respectively. But turgor alterations in
occasioned also by non-photeolic changes in the wa
cells and the neighboring tissues, and stomatal mc
not always to be explained by reference to alteratiᵢ
Motile stomata frequently show closing movements
advanced stages of incipient drying.

"It is clear that outward diffusion of water
stomatal chambers must be greatly retarded when
are closed and when the vapor tension of the outer
same time greatly retard that diffusion. An idea ᴄ
in which motile stomata may alter throughout a clᴇ
from the following approximate averages of length
elliptical cross-sections of stomatal openings on t
mature leaf of August lily. Dimensions are given
millimeter. "

Hour of observation	4:30 AM	5:30 AM	10:00 AℲ
Length [of opening]	1.4	3.6	6.5
Width [of opening]	.6	1.1	1.3
Average elliptical perimeter	2.0	4.7	7.8

"When incipient drying of the leaf tissues induces sufficient closing movement in motile stomata, stomatal transpiration is thereby retarded in another way. Thus stomata frequently attain their maximal degree of opening for the day, and begin to close, long before light intensity begins to decline. This early closure, which is not photeolic, appears to be of general occurrence, but it is naturally most pronounced when transpiration is unusually rapid or when the leaf is inadequately supplied with water. Motile stomata generally begin to close in the daytime long before wilting becomes apparent."

24. STUDIES OF EVAPORATION AND TRANSPIRATION UNDER CONTROLLED CONDITIONS

Emmett Martin. Carnegie Inst. Wash. Pub. 550. 1943. (Reprinted by permission.)

Helianthus annuus (sunflower) and Ambrosia trifida (ragweed) were grown in cans holding about 1 kilogram of sandy loam at field capacity. The correlation between rate of transpiration and relative humidity in darkness in calm air was investigated at temperatures of 27, 38, 49° C. The youngest plants tested were 34 days old; the oldest, 84.

"For young plants, the relation between transpiration rate and relative humidity was approximately linear at all three temperatures, although at 49° there was a tendency for the rates at low relative humidities to fall below the values expected from the straight-line relation.

"For older plants, the rate of transpiration at 27° was less than for the young ones, but the relation with relative humidity still appeared to be linear. At 38°, however, the older plants showed lower rates of water loss than the young ones only for relative humidities below 50 per cent. Above this point, no difference was detectable. At 49° no difference between young and old plants was observed at any relative humidity. These differences between young and old plants are probably due to dependence of permeability of protoplasm on age and temperature.

"The influence of wind on rate of transpiration increased considerably with temperature, probably because of an increase in the cuticular component of transpiration. Exposure to wind of 250 cm/sec.

at high temperatures and low relative humidities sometimes in closure of the stomata in less than 3 minutes."

"The ratios of nighttime to daytime transpiration r. Helianthus annuus under constant conditions in darkness we: to be 0.64, 0.67, and 0.91 at air temperatures of 27°, 38°, a respectively, indicating a reduction in the regulatory power stomata at high air temperatures. This behavior is probabl an increase in the cuticular component of transpiration at h temperatures.

"The maximum temperature depression of leaves o these experiments was 20° C. Ordinarily transpiration pro cooling of only 10° C. or less. The depression of the tempe the leaves below that of air for a given rate of transpiration leaf area decreased with increasing leaf temperature. This could be explained as due to an increase in the wetness of th surface, which indicates an increase in the permeability of l and epidermal cell walls to water.

"The influence of visible radiation on the transpirat single attached leaves of Helianthus annuus is linear. The a influence of radiation is apparently about 25 percent greater be expected as a result of its effect in raising the leaf temp Presumably radiation has an effect on permeability of proto pendent of temperature."

25. ECOLOGICAL ASPECTS OF TRANSPIRATION

   I. Pike's Peak region: climatic aspects.
   II. Pike's Peak and Santa Barbara regions: edaphic and aspects.

Charles J. Whitfield. Bot. Gaz. 93(4): 436-452 and 94 (196. 1932. (Reprinted by permission.)

Sunflower, wheat, and corn were used as phytomete climatic formations to obtain a measure of transpiration un climatic factors.

The plants were grown in water-proofed cylinders and 5.5 inches wide. The plants were watered, and the cont

weighed and sunk in ground to the same level as native vegetation.
The containers were left in ground for 2 - 4 days and then reweighed.

"In all cases transpiration decreased with increased altitude.
This was true for all species used in the different series. Of the
various climatic conditions measured, evaporation, soil and air temperatures, and saturation deficits decreased with increased altitude.
On the other hand, the approximate values of the maximum intensity
of radiant energy on a clear day were 1.48, 1.58, and 1.64 cal. per
cm. ^{2} per min. for plains, montane, and alpine respectively, showing
increased values with increased altitude. However, the total energy
over a season might reverse these readings because of the more cloudy
conditions in the alpine region. No data are at present available on this
problem. Relative humidities, rainfall, and holard are usually greater
at the higher elevations. Wind is constantly highest in the alpine region,
with the plains next, and the montane zone ordinarily the lowest."

"A close relationship exists between transpiration, air temperature, and relative humidity, and between transpiration and a combination of these two factors. The average air temperature, relative
humidities, and transpiration of the natural groups approximates a
straight line. This relationship is strengthened by the fact that the
average air temperatures, relative humidities, and transpiration at the
main climatic stations indicate a straight line. While the results approximate a straight line between transpiration, air temperature, and relative
humidity, such a correlation does not exist between transpiration, saturation deficit, and evaporation. The transpirational points which fall naturally on an approximately straight line with air temperature and relative
humidity do not do so with saturation deficit and evaporation. These results
show that the two curves of transpiration and saturation deficit do not run
parallel. In addition, the mathematical ratios of transpiration and saturation deficit are irregular. The fact that the points do not approximate a
straight line, as well as the irregularity of the ratios, indicates that
saturation deficit does not bear so close a relation to transpiration as do
air temperature and relative humidity. "

"Experiments to determine the effect of various soil temperatures
on transpiration were conducted at the Alpine laboratory during the
summers of 1930 and 1931. Three conditions, cold, intermediate, and
warm, with respective temperatures of 36°, 51°, and 113° F., were used
for the initial experiment. Thirty standardized sunflower plants were
divided into three groups of ten each, weighed and placed in the conditions already described. When reweighed, the batteries showed the
following losses: cold, 19.0; intermediate, 46.2; and warm, 53.3 gm.

The results show a rapid rise of the transpirational curv
comparatively low increase of soil temperature, and a fl
of the transpirational curve with high soil temperatures.
indicate, as does the initial experiment conducted in the
that when soil temperatures get above 40° F., they are n
ant in influencing transpiration as they are below this ter

"In order to check the relation of different soil
transpiration, sunflower plants were grown in a uniform
standardized. The plants were then divided into three gr
set allowed to approach the wilting coefficient (or a low
another brought to medium content, and the third given a
The experiment was conducted for a 5-day period, at the
time water contents were figured on the basis of dry wei
osmotic values were determined.

"Water loss apparently increases with the water
the optimum conditions are reached. In this experiment
water content gave a water loss below that of the mediun
doubtedly to poor aeration. Osmotic values showed a cor
with increased soil moisture."

26. DAILY TRANSPIRATION DURING THE NORMAL GI
PERIOD AND ITS CORRELATION WITH THE WEA1

Lyman J. Briggs and H. L. Shantz. Jour. Agr. Res.
7(4): 155-212. 1916.

At Akron, Colorado, plants of 22 crops were gro
nized-iron pots containing about 115 kilograms of soil.
tight-fitting covers with openings for the stems. Loss o
limited to transpiration. The pots were weighed daily du
growth period, and water was added daily to insure an ac

"...The direct solar radiation received by the p
is usually not sufficient to account for the observed tran
the midday hours. In some of the small grains the ener
through transpiration is twice the amount received dire
[This indicates the importance of indirect radiation.]

"The march of transpiration due to changes in t
(change in the transpiration coefficient) may be express
of the daily transpiration to the daily evaporation if we

to constitute a perfect summation of the weather conditions determining transpiration. The transpiration of the annual crop plants (aside from fluctuations due to weather) rises to a maximum a little beyond the middle of the growth period and then decreases until the plants are harvested. Perennial forage crops such as alfalfa increase steadily in transpiration to a maximum at or near the time of cutting. Various crops show their individuality by departing more or less from these types.

"The transpiration coefficient of many of the crops increases exponentially during the early stages of growth. Sudan grass, for example, doubled its transpiration coefficient every four days during the early growth period. Alfalfa throughout practically the whole period between cuttings doubled its transpiration every eight days."

"The correlation has been determined between the various physical factors of environment and the transpiration of the different crops, considered both individually and as one population. The correlation coefficients in the latter case for the season of 1914 are as follows:

"Transpiration with radiation, 0.50 + 0.01; with temperature, 0.64 + 0.01; with wet-bulb depression, 0.79 + 0.01; with evaporation (shallow tank), 0.72 + 0.01; with evaporation (deep tank), 0.63 + 0.01; and with wind velocity, 0.26 + 0.01."

27. THE TRANSPIRATION OF DIFFERENT PLANT ASSOCIATIONS IN SOUTH AFRICA

Part·IV--Parkland; forest and Sour Mountain - grassveld; large Karoo bushes.

M. Henrici. Union So. Africa Dept. Agr. and Forestry Sci. Bul. 244. Pretoria, 1946. (Reprinted by permission.)

Transpiration was measured by weighing material cut from trees or plants.

"From a number of preliminary experiments it was observed that the transpiration of woody branchlets was constant for about 6 minutes after their removal from the tree; after the 9th minute, however, their transpiration dropped rapidly, corresponding to the closing of the

stomata (Henrici 1940). No increase in transpiration after
was observed. Gramineae showed a slowing down of the t
tion from about the 5th minute after they had been cut. It
fore decided to take the reading of the first 3 minutes as th
the transpiration value....Within a few seconds after cuttin
were on the balance ready for weighing, which is importan
method. "

"In an investigation on transpiration in Europe the
soil moisture would scarcely come into consideration. In
soil moisture is a most important factor. All xerophytes
so far investigated--succulents excluded--have a high tran
power if exposed to the sun, so long as the soil moisture i
In misty weather or rain no transpiration takes place. "

"Of special interest is the effect of wind. Even in
is still a controversy as to its effect on transpiration. In
it never has an accelerating effect on the transpiration of
contrary, the stomata of all tree foliage close as soon as t
wind and the leaves therefore restrict transpiration almos
On grass the effect is different. An accelerating effect is
this does not last long, as after a while the aperture of the
decreased and the transpiration eo ipso diminishes. "

"Last, but not least, temperature and light have to
as external factors influencing transpiration. There is no
all the investigated associations the change from darkness
transpiration, but a quarter of the intensity of sunlight is
ensure maximal transpiration. After 9 to 10 a. m. no furt
transpiration, due to stronger light, could be recorded on c

Regarding the daily trend of transpiration: "The m
piration is not correlated with either the maximum light ir
the maximum temperature, or with the minimum humidity
wet days or early in spring. On the contrary, nearly all c
have two maxima for transpiration--one before 11 a. m. a
after 3 p. m. The midday depression of the transpiration
observed in South Africa long before anything about it was
Europe. The fact that it is not regularly observed in earl
the soil is really saturated, suggests that it is an automat
measure of the plant against excessive transpiration; in m
stomata are not fully open during the midday depression,
so much closed that they could be the only reason for the
piration. "

"The noon depression of transpiration is doubtlessly connected with the decrease in free water in the leaves which is only replaced later in the afternoon or in the evening."

"It is clear that the factors restricting transpiration are very different in the various climates. In the Sour Mountain-grassveld or in the 'Ndema forest of the Drakensberg, the very high relative air moisture, often mist, causes 0 values, especially early in the morning. At Fauresmith the low moisture inhibits transpiration, this being particularly noticeable in the afternoon of hot summer days. In winter the temperature early in the morning usually depresses transpiration, presumably not as such, as radiation is strong enough, but by preventing a regular supply of water from the roots. In extreme cases real dormancy of the trees can be observed.

"If, broadly speaking, the potential transpiration (not the actual) of a tree in the Orange Free State is the same as for the same species or a near related species in the Drakensberg, it is eo ipso understandable that the espacement of trees in the dry climate must be very great."

28. TRANSPIRATION OF SOUTH AFRICAN PLANT ASSOCIATIONS

II. Indigenous and exotic trees under semi-arid conditions.
M. Henrici. Union So. Africa Dept. Agr. and Forestry Sci.
Bul. 248. Pretoria, 1946. (Reprinted by permission.)

Henrici measured transpiration by determining the loss in weight of cut branches during a three-minute period beginning one minute after the branches were cut.

"Even on clear days, the light proved too weak after 5 p.m. for any appreciable transpiration, although the stomata were sometimes wide open. It seems that the transpiration of this botanical formation is more or less limited to nine hours of the day. This is readily understandable for the moist valley itself, but it is rather interesting for the slopes and ridges where the trees simply stopped transpiration about 5 p.m. with the disappearing, not setting sun."

"...Harder (1935) still considers the possibility, in an arid region, of wind having a beneficial influence on plant life in decreasing the transpiration by lowering the temperature of the plant and by causing

the closing of the stomata. Firbas (1931) points out that the fi
of wind always accelerates the transpiration which, in the case
hardy plants, decreases after the primary increase and may th
fall below the value obtained in calm air.

"To the author's mind these two points govern the trans
of the trees in the semi-arid region of Klapperkop. If the winc
duration, there is certainly no increasing effect, but with sing
there may be a momentary increase."

"Does the transpiration curve of the investigated trees
temperature, air moisture and light intensity? The question h
been answered in the affirmative for grasses at Pretoria (Mes
where the lowest air moisture and highest temperature appare
cided soon after 12 noon with the highest transpiration calculat
percentage of the water content. The soil moisture was excep
high, being rather like laboratory than veld conditions. In the
investigation, which was made during the rainy season of a ver
year, the soil moisture was certainly as favourable as it could
be, although it is fully realized that the water content of shale
hardly exceed 20%, and that the soil of Fountains Valley itself
humus, 25%."

"During the whole investigation, it only once occurred
highest transpiration coincided with the lowest relative moistu
highest temperature, although at times the maxima are near th
of the highest temperature in the afternoon. It may be safely (
that in a semi-arid climate, on porous soil like shale, the trar
of trees never follows temperature over a certain range, is no
tional to the increasing light, and certainly does not follow the
relative moisture. On the contrary, the maxima are found not
time of the strongest light, but earlier or later. Some of the
are even found at the time of the highest air moisture. One m
whether transpiration only follows light, temperature and air
to a certain degree as, of course, there is no questioning the
light and temperature greatly increase the water output. For
the transition from night to day, the change in the evening fror
darkness, and even from sunlight to diffuse light, brings out t
changes in the water output.

"It is quite possible that light, especially when added t
temperature, becomes too much for the trees, and that they th
restrict their water output about noon. The sudden drop in trr
tion when the light gets weaker towards 5 p.m., shows how m

trees are affected by light, and not so much by temperature in the range of the investigation. The result is the graph with the decrease over the noon period."

"The author is convinced that the decreased transpiration about noon was partly due to an incipient drying or at least a water deficit (Maximov, 1929; Stocker, 1929) at that time, though no direct determination could be made during the veld experiment. Later in the afternoon the water intake apparently compensates for the loss of water by transpiration. The behaviour of the stomata is in close accordance with Stalfelt's findings (1929) that even in good light stomata close when water reserves are sinking. Even when external factors and the apertures of stomata are made responsible for the rate of transpiration, there must still be some internal factor of the plant which regulates it, otherwise, there would not be found, as in many of the investigated trees with fully open stomata, a very small transpiration at 5 p.m. on bright days."

"It is clear that our natural forest in the surroundings of Pretoria transpires well within the rainfall. Plenty of water is available for the undergrowth which considerably decreases the run-off of the water. As regards the exotics it is in most cases not the transpiration as such which preclude plantations; the espacement and size of the trees will be the deciding factor.

"If a plantation is not thinned out in time, a point may be reached when, even for conifers, there is not enough water in the soil to supply all the trees, and some of them must consequently die. This is what happened in plantations in the eastern Transvaal...where in 1933 about half the trees died. The espacement before the disaster was only three to four feet; after the dying off of the trees it was six to eight feet. The surviving trees withstood a further drought in 1937 perfectly well."

29. RELATIVE TRANSPIRATION OF CONIFEROUS AND BROAD-LEAVED TREES IN AUTUMN AND WINTER

J. E. Weaver and A. Mogensen. Bot. Gaz. 68(6): 393-424. 1919. (Reprinted by permission.)

At the University of Nebraska, transpiration of ponderosa pine, jack pine, lodgepole pine, white fir, Douglas-fir, white elm, bur oak, and soft maple seedlings was measured. Seedlings were grown in pots placed outdoors. Pots were weighed daily and water added. Comparisons were mainly on a relative basis.

The seedlings with greatest leaf area had the highest water losses. Relative high transpiration rates in autumn were followed by a rapid reduction to very low winter rates. For ponderosa pine, for instance, "the entire period from Sept. 24 to Oct. 16 is character- ized by relatively high transpiration losses after which there is a decided falling off. On Oct. 11 the stomata were found to be closed. The midwinter transpiration rates are exceedingly small....A general but slow rise during the cold month of April may be noted, with a shar increase following the milder weather in May." This general trend hel for both conifers and hardwoods.

Summary:

"1. Autumn transpiration losses from conifers are just as great as or even greater than those from broad-leaves.

"2. The decrease in water losses from broad-leaved trees resulting from defoliation is gradual, and not greatly unlike the decrease sl in the transpiring power of conifers.

"3. Winter transpiration losses from conifers are only 1/55 - 1/251 a great as those in autumn.

"4. The increased losses of broad-leaved trees in spring occasioned foliation are in proportion to the leaf areas exposed, and are clos controlled by weather conditions, but in the main are similar to increased losses of conifers.

"5. Winter transpiration losses from conifers are scarcely greater t those from defoliated stems of broad-leaved trees."

30. TRANSPIRATION RATES OF SOME FOREST TREE SPECIES DU THE DORMANT SEASON

Theodore T. Kozlowski. Plant Physiol. 18: 252-260. 1943. (Reprinted by permission.)

Kozlowski compared the winter transpiration rates of white oa yellow-poplar, sugar maple, cherry laurel, eastern white pine, and loblolly pine. Three-year old seedlings were grown in buckets of sandy loam. Soil moisture was at field capacity at the beginning of the record, and water losses were replaced at the end of the experimental periods.

"...No great difference was observed in foliar transpiration of conifers and stem transpiration of deciduous species on a unit area basis [of leaf surface or stem area].

"The October-November average rate for loblolly pine and cherry laurel was more than twice the average maximum for December and January; for white pine the October-November average rate was more than three times the December-January maximum.

"Independent comparisons of the evergreen versus deciduous species and of pines versus cherry laurel indicated highly significant differences on a unit area basis. During December and January the weekly transpiration rates of cherry laurel were approximately from 2 to 4 times as great as those of either of the pines. Comparisons of rates of loblolly pine versus eastern white pine, white oak versus yellow poplar and sugar maple, and yellow poplar versus sugar maple indicated no real statistical differences in transpiration of these species during the dormant season.

"A decrease in soil temperature was found to decrease the transpiration rates of loblolly pine and eastern white pine. This is in general agreement with results obtained by other investigations with herbaceous materials. Transpiration of loblolly pine was reduced more than that of eastern white pine over the temperature range between 17° and 0° C. The difference in behavior of the two species is probably caused by differences in inherent protoplasmic qualities. The greater reduction in absorption by loblolly pine in cold soil might be a factor in its inability to survive in colder regions."

31. SOIL WATER IN RELATION TO TRANSPIRATION

V. M. Spalding. Torreya 5(2): 25-27. 1905.
(Reprinted by permission.)

"In a recent article by the writer on the creosote bush in its relation to water supply, the statement was made that the amount transpired appears to stand in direct relation to the amount of water available in the soil in which the plant is growing. Further observations on this and some other desert plants not only confirm this view but go to show that water in

the soil is a controlling factor, and that even as effici
light may, in comparison, take quite secondary rank.

"The plants employed were seedlings of the cr
(Covillea) and palo verde (Parkinsonia Torreyana and
growing in cans and supplied with measured quantities
stated intervals. The rate of transpiration was deter
the plants under a bell-jar, with suitable precautions
absorption or escape of water vapor, the amount of w.
being derived from readings of a hygrometer."

"Beginning with the palo verde, two sets of pla
as a check on the other, were used. August 11, the p
well watered the day before, the rate of transpiration
The following day, August 12, the plants meantime ha
water, but having been treated precisely as before, a
other controllable conditions, the rate of transpira tio
only 52.6 and 38.5 per cent, as high as it was on the]
result apparently attributable to nothing else than the
quantity of water in the soil in which the plants were ε

"The same plants were again placed under obs
18, having been given no water since August 15. Exte
were favorable to transpiration, full sunlight, a fresh
high temperature. At 11:40 A. M., after the rate of ·
noted, number 1 was given one ounce, and number 2 t
water. At 1:15 P. M., the rate of transpiration of nu
be the same as at the time of the preceding observatic
number 2 was twice as great. At 4:00 P. M., observ
made, and at this second afternoon reading it was fou
was transpiring twice and number 2 four times as raμ
of the forenoon observation."

"Experiments with Covillea gave even more s·
September 5, the transpiration of two plants, designa
determined in the forenoon between 11 and 12, and ag
noon between 3 and 4 o'clock. Number 2 was given th
at 12:20, none being given to number 1. At the time ·
observation it was found that number 2 was transpiriι
times as rapidly as it was before the water was giver
1, which was not watered, was transpiring only one-f
it was in the forenoon."

"It was found that exposure to bright sunlight was uniformly
followed by accelerated transpiration, whenever the plant under observation had a full supply of water, but that otherwise such acceler-
ation did not take place."

"It is noteworthy that plants which had all along received a
meagre supply of water were nevertheless in a position to transpire
rapidly when once a full supply of water was furnished them, while
plants which from the beginning had received a very large amount of
water showed promptly a marked lowering in rate of transpiration
when the water supply was reduced."

"With so complicated a problem general statements may well
be made with extreme caution, but the evidence in the present case
is sufficient to show that in studies of transpiration it is altogether
unsafe to attempt to estimate any other factors whatever without
taking due account of water in the soil."

32. INFLUENCE OF SOIL MOISTURE CONDITIONS ON APPARENT PHOTOSYNTHESIS AND TRANSPIRATION OF PECAN LEAVES

A. J. Loustalot. Jour. Agr. Res. 71(12): 519-532. 1945.

Loustalot grew seedlings in crocks of coarse sand and silt
loam, and measured photosynthesis and transpiration as soil dried.
The sand had a field capacity of 6 percent and a wilting coefficient of
1 percent; silt loam 34 and 12.2 percent respectively. During drying
from July 18 to the morning of July 22, the soil moisture of sand
dropped from about field capacity to 1.5 percent above the wilting point.
The transpiration rate dropped to two-thirds of normal, but photosynthesis was not appreciably affected. During the afternoon of July 22 the
rate of transpiration dropped sharply and photosynthesis was only two-
thirds of normal.

Silt loam drying started September 7. By September 16, with soil
moisture at 18 percent, there were small decreases in both photosynthesis and transpiration. By September 21, with soil moisture close to
wilting point, transpiration dropped to one-half normal, photosynthesis
to one-fifth normal.

"The amounts of reduction in the rates of photosynthesis and
transpiration of leaves of pecan seedlings subjected to drought were

closely correlated with the proximity of the soil moisture to point as well as with the atmospheric conditions during the periods of moisture shortage. Under conditions highly favo moisture evaporation, as in the afternoons, photosynthesis ceased when the soil moisture was at the wilting point or sl but the reduction in transpiration was considerably less. U less favorable for evaporation, as in the mornings, the tra rates were reduced to a greater extent relative to the reduc rates of photosynthesis, although the actual reduction in tra usually somewhat less than that under conditions highly favo evaporation.

"The rate of recovery in photosynthesis and transpi activity from the effects of drought was usually very rapid c first day or two after termination of the drought, but severa more were required before the rates reached normal or the

33. INFLUENCE OF SOIL MOISTURE ON PHOTOSYNTHES RESPIRATION, AND TRANSPIRATION OF APPLE LE.

G. William Schneider and N. F. Childers. Plant Phys 16: 565-583. 1941. (Reprinted by permission.)

"The results presented in this paper trace the daily a gradually drying soil on apparent photosynthesis, respira transpiration of small apple trees as the soil in which they ing gradually dried to approximately the wilting percentage.

"Before wilting was evident, there were marked red apparent photosynthesis and transpiration and an increase i tion; in one case there was a 55 per cent reduction in photo 65 per cent reduction in transpiration, and a 62 per cent in respiration. Stomata at this time appeared to be completel

"When water was applied to the soil in which wilted were growing, the leaves usually attained turgidity within tl hours, depending upon their degree of wilting. They did no recover to their original relationships with the controls in j sis and respiration before two to seven days after the wate piration usually recovered about the same time as photosyn slightly earlier."

C. W. Thornthwaite. Unpublished manuscript. Undated. (By permission.)

"In 1894, a Division of Agricultural Soils was created in the Weather Bureau with Milton Whitney as Chief. This Division was expected to continue 'the study of rainfall and temperature after they enter the soil, and to keep a continuous record of the moisture and temperature conditions within some of the most important types of soil in the country.' " (Whitney, 1894).

The author cites data indicating that "...There is little difference in the proportionate amounts of water obtained by different types of plants from different depths of the soil. The percentage of water obtained from various depths of the soil is comparable to the percentage of feeding roots at different depths and depends upon the type of soil rather than the kind of vegetation."

"Investigations should be undertaken to determine the actual amount of water readily available to different types of vegetation growing in different soils. The few data that relate to the problem indicate that the frequency distribution of available moisture will range from about 2 inches to the neighborhood of 6 or 7 inches, with the most frequent value approximately 4 inches. These values may need to be revised upward or downward as additional observations are made. Nevertheless, it is to be expected that the amount of water available to plants in agricultural soils will prove to be far less variable than conventional studies of soil moisture would indicate.

"It has already been emphasized that plants are not absolutely limited to the equivalent of 4 inches of moisture in the soil. Their roots will grow into moist soil and get additional moisture. Such moisture enters the plant too slowly to aid effectively in satisfying its needs and transpiration is very small. The plant wilts daily and soon enters the stage of permanent wilting. Growth is retarded or stopped. It is often overlooked that root growth also stops, when food production stops in the green aerial parts. Thus, the more a plant may be in need of water, the less able it is to extend roots to get it.

"It is well known that plants can utilize the soil moisture from considerable depth in the event of a moisture deficiency in the surface layers. What is not generally appreciated is that this water utilization is at a much slower rate than the optimum. Life may be preserved in

the plant but its development is seriously arrested
Miller (1938) states that transpiration under these
only one-twentieth to one-thirtieth of that which pr
optimum amount of moisture is present."

"When there is no deficiency of moisture in
roots are able to absorb the moisture freely, the r
transpiration are governed by atmospheric factors
represent the optimum water loss for the vegetatio
particular locality and may be designated the poten
If, on the other hand, soil moisture is deficient, tl
condition of the atmosphere diminishes and that of
increases. The actual transpiration is influenced
centration in the soil as well as by atmospheric fac
conditions it is almost everywhere less than potent

35. SOIL MOISTURE AND EVAPORATION

R. K. Schofield. Internatl. Cong. Soil Sci.
1950.

"...Regarding their transpiration plants ar
mercy of their physical environment. This import.
years obscured by the use of the term 'transpiratic
ratio of the water transpired by a plant to the amou
duced by photosynthesis. There is, in fact, no qua
between transpiration and the production of dry ma
clearly shown by the results of recent experiments
[1948] at Rothamsted in which measurements were
tion from several patches of short-cut grass. One
heavy dressing of fertilizer and yielded three time
grass cuttings than two others of equal size which
lizers. The evaporation from (and hence the trans
lized grass was, however, no greater than that of

"The principal reason for the seasonal vari
relating maximum evaporation (transpiration) fron
from open water] appears to be the closing at night
leaves which interrupts transpiration.

| Sept and Oct. | 13 | 0.7 |
| May to Aug. incl. | 16.5 | 0.8 |

"...Up to a certain value, C, of the soil moisture deficit, evaporation goes on at the maximum rate. If after this deficit has been built up, evaporation continues to exceed rainfall it will fall more and more behind the maximum evaporation as the deficit increases."

" 'C' of course depends on the root habits of the plants as well as on the physical properties of the soil. If the depth of rooting can be increased C will be increased, and full crop growth will continue proportionately longer in a dry season. Here is yet another desirable property for which it may be possible to select and breed plants. Soil cultivation and manurial treatments appear more likely to increase C by increasing root penetration than by increasing the water-holding power of the soil."

36. THE ROLE OF VEGETATION IN METEOROLOGY, SOIL MECHANICS AND HYDROLOGY

H. L. Penman. Brit. Jour. Appl. Phys. 2:145-151. 1951.

"The important effects of plant transpiration are found where deficits can be built up. As the deficit increases the soil reaches a state of dryness at which extraction of water by the plant ceases to be easy, the transition being sharp in some soils gradual in others, i.e. there is a value of the deficit beyond which the transpiration rate no longer is as great as the potential rate calculable from weather data."

"The calculation of soil moisture deficit from weather data is only valid when there is adequate water available to meet the potential demand...The normal condition in large areas of the world...is that summer water supply is not sufficient to maintain maximum transpiration and growth. The plant's first main response to water shortage is to extend its root system, so tapping a deeper layer of soil, but there

EVAPO-TRANSPIRATION

37. EVAPORATION IN NATURE

E. L. Stone, Jr. Jour. Forestry 50(10): 746-747. 1952.
(Reprinted by permission.)

"...It seems proper to call attention to the very pertinent work of the Rothamsted investigators, H. L. Penman and R. K. Schofield. Although not unknown in this country, this work is yet to have the impact which it surely must upon the thinking of all concerned with 'evaporation in nature'. Much of its significance lies in the very simple basic views of evapo-transpiration from land surfaces that emerge, together with a fresh approach to estimation of its magnitude."

"Penman and Schofield approached the problem of evaporation of water from any surface, including plant tissue, as a physical process accountable in terms of an energy balance. It was at once obvious that the only significant source of energy is direct solar radiation; in the case of a free water surface, evaporation accounts for some 40 percent of the total radiation received. Accordingly the annual variation in evaporation follows the solar radiation cycle closely and such correlation with mean periodic air temperature as may occur is due to their dependence on a common cause. Thus an analysis of some Rothamsted records showed little correlation between mean daily temperature and daily evaporation although the monthly means were correlated. Annual evaporation losses vary with latitude, approximating 15 inches from a free water surface in Ontario and 60 inches in the tropics.

"Since water absorbs very nearly all of the high energy radiation falling upon it, losses from a water surface represent the maximum possible values for natural evaporation. Measured losses from sod adequately supplied with moisture were approximately 80 percent of those from free water during the May-August period, with the lesser amount attributable in part to a greater reflectance of incoming radiation by the leaf surfaces. From other evidence it appeared likely that, so long as soil moisture was not limiting, deciduous forest and sod differed very little in the amount of water transpired. Nor did a trebling of dry matter production, brought about by fertilization of certain sod areas, increase water losses.

"Some of the implications for forest conditions are evident: Where soil moisture is restricted, equivalent areas of deep soil supporting unlike species or vegetation types may differ markedly in water loss because of differences in rooting depth. When moisture is not limiting, however, losses from the various areas will probably be of similar magnitude and close to a fixed maximum, with such differences as do occur associated with reflectance of short wave radiation. When so supplied, the important factor is simply the radiation absorbed per unit land area (less losses to conduction and back radiation) and once full coverage of the area is obtained considerations of total leaf surface and height become of minor importance. Likewise air temperature, wind movement, and atmospheric humidity, within the usual ranges occurring during the growing season, are of much less significance than radiation in bringing about the vapor pressure differences on which evaporation depends.

"The consequences of this work challenge a number of beliefs and textbook generalities, some of them already suspect or discarded by watershed investigators. Thus evaporation measured from isolated trees or portions thereof cannot be related, even theoretically, to losses from an area of forest vegetation nor can the results of conventional pot cultures. 'Transpiration ratios' are likewise meaningless. Definite limits are set on evapo-transpiration losses from any area and the alleged superiority of one species over another is unlikely except as attributable to season of growth or rooting depth and habit. The separation of evaporation and transpiration on vegetated areas becomes of doubtful utility inasmuch as they cannot proceed independently; for the same reason the significance of canopy interception during the growing season may prove less than has been imagined."

38. NATURAL EVAPORATION FROM OPEN WATER, BARE SOIL AND GRASS

H. L. Penman. Proc. Roy. Soc., A, 193: 120-145. 1948.

"Evaporation from bare soil involves complex soil factors as well as atmospheric conditions: transpiration studies add to these further important physical and biological features, for a plant's root system can draw on moisture throughout a considerable depth of soil, its aerial parts permit vapour transfer throughout a considerable thickness of air, and its photo-sensitive stomatal mechanism restricts this transfer, in general, to the hours of daylight. A complete survey of evaporation from bare soil and of transpiration from crops should take account of all relevant factors, but the present account will be largely restricted to consideration of the early stages that would arise after thorough wetting of the soil by rain or irrigation, when soil type, crop type and root range are of little importance."

"Two theoretical approaches to evaporation from saturated surfaces are outlined, the first being on an aerodynamic basis in which evaporation is regarded as due to turbulent transport of vapour by a process of eddy diffusion, and the second being on an energy basis in which evaporation is regarded as one of the ways of degrading incoming radiation. Neither approach is new, but a combination is suggested that eliminates the parameter measured with most difficulty--surface temperature--and provides for the first time an opportunity to make theoretical estimates of evaporation rates from standard meteorological data, estimates that can be retrospective."

"...For the combined estimate there must be known, mean air temperature, mean dewpoint temperature, mean wind velocity; and mean daily duration of sunshine."

"The evaporation rate from continuously wet bare soil is 0.9 that from an open water surface exposed to the weather conditions in all seasons.

"The corresponding relative evaporation rate from turf with a plentiful water supply varies with season of the year. Provisional values...for southern England are:

Midwinter (November - February)	0.6
Spring and autumn (March - April Sept. - Oct.)	0.7
Midsummer (May - August)	0.8
Whole year	0.75"

For the bare soil the water table was 5 inches below surface, for turf 16 inches.

39. THE MAGNITUDE AND REGIONAL DISTRIBUTIOI
LOSSES INFLUENCED BY VEGETATION

Joseph Kittredge, Jr. Jour. Forestry 36(8): 775-
(Reprinted by permission.)

Sources of data were isohyetal maps of rainfall
author subtracted runoff from rainfall to obtain estimat
by forest regions. Annual water loss:

Eastern regions	Inches	Western re:
Longleaf-loblolly-slash pine	30-40	Pacific Dou
River bottom hardwoods and		
cypress	30-40	Redwood
Oak-pine	25-35	Sugar and p
		pine
Oak-chestnut-yellow poplar	20-30	Western la:
		tern wh
Oak-hickory	20-30	Spruce-fir
Tall grass	20-30	Ponderosa
Birch-beech-maple-hemlock	15-20	Short grass
White-red-jack pine	15-20	Lodgepole ɪ
Spruce-fir	10-20	Pinon-junip
		Chaparral
		Sagebrush
		Desert shrɪ

In the East, "the regions thus form a series in
are at a maximum in the South, where the temperature
and the growing season longest, and in which they decr
sively toward a minimum in the North. In general the
losses of the different types follows the same sequence
growth rate of well-stocked forest stands."

"...In the West the progression from South to
disturbed by the low precipitation which prevails over
and strictly limits the losses... For the forest types th
higher to lower water losses is again the progression fɪ
less rapid growth, whether the differences in growth a
tude and temperature or by deficient moisture."

40. AGRICULTURAL HYDROLOGY AS EVALUATED BY MONOLITH LYSIMETERS

L. L. Harrold and F. R. Dreibelbis. 149 pp. U. S. Dept. Agr. Tech. Bul. 1050. 1951.

"This bulletin is in the nature of a progress report on the lysimeter investigations carried on at the North Appalachian Experimental Watershed near Coshocton, Ohio, to 1949.

"The hydrologic data were obtained from eleven monolith lysimeters, each 0.002 acre in area and 8 feet deep, three of which were weighed automatically. The features of the installations, some of which are unique, are described. Records of precipitation, runoff, and percolation are presented for each lysimeter. Weight records provided data for determination of condensation-absorption of moisture from the atmosphere, evapo-transpiration of soil moisture, and changes in storage of soil moisture.

"The amount of moisture condensed and absorbed from the atmosphere was fairly large, amounting to over 6 inches of water annually. Of the water added to the soil, precipitation accounted for 81 percent and condensation-absorption 19 percent. From 80 to 85 percent of the soil-moisture depletion was due to evapo-transpiration. Percolation accounted for the remainder. Different crops had strikingly different effects on seasonal evapo-transpiration rates. Wheat and meadow crops depleted soil moisture most rapidly in May and June. Corn used water at high rates in July and August. Cultivating the cornland had a noticeable effect on evapo-transpiration, and the effect of hay cutting was still more marked."

"As expected, the growing season had the greatest ET [evapo-transpiration]. June and July were the months of highest water demand by plants. The 6-year average daily ET for June ranged from 0.179 to 0.203 inch, and the July range of ET values was not materially different. The variation between the 6-year average ET values for the three lysimeters, for the different months is small. Average ET values for all lysimeters, however, had a definite trend throughout the year similar to that for air temperature. Average ET for the warm months was from three to four times as much as that for the cooler months. This difference reached a ratio as high as six to one in some years."

"Water removal from soil pores was more rapid when the land was in good sod and wheat than when it was in poor sod or corn. Conse-

quently, poor sod and cornland were less capable of absorb
storm rainfall at this season [May]. In July and August, ho
ET from cornland was very great, especially in 1949, and t
of soil-moisture depletion was high."

"...Moisture on vegetation from the preceding night
from rain or dew, had no noticeable effect on evapo-transpi
Evapo-transpiration was no greater from wet vegetation tha
dry... When there was little or no dew, larger quantities of
were used in the ET process. In other words, the evapo-tr
from vegetated land following nights of little or no dew was
transpiration. Furthermore, sizeable quantities of the ET
moistened by CA [condensation-absorption] must have been
Dew fall, or absorption of water by the soil, or both, have,
a soil-moisture conservation value."

"A study of the crop use of water by semimonthly pe
that the rate of water consumption varies during the growin
Depletion of soil-water supplies by corn, for example, was
July. The usage exceeded the normal rainfall in both the fi
second half of this month 9 out of 10 times."

"Wheat used water most heavily in late May and earl
Use exceeded normal rainfall from the middle of April thro
first half of June. Wheat usually removed water from soil
evapo-transpiration faster than cornland prior to June 15.
for wheat diminished after June 15. This is the ripening pe
lowing wheat harvest and the removal of straw, the new me
timothy, red clover, and alfalfa put on rapid growth. Wate
tion increased somewhat in late July and early August.

"Water use by first-year meadow.... Vegetation was
red clover, timothy grass, and small alfalfa plants. Water
was greatest in the first half of June. Hay cutting in the las
June decreased the demand for water. The second cutting c
in August 1947 had no noticeable effect on the semimonthly
cutting of August 1943 was effective in reducing ET-CA wat
Hay cutting may result in striking changes in daily evapo-t
and condensation absorption...."

"These data show that the net soil—moisture depletion (ET-CA) is fairly rapid on land with a good meadow crop... ranging from 0.16 to 0.38 inch per day for a vigorous and full-grown legume-grass meadow. The removal of most of this growth decreased the daily loss of soil moisture by at least one-half. The reduction may reach two-thirds, as in June 1947."

41. CONSUMPTIVE USE OF WATER. A SYMPOSIUM.
    Forest and range vegetation.

L. R. Rich. Trans. Amer. Soc. Civ. Engin. 117: 974-987. 1952. (Reprinted by permission.)

"Data from studies at Sierra Ancha Experimental Forest [central Arizona] have shown small differences in consumptive use between areas kept bare of vegetation and areas in various types of vegetation, and no significant differences in consumptive use between a watershed with good and poor grass cover. Decreased use on bare watersheds is largely hypothetical since no watershed can be kept bare of all vegetation; some plants will grow under extremely adverse conditions. If perennial grasses are depleted through abuse, inferior species such as annuals, weeds, and half-shrubs usually replace the original vegetation. If chaparral species are burned out, the sprouting varieties readily recover and may even increase.

"From a water-yield standpoint, the major question is not whether watersheds bare of vegetation would yield most water, but the type of vegetation that interferes least with water yield and still controls sediment. In the Southwest, with very few exceptions, grasses are dormant and do not use water during the winter water-yielding period. Water use is largely confined to the summer period when there is no surplus water, and surface evaporation alone could consume all the precipitation . During the winter, high transpiration by half-shrubs, winter annuals, and evergreen shrubs, all characteristic of a deteriorated vegetation, tends to reduce water yield. Total annual use by grasses is slightly higher than evaporation from bare ground but lower than losses from a deteriorated vegetation. The benefit of well-developed and well-maintained plant cover in checking soil erosion and sedimentation of reservoirs outweighs by many times the value of the slight amount of water it uses.

"Consumptive use of water by forest and range veget‹
pends primarily on climatic and watershed conditions and on ‹
ability. There are few areas in the West where unlimited wat‹
able for full potential consumptive use. The ability of plants t
use water is greatest in summer and least in winter. . . . Actual
pendent on growing conditions when moisture is available. Tl
bility of moisture depends on distribution of precipitation and ·
held in the soil for use during drought periods. Water use in ti
grassland zone was 92% of the precipitation for perennial gras
for winter annuals, and 89% of the precipitation lost from ba
evaporation. Consumptive use of water in the chaparral zon‹
of the precipitation for grasses, 84% of the precipitation for
and 78% of the precipitation lost from bare soil by evaporati
on watersheds in the mixed grassland-chaparral zone varied
to 95% of the precipitation on 4 adjacent watersheds, depend
type and depth of soil. Water on forested watersheds has va
77% to 90% of the rainfall, depending on depth of soil and sl‹

42. EFFECT OF REMOVAL OF FOREST VEGETATION UP
YIELDS

M. D. Hoover. Trans. Amer. Geophys. Union. Part V
1944. (Reprinted by permission.)

"The location of the experimental watersheds is the (
Experimental Forest, a field-laboratory of the Southeastern
Experiment Station. It is situated in the Nantahala Range of
Appalachian Mountains near Franklin, North Carolina. Thi‹
in the zone of maximum precipitation in the eastern United S

"Natural vegetation is composed of deciduous trees v
ant shrubs and minor vegetation. Because of the favorable ‹
plant-growth is rapid and but rarely checked by summer drc

"Under these conditions, vegetation has a maximum
for transpiration. Soil-moisture seldom reaches the wilting
does not become a limiting factor for growth of established ;

"The plan of this experiment was to compare water-
watershed before and after removal of forest. Before cuttir
cient stream-flow and climatic records were to be obtained
normal relationship between rainfall and runoff could be est
In addition, another similar drainage-area was to be left un
as a control."

"All woody vegetation on [the treated] watershed was cut
between January 6 and March 31, 1941. To prevent disturbance
to the soil, trees were left where they fell and no material was removed.
Tops and limbs were lopped to lay close to the ground and scattered uniformly over the soil-surface. Because of the large proportion of evergreen rhododendron and mountainlaurel, the resulting slash-cover formed
a loose mulch over the area. This protected the soil and maintained
infiltration-capacity. It also served to protect the soil from drying by
sun and wind, preventing any great increase in evaporation after cutting.
An isolation strip 50 feet in width around the watershed-boundary received
identical treatment."

"The cumulative runoff from April 1 to March 31 was 17.29
inches for 1941-42 and 13.26 inches for 1942-43. The value for the first
year was obtained under minimum plant-cover and should approximate
the transpiration that would have occurred under natural forest-cover.
A lower value occurred in the second year because the sprouts had a
longer average opportunity for growth."

"The maximum monthly increase to runoff of 2.84 inches occurred
in July 1941. In this month rainfall was 10.80 inches but preceding rainfall was deficient. Under forested conditions the heavy rainfall in July
was barely sufficient to satisfy soil-moisture deficiencies and but little
runoff resulted. However, on the treated area, because of the lower
drain on soil-moisture, the rains during July made large contributions
to ground-water storage resulting in a greatly increased runoff for this
and the following months. The normal period of soil-moisture replenishment is in the fall. In both 1941 and 1942 the rapid replenishment of soil-
moisture on the treated area resulted in a large increase to runoff during
this time."

43. DISPOSITION OF RAINFALL IN TWO MOUNTAIN AREAS OF CALI-
FORNIA

P. B. Rowe and E. A. Colman. 84 pp. U. S. Dept. Agr. Tech. Bul.
1048. 1951.

Evapo-transpiration losses were determined in woodland chaparral
and ponderosa pine in central California, and under brush cover in southern California.

"A large part of the annual evapo-transpiratio
during the spring and summer dry season. The loss
period ranged from 56 percent of the total annual loss
chaparral to 76 percent of the annual loss in the pond
water thus lost included all water stored in the soil fl
to slightly below wilting point and in addition all wate
by the infrequent late spring and summer rains. Und
of winter rain and summer drought typical of these al
evapo-transpiration loss was more strongly influence
water storage capacity of the soil than by any other fa

"Water-loss rates decreased markedly at abo
each soil layer reached wilting-point storage. This v
vegetated plots studied. It seems probable, from thi
the woody vegetation on these plots can draw little if
from the soil than can herbaceous plants, which are l
from lack of water when soil moisture has been deple
point. Yet the woody plants involved in this study do
exposed to soil having less than wilting-point storage
time (nearly 3 1/2 months for 1940). During such per
cannot obtain any significant quantity of water from tl
appreciable amount available to them from the underl
fore it must be assumed for the present that these pl
summer drought by entering some type of dormant st
tion is supported by the observation that chaparral sh
confined in lysimeters at San Dimas have survived ev
ture content of the confined soil remained below the v
long as 5 months at a time."

"Removing the vegetation, trenching, and ma
surface on plots in the woodland chaparral, ponderos
Dimas chaparral eliminated all interception and tran
Surface runoff and soil erosion were greatly increase
loss of water from the soil was reduced. As a resul
evaporative losses there was a greater carryover of
plots from one year to the next than was found on the
natural plots. During each summer the bare soils lc
quantities of water from all depths, but drying was m
less complete in the deeper soil layers. Thus the pl
entered each rainy season with a proportionately gre
water than did those with shallow soil."

- 68 -

"...It appears that denudation is more effective in reducing evaporation losses from deep than from shallow soils, and from soils protected from full insolation than from those exposed to sun and wind."

"...Increases in usable water yield can possibly be achieved in this area if soils are deep, by reducing interception and evapo-transpiration losses, but only if surface runoff and soil erosion can be controlled."

44. ESTIMATING THE MOISTURE CONTENT OF THE 0- TO 6- INCH SOIL HORIZON FROM CLIMATIC DATA

A. W. Zingg. Soil Sci. Soc. Amer. Proc. 8: 109-111. 1944.
(Reprinted by permission.)

"Measurements of soil moisture were secured under various crops during the 3-year period 1934-36 on the Shelby loam soil at the Soil Conservation Research Station at Bethany, Mo. During this period soil moisture determinations were made under crops of corn, small grain, meadow, and bluegrass on 62 occasions."

Average results of triplicate soil samples taken in the 0 - 6 inch horizon were used in this analysis.

"As an initial approach to the problem, triplicate field moisture determinations on a given date for each of the crops of corn, small grain and meadow were averaged. These crops represent a 3-year crop rotation commonly grown in the soils region and the average moisture figure on a given date would, therefore, represent rotation average moisture."

"Rotation moisture was correlated with various combinations of temperature and rainfall, and time of occurrence thereof. Lapse of time required for rainfall to accumulate to a total of 1/2 inch, and finally, a factor for estimating the moisture content at a prior date were added in an attempt to further refine such relationships. Data for the winter months were excluded to avoid snow and freezing temperature."

"After determining the relationship of factors resulting in the most practical equation for estimating moisture under the crop rotation, equations containing these identical variables were derived for moisture under individual crops."

Total precipitation in inches for 21 days
determination and mean 21-day temperature in
determination, and the number of days required
1/2 inch prior to the time of moisture determin.
practical equation for estimating soil moisture.

From graphs of actual and predicted mo
four crops: "Reasonable agreement between th
and the ability of the estimating equations to sho
crops, and high or low moisture content for all
illustrated. A seasonal trend in moisture, wher
in July, is also apparent."

45. DETERMINING WATER REQUIREMENTS I
    FROM CLIMATOLOGICAL AND IRRIGATIC

Harry F. Blaney and Wayne D. Criddle. U.
SCS-TP-96. 1950.

"Consumptive use (evapo-transpiration
of water used by the vegetative growth of a give
and building of plant tissue and that evaporated
or intercepted precipitation on the area in any s
the given area."

The authors discuss factors that affect
precipitation, temperature, humidity, wind move
latitude, available irrigation water supply, soil
pests and diseases.

"...Consumptive use of water is affecte
dent and related variables; and of the climatic f
growth, temperature and precipitation undoubte
influence. Furthermore, records of temperatu
far more universally available throughout the w
data for other factors. The actual hours of sun
portant part in the rate at which plants grow an
sunshine records are not generally available.
hours for each day are available for all the lati
may be used in place of the actual data. Althou
these may be misleading in areas where heavy
exists during a large part of the year, tempera

such a condition. Humidity records, if available, may also be used as a correction. [Blaney, Morin, and Criddle, 1942]

"Disregarding the unmeasured factors, consumptive use varies with the temperature, daytime hours, and available moisture (precipitation, irrigation water or natural ground water). By multiplying the mean monthly temperature (t) by the monthly percent of daytime hours of the year (p), there is obtained a monthly consumptive-use factor (f). It is assumed that the consumptive use varies directly as this factor when an ample water supply is available. Expressed mathematically, $U = KF$ = sum of kf where

U = Consumptive use of crop (or evapo-transpiration) in inches for any period.

F = Sum of monthly consumptive-use factors for the period (sum of the products of mean monthly temperature and monthly percent of daytime hours of the year).

K = Empirical consumptive-use coefficient (irrigation season or growing period).

t = Mean monthly temperature, in degrees Fahrenheit.

p = Monthly percent of daytime hours of the year.

$f = \dfrac{t \times p}{100}$ = monthly consumptive use factor.

k = Monthly consumptive-use coefficient.

$u = kf$ = monthly consumptive use in inches."

"The consumptive use factor (F) for any period may be computed for areas for which monthly temperature records are available. Then by knowing the consumptive-use coefficient (K) for a particular crop in some locality, an estimate of the use by the same crop in some other area may be made by application of the formula $U = KF$."

The authors calculated K from $K = \dfrac{U}{F}$, having obtained values of U from previous investigations. K varies somewhat because of different conditions (soils, water supply and method) in various studies. For normal conditions it is as follows:

- 71 -

	Length of	Consu
'rop	growing season	coeffi
	or period	(

lfa	Between frosts	0. 80 t
ιs	3 months	. 60 t
ι	4 months	. 75 t
ɔn	7 months	. 60 t
:	7 to 8 months	. 80
ns, small	3 months	. 75 t
ns, sorghums	4 to 5 months	. 70
ιard, citrus	7 months	. 50 t
ιard, walnuts	Between frosts	. 70
ιard, deciduous	Between frosts	. 60 t
.ure, grass	Between frosts	. 75
ːure, Ladino clover	Between frosts	. 80 t
ιtoes	3 1/2 months	. 65 t
:	3 to 5 months	1. 00 t
ιr beets	6 months	· . 65 t
ιatoes	4 months	. 70
ck - small	3 months	. 60

The lower values of K are for coastal areas, the higher
areas with an arid climate。

"The coefficients so developed are used to transpos⸱
ptive-use data for a given area to other areas for whicl
ιatological data are available. The net amount of irrigᴇ
ːssary to satisfy consumptive use is found by subtractin
precipitation from the consumptive water requirement
ving or irrigation season. This ꞥet requirement, divid
gation efficiency, indicates the seasonal irrigation requ
crop。"

AN APPROACH TOWARD A RATIONAL CLASSIFICAϹ
CLIMATE

C. W. Thornthwaite. Geog. Rev。 38(1): 55-94. 1948.
(Reprinted by permission。)

"We cannot tell whether a climate is moist or dry Ꮟ
precipitation alone。 We must know whether precipitati

or less than the water needed for evaporation and transpiration. Precipitation and evapotranspiration are equally important climatic factors."

"The vegetation of the desert is sparse and uses little water because water is deficient. If more water were available, the vegetation would be less sparse and would use more water. There is a distinction, then, between the amount of water that actually transpires and evaporates and that which would transpire and evaporate if it were available. When water supply increases, as in a desert irrigation project, evapotranspiration rises to a maximum that depends only on the climate. This we may call 'potential evapotranspiration', as distinct from 'actual evapotranspiration'.

"We know very little about either actual evapotranspiration or potential evapotranspiration. We shall be able to measure actual evapotranspiration as soon as existing methods are perfected. But to determine potential evapotranspiration is very difficult. Since it does not represent actual transfer of water to the atmosphere but rather the transfer that would be possible under ideal conditions of soil moisture and vegetation, it usually cannot be measured directly but must be determined experimentally. Like actual evapotranspiration, potential evapotranspiration is clearly a climatic element of great importance. By comparing it with precipitation we can obtain a rational definition of the moisture factor."

"Although the various methods of determining potential evapotranspiration have many faults and the determinations are scattered and few, we get from them an idea of how much water is transpired and evaporated and how much would be if it were available. We find that the rate of evapotranspiration depends on four things; climate, soil-moisture supply, plant cover, and land management. Of these the first two prove to be by far the most important.

"Some scientists have believed that transpiration serves no useful purpose for the plant. We now understand that transpiration effectively prevents the plant surfaces that are exposed to sunlight from being overheated. Most plants require sunlight for growth. The energy of the sun combines water and carbon dioxide in the leaves into food, which is later carried to all parts of the plant and used in growth. This process, which is called 'photo-synthesis', is most efficient when the leaf temperatures are between 85° and 90° F. But a leaf exposed to direct sunlight would quickly become much hotter if the energy of the sun were not disposed of in some way. The surface of dry ground may

reach a temperature of 200° F.; temperatures higher thar
been measured one-fourth of an inch below the ground sur
plant is admirably designed to dissipate heat, the leaves b
fins of a radiator, and some of the excess heat is conduct
jacent air and carried away in turbulence bodies. In this
heated. But some of the excess heat energy is utilized in
to change water from a liquid into a vapor. Most of the h
tion must come from the plant. Thus, the greater the int
shine, the greater will be the tendency to overheating, an
will be the transpiration of a plant exposed to it, if water
the process. Transpiration is a heat regulator, preventir
excesses in both plant and air. Dew formation at night is
this process and tends to prevent low temperature extrem
heat released goes mainly to the plant. Both transpiratio
are related to temperature in the same way.

"Atmospheric elements whose influence on transpi
studied include solar radiation, air temperature, wind,
humidity. These factors are all interrelated. Although s
the basic factor, there seems to be a closer parallelism
temperature and transpiration. The temperature of the t
is most closely related to the rate of transpiration.

"Transpiration and growth are both affected in the
variations in soil moisture. Both increase with increase
water in the root zone of the soil, to an optimum. Above
both are less, presumably because of poor aeration of the
results in a lack of oxygen to supply the roots and an exc
dioxide [Loustalot, 1945; Schneider and Childers, 1941].
hand, as water in the soil increases above the optimum f
evaporation from the soil surface also continues to increa

"We do not yet know how much we may increase o
piration by varying the type of plants or by modifying the
Since transpiration regulates leaf temperature, and since
reach their optimum growth at about the same temperatu
cannot change it very much except by reducing the densit
cover and thus wasting a part of the solar energy. If all
is removed from a field, there will be no transpiration.
the root zone of the soil is well supplied with water, the
transpired from a completely covered area will depend m
of solar energy received by the surface and the resultant
on the kind of plants."

"...Determinations have shown that potential evapotranspiration is high in the southern part of the United States and low in the northern part and that it varies greatly from winter to summer. From observations...it has been found that when adjustments are made for variation in day length, there is a close relation between mean monthly temperature and potential evapotranspiration. Study of all available data has resulted in a formula that permits the computation of potential evapotranspiration of a place if its latitude is known and if temperature records are available. The formula is given, and its use described...."

47. MANUAL OF EVAPOTRANSPIRATION

John R. Mather. The Johns Hopkins University Laboratory of Climatology, Micrometeorology of the Surface Layer of the Atmosphere. Supplement to Interim Report No. 10. April 1, 1950, to June 30, 1950. (Reprinted by permission.)

"Values of evapotranspiration may be obtained for short periods of time with the use of empirical formulae. It has long been realized that a number of meteorological factors are important in determining the amount of evapotranspiration that will occur. These factors include the temperature of the soil which in turn depends on the insolation and air temperature; in addition, evapotranspiration depends on the dew point temperature of the air, the wind velocity, and the roughness of the ground. Although many attempts have been made to find a relation between evapotranspiration and one or more of these factors, no definite relationship has emerged. The reason for this is that the amount of evapotranspiration depends also on the amount of moisture in the soil available to be utilized. Thus during drought conditions, evapotranspiration will be far different than during moist conditions. Although various expressions have been worked out relating evapotranspiration to meteorologic factors and the number of days since the last rain no good relationship has been found.

"In an effort to obtain a clearer understanding of evapotranspiration, Thornthwaite (1948) suggested the concept of potential evapotranspiration. Potential evapotranspiration is defined as the amount of water which will be lost from a surface completely covered with vegetation if there is sufficient water in the soil at all times for the use of the vegetation. Adequate moisture can be insured if the soil moisture tension is always maintained below 100 mb. The moisture in the soil would then be above the field capacity.

"Potential evapotranspiration offers an approach toward the
understanding of the role of moisture in climate. The climatic
elements that are usually measured— temperature, precipitation,
humidity, pressure, and wind—do not by themselves equal climate.
Evaporation or evapotranspiration, which is the transfer of moisture
from the soil and plants to the atmosphere, is as much an element of
climate as is precipitation, yet it is seldom included in the measurements of climatic stations. Because instruments are available to give
us measurements of precipitation easily, we know its distribution over
the earth through time reasonably well. The lack of adequate instruments or techniques to measure the movement of water from the earth
to the atmosphere has sorely hampered our knowledge of the evaporation
phase of climate.

"It is recognized that precipitation by itself does not indicate
whether a climate is moist or dry. Only the relation between the water
need of a place and the amount of precipitation will indicate this. Thus
the yearly distribution of evapotranspiration is a climatic factor as important as precipitation. Actual evapotranspiration, as has been shown,
depends, among other things, on the amount of water available in the
soil. During drought periods evapotranspiration will practically cease
so that it is difficult to obtain a picture of true water-need."

"With the further use of evapotranspirometers several problems
demand immediate solution. It was stated before that potential evapo-
transpiration depends on climate, soil moisture supply, plant cover, and
land management. In order to evaluate its dependence on the climatic
factors its relation to the other three factors should be investigated. Work
already in progress indicates that potential evapotranspiration is independen
of land management practices and kind of crop cover within a wide range of
economic crops. As long as the soil moisture tension is less than 100 mb.,
potential evapotranspiration is also independent of soil type or structure.
Some dependence on the density and stage of growth of the crop has been
found."

48. REPORT OF THE COMMITTEE ON TRANSPIRATION AND EVAPO-
    RATION, 1943—1944

    Contribution by C. W. Thornthwaite. Trans. Amer. Geophys. Union
    Part V: 686-693. 1945. (Reprinted by permission.)

    Presents an equation for calculating potential evapo-transpiration
and shows close correspondence between predicted and actual runoff.

"There are still many unanswered questions in my mind, but space here does not permit me to state them. An implied conclusion of this study is that evapotranspiration is independent of the character of the plant-cover, of soil-type, and of land-utilization to the extent that it varies under ordinary conditions. This conclusion is contrary to currently accepted notions concerning evapotranspiration-losses and I am reluctant to accept it myself; however, I have not yet found a reason for denying it. Of course, it is clear that the evapotranspiration regime can be altered. Destruction of vegetation will eliminate transpiration and mulching of bare soil will greatly reduce evaporation. But such practices are not carried out on a wide scale and surely could not modify the evapotranspiration map of the United States."

49. EVAPOTRANSPIRATION ESTIMATES AS CRITERIA FOR
   DETERMINING TIME OF IRRIGATION

   E. H. M. van Bavel and T. V. Wilson. Agr. Eng. 33(7):
   417-418, 420. 1952. (Reprinted by permission.)

   "On drained land, water is lost by evaporation from the land surface and by transpiration of the vegetative cover. The combined process, known as evapotranspiration, accounts for all water losses while irrigation and rainfall minus runoff account for all additions. It follows that, if the quantities of water added to the soil were known and if the total daily evapotranspiration amounts were known, a simple bank account procedure could be used to determine the readily available supply of water in the soil at any time. When this supply appears to be reduced to zero, or nearly so, the need for irrigation is indicated."

   "The use of the outlined principle entails two main sources of error. In the first place, it is not always easy to determine what proportion of rainfall infiltrates to replenish soil moisture, particularly in the case of high-intensity rainstorms. Fortunately, in the case of heavy rains exceeding the available water-storage capacity of the soil, runoff is oftentimes not too significant because it would represent surplus water, whether it entered the soil or not. However, there are some summer rains that fall at high intensities for short duration which, even on dry soils, exceed the maximum infiltration rate.

   "Secondly, one must have a reasonably accurate estimate of evapotranspiration losses. Considering the fact that such losses are determined by such variable factors as the radiant energy from the

- 77 -

sun, the wind velocity, the temperature of the air, the relativ
humidity of the air, the temperature of the leaf or ground surj
and the density of   cover, it seems that computing evapo·
transpiration might be a complicated process, the results of v
have small general value. Actually, experience has shown, a
be discussed later, that evapotranspiration values computed j
long-time averages of the cited meteorological factors have co
able validity. This is possibly due to the fact that the daily va
if considered over several days, very nearly cancel each othe
must also be remembered that the estimate of evapotranspirat
not be any more accurate than the other elements of the compt
such as the rainfall record, error due to runoff, and variable
moisture characteristics."

    "An important question is, whether the evapotranspira
of different crops is different under similar conditions. So fa
provided an ample moisture supply is in the soil, few experim
data would support this view. More research should be done t
this point. The work reported here is based on the assumptio
closed vegetative cover under equal meteorological conditions
of the soil water supply with equal rapidity, regardless of bot;
composition.

    "The practical application can probably best be brough
by an example here arbitrarily taken to be a field of tobacco o
coarse sandy loam at Raleigh, N. C., in July. The moisture
istic of a Ruston coarse sandy loam shows that at field capaci
at 50 cm. tension) the moisture content is 12.5 percent. If th
allowable tension is taken at 800 cm. of water (unpublished da
this value, but its exact magnitude is not important for this di
the corresponding moisture content is 4.4 percent. The maxi
ful capacity is then 8.1 percent and for a rooting depth of 12 i
bulk density of 1.55, this represents 1.50 in. of water.

    "According to Thornthwaite, the daily evapotranspirat
Raleigh area is 0.21 in. in July. Therefore, as indicated in
subtracting evapotranspiration from the supply and by adding
irrigation to the supply, a running record of available soil mc
be kept.

Table 1. Example of a Soil Moisture Account
Ruston coarse sandy loam

Volume weight, 1.55
Moisture content (weight basis) at field capacity = 12.5 per cent
Moisture content at maximum allowable tension = 4.4 per cent
Useful moisture range 8.1 per cent

Volume of water in useful range (12 in. of soil) 1.50 in.

Date	Evapotranspiration	Precipitation	Irrigation	Supply
July 1	0.21	1.80		1.50
2	0.21			1.29
3	0.21			1.08
4	0.21			0.87
5	0.21	0.26		0.92
6	0.21			0.71
7	0.21			0.50
8	0.21			0.29
9	0.21			0.08
10	0.21		1.50	1.37
11	0.21			1.16
12	0.21	1.06		1.50

"When a need for irrigation is indicated as on July 10 in Table 1, 1.50 in. should be applied. Rainfall is also added to the supply, except when the total exceeds 1.5 in. If 1.5 in. is exceeded, the additional water is not added to the supply because it becomes unavailable to the plants by storage beyond the root zone or by drainage to the water table.

"In order to use the evapotranspiration approach, the rooting depth of the crop, the moisture characteristic of the soil, the moisture-tension tolerance of the crop, evapotranspiration rates, and a record of rainfall will have to be known."

ROOTS AND SOIL MOISTURE

50. RATE OF LEAF ELONGATION AS AFFECTED BY THE
INTENSITY OF THE TOTAL SOIL MOISTURE STRESS

C. H. Wadleigh and H. C. Gauch. Plant Physiol. 23(4):
485-495. 1948. (Reprinted by permission.)

"It is difficult to evaluate the moisture stress to which a
plant is responding when grown on a given mass of saline soil. This
moisture stress which is conditioning the entry of water into the roots
will be largely affected by five variables (a) the variation in salt distribution within the soil mass and its consequent effect on the variation
in the osmotic pressure of the soil solution at a given moisture content;
(b) variation in osmotic pressure in relation to change in moisture
content; (c) variation in moisture tension in relation to moisture content; (d) variation in moisture content within the soil mass at a given
time; and (e) variation in total water content of the soil mass with
time, i.e., over an irrigation interval."

Rates of leaf elongation of cotton plants grown in steel drums
containing 100 lbs. of loam were found to vary with moisture stress,
a parabolic relation in which cessation of growth was mathematically
calculated to occur consistently close to 15 atmospheres stress.

51. DISTRIBUTION OF SOIL MOISTURE UNDER ISOLATED FOREST
TREES

H. A. Lunt. Jour. Agr. Res. 49(8): 695-703. 1934.

"...The whole root system is involved in moisture absorption
and not any one particular portion. This is not true of nutrient absorption, since the bulk of the available plant nutrient supply, particularly nitrogen, is to be found comparatively close to the surface, and
therefore most of the feeding roots are located in that portion of the
profile. This is particularly true in northern forests where there is a
considerable accumulation of duff on the forest floor. In other words,
the tree may obtain most of its nutrients from the surface 10 or 12
inches, but except in wet soils it must draw also upon the subsoil to
meet its moisture needs."

52. ROOT DEVELOPMENT AND SOIL MOISTURE

John P. Conrad and F. J. Veihmeyer. Hilgardia 4: 113-1?
1929. (Reprinted by permission.)

"...Moisture under rows of grain-sorghum plants is app
extracted in successive zones and the extraction is progressive
ever no material additions of moisture occur during the growing

"The percentages for relative wetness expressed as rati
soil-moisture contents to their respective moisture equivalents.
be used to indicate the development of roots, and the results of
moisture samples, taken at proper times, indicate with a fair d
of accuracy the presence or absence of roots of plants growing ·
the soil tested."

"A correlation has been shown to exist under conditions
this study between the amount of roots and the extent to which th
soil has been dried by root activity. The writers reason that if
soil is wet at the beginning of the growing season to the full dep
to which roots of plants would normally penetrate, subsequent a
of water by rain or irrigation, unless adverse conditions for gr·
are brought about thereby, can have but little influence on the e·
of the root system developed."

53. RANGE OF SOIL-MOISTURE PERCENTAGES THROUGH V
PLANTS UNDERGO PERMANENT WILTING IN SOME SOIL
SEMIARID IRRIGATED AREAS

J. R. Furr and J. O. Reeve. Jour. Agr. Res. 71(4): 149-

"...The first permanent wilting point is marked by perr
wilting of the basal leaves [sunflower as test plant], and the low
of the range, the ultimate wilting point, is marked by complete
wilting of the apical leaves."

"At soil-moisture percentages near or in the wilting rar
low rate of water loss from the plant had an appreciable effect ·
osmotic pressure of the sap and upon the turgor of the plant. A
in soil moisture from field capacity to the first permanent wilti
caused, in plants in dry air, an increase of 5 atmospheres in th
pressure of the sap and, in plants in humid air, an increase of

atmospheres. The changes in osmotic pressure of plants in humid air
indicate that the diffusion-pressure deficit of the plant was somewhat
less than 9 atmospheres at the first permanent wilting point and about
22 atmospheres at the ultimate wilting point."

"From extensive field work relating to irrigation problems,
there has been formulated the following picture of the typical pattern
of root distribution and the sequence of events in the extraction of water
from soil initially wet to soil at the ultimate wilting point. While the
distribution of roots varies greatly with species and soil, the concentration of absorbing roots is typically greatest in the upper part of the
root zone and near the base of the plant and decreases with soil depth
or distance from the plant. Extraction of water is most rapid in zones
of highest root concentration and most favorable conditions of temperature, aeration, and other environmental factors. When the moisture
content in the zone of highest root concentration has been reduced to
the first permanent wilting point, extraction in this zone does not cease,
but the rate falls off sharply and the total water absorption rate of the
plant decreases. As the total absorption rate and the turgor of the plant
decrease, the diffusion-pressure deficit of the root system as a whole
increases, the soil-moisture percentage is lowered into the wilting
range progressively in zones of lower and lower root concentration , and,
finally, as the severity of wilting increases, the soil-moisture percentage
is reduced to the ultimate wilting point progressively from zones of highest root concentration to zones of lower root concentration. By the time
the plant in the field dies as a result of desiccation, the soil-moisture
percentage in a large part of the root zone may have been reduced to
the ultimate wilting point. Soil at the extremities of the root system,
however, may still be well above the first permanent wilting point. The
plant dies, not because water absorption has absolutely ceased, but
because the rate of absorption finally lags too far behind the rate of loss
to support life."

54. PEAR ROOT CONCENTRATION IN RELATION TO SOIL MOISTURE
EXTRACTION IN HEAVY CLAY SOIL

W. W. Aldrich, R. A. Work, M. R. Lewis. Jour. Agr. Res. 50(12):
975-988. 1935.

In a pear orchard (trees 30 feet apart) in the Rouge River Valley
of Oregon the authors took soil samples in one-foot increments with a
King tube. They then determined the rate of soil-moisture extraction

during summer, when soil moisture decrease was rapid
by rainfall or irrigation for one or more periods. The a
ture extracted was expressed as a percentage of the su
decrease for all depths. The relative root concentration
as a percentage of the sum of concentrations.

Depth (ft.)	Average extraction (percent)	Relative (1
0 - 1	34	
1 - 2	28	
2 - 3	22	
3 - 4	16	

r :   0.98   ±   0.01

55. A NEW TECHNIQUE FOR STUDYING THE ABSORP'
MOISTURE AND NUTRIENTS FROM SOIL BY PLAN

Albert S. Hunter and Omer J. Kelley. Soil Sci. 62:
1946. The Williams and Wilkins Co. (Reprinted by

"A new technique was developed for the study of t
plant root systems in the absorption of moisture and nutr
Guayule and alfalfa plants were grown in columns of soil
divided into 8-inch sections by means of tar-paraffin mer
were permeable to roots but which restricted the movem
and nutrients between adjoining sections. Each 8-inch se
column was provided independently with a tensiometer a
block, for the continuous measurement of soil moisture
perforated plastic tube for the addition of water and nutr
aeration. Water and nutrient elements could be added at
section of the columns."

"Studies of the moisture tension changes in the se
as absorption by the plant roots dried the soil column fr
low tension, were made with both boxes in which guayule
For a period of several weeks (3 weeks in the case of bo
for box 3) the moisture tension throughout the soil colum
at relatively low values (usually below 350 cm. of water
roots in all sections might be capable of normal absorpt
this, all sections were irrigated to tensions of about 100

Changes in moisture conditions in the several sections of the soil
column were then measured by the instruments at frequent intervals
(hourly, at first) as the plants lowered the moisture content to the
wilting point range."

"In general, at a given time the tensions ranged from higher to
successively lower values from top to bottom of the soil column
These data indicate that the roots extracted moisture held in the topsoil
at fairly high tensions, while moisture was available at lower tensions
in the subsoil The same general behavior was noted with alfalfa. The
greatest concentrations of roots were present in the upper part of the
soil column, as is true under normal growing conditions. It is probable that this was an important factor.

"The following experiment was made to study the growth of
guayule plants having water available to their roots in the subsurface
layers of soil but with the surface soil at the permanent wilting percentage. After a period during which the whole soil column was kept at
tensions of less than 1 atmosphere, no further irrigations were made,
and during a period of about 3 weeks, moisture was removed by the
plants from the six upper sections to such an extent that the Bouyoucos
block resistances were in the range of 400,000 to 800,000 ohms. The
plants were then defoliated, and during the next six weeks water was
applied to only the five lower sections. With the top 32 inches of their
root systems in soil having a moisture content of less than the permanent wilting percentage, the guayule plants put forth new leaves and
continued to grow. Alfalfa behaved similarly, putting forth new shoots
while the top 32 inches of soil was at or below the wilting percentage.
Growth, however, was considerably less luxuriant than when moisture
was available to the plants throughout the soil columns."

56. THE INTEGRATED SOIL MOISTURE STRESS UPON A ROOT
     SYSTEM IN A LARGE CONTAINER OF SALINE SOIL

     C. H. Wadleigh. Soil Sci. 61: 225-238. 1946. The Williams
     and Wilkins Co. (Reprinted by permission.)

     "The tension on the soil moisture at field capacity lies within
the range of 0 1 — 0.4 atmosphere [Richards and Weaver, 1944] .
Values for the soil moisture tension over the wilting range may vary
widely, but the 15-atmosphere-percentage almost invariably falls
within this range [Richards and Weaver, 1944] . Even though this

wide diversity in tension is found over the range of ava
it is frequently considered to be equally available throu
The tenability of this postulate rests with the hyperbolic
relationship between soil moisture percentage and tensi
and Edlefsen, 1937; Wadleigh and Ayers, 1945]. In oth
soil moisture tension in most soils does not exceed 1
until most of the available water is removed, but a trer
in tension takes place with removal of the last portion o
water.

"It is known that there are two groups of forces
the decrease in the free energy of soil moisture: (a) th
(hydrostatic, gravitational, adsorptive) which induce a 1
soil water; and (b) the osmotic forces due to dissolved s
solution [Bodman and Day, 1943, Day, 1942, Richards a
As Richards and Weaver [1944] have pointed out, most
the free energy of soil moisture have given inadequate t
osmotic effects. They reported that soil moisture rete
derived from tensiometers or from pressure-plate [Ric
1943] or pressure-membrane [Richards, 1941] apparatu
mented by a determination of the osmotic pressure of th
order to arrive at an evaluation of the energy relations
ture. Their studies indicated that the summation of the
of the soil solution plus the moisture tension at the perr
percentage for many different soils covered a narrower
the corresponding values for moisture tension alone, W
[1945] related this summation of the two groups of forc
of beans in saline soil. For convenience they designate
ed summation as the 'total soil moisture stress'.

"Thus, letting S designate the total stress in a

$$S = T + \pi \quad (1)$$

where T designates the soil moisture tension in atmos
osmotic pressure of the soil solution in atmospheres.
Obviously,

$$T = f(Pw) \quad (2)$$
$$\pi = f(Pw) \quad (3)$$

where Pw is the soil moisture percentage. It is eviden
as a plant removes water from a soil, the water stress
system is continually increasing. Since this stress ca
constant, the rate of change of stress with time should

"The situation is further complicated in a saline soil by the fact that it is not possible for liquid water to move into and through a soil without carrying soluble salts with it. Consequently, if the container of soil is surface-irrigated, as is usually the case, solutes will tend to accumulate in the lower strata of soil. This means that at constant soil moisture percentage throughout the soil mass there would be an increase in the osmotic pressure of the soil solution from the surface downward [Ayers, Wadleigh, and Magistad, 1943]. Under such a condition, even though the roots thoroughly permeate the soil mass there will be an unequal removal of soil moisture from different strata, the rate of removal being lowered as osmotic pressure of the soil solution increases. This is in accordance with observations from controlled experiments showing that rate of water absorption by plant roots is inversely related to the osmotic pressure of the substrate [Eaton, 1941; Hayward and Spurr, 1943, 1944; Long, 1943]. Such data enhance the validity of the main assumption made in the following presentation. That is, it is assumed that as water is removed from the soil mass, the total stress on the water being absorbed at a given time tends to approach uniformity in all the various strata of soil, even though the components of the total stress--osmotic pressure and tension--vary considerably among these strata."

"The primary assumption...was that the soil moisture stress over the various portions of the absorbing surface of the root system tends to approach uniformity. This assumption is fully in accordance with the second law of thermodynamics, in that it is assumed that the plant will not absorb water at a higher energy level if water is available at a lower energy level when the system is at equilibrium. But the tenability of this postulate is conditioned by the degree of constancy of the diffusion pressure deficit within the innumerable absorbing cells over the root system, and the degree to which equilibrium in this force is maintained among these cells. Furthermore, a growing plant is never in equilibrium with its environment. It is probable that the status of the two conditions pertaining to the absorbing forces of the roots deviates appreciably from the ideal. Variations in moisture stress over the root system may, to a slight degree, be concomitantly associated with variations in diffusion pressure deficit of the absorbing cells. Furthermore, this pressure deficit of the water in an absorbing cell is usually high enough that an appreciable variation in the pressure deficit of the external water is possible without exceeding that in the cell. Differences in magnitude of this diffusion pressure gradient would be reflected by variations in rate of absorption. In the final analysis, these variations in rate of absorption over the different parts of the roots system which were brought about by variations in the soil moisture stress would effectively tend to bring about uniformity in the external stress at a given time."

57. AN ESTIMATION OF THE VOLUME OF WATER ABLE BY ROOT EXTENSION

Paul J. Kramer and T. S. Coile. Plant Physiol
1940. (Reprinted by permission.)

"Modern views on plant-soil moisture relatio
importance of continuous elongation of roots into new
soil as an important factor in making water available
long believed that as roots absorb the available water
particles with which they are in direct contact more
available by capillary movement from more distant s
Investigations conducted during the past 15 years ind
belief is not true, capillary movement of water towar
so slow under average field conditions that it is of ne
A brief discussion of soil-water relations will show
ment is relatively unimportant.

"If a limited amount of water is applied to a l.
soil a part of the soil will be wetted uniformly while
is unaffected. This situation is often observed after
wetted the upper three or four inches of soil, leaving
demarcation between the moist soil and the dry soil k
ment of water in a dry soil can occur only from large
laries. As Puri (1939) has indicated, water will mov
to a dry soil only if the dry soil contains some capill.
the largest ones in the wet soil which are full of water
from wet soil to dry soil continues until contact is br
smaller capillaries of the dry soil which are respons
ment. When continuity of the liquid phase is thus bro
of water ceases. The swelling of certain types of so
wetted is also of importance in inhibiting water move
the size of capillaries to a diameter so small that wa
readily pass. The amount of water held by capillarit
had been uniformly distributed by gravitational and c.
usually termed the field capacity (Veihmeyer and Her
Under field conditions well-drained soils are usually
their field capacity a day or two after being wetted by
Shantz (1927) stated that on theoretical grounds the c.
of water from moister to drier regions in soils at or
pacity would be very slow and this was experimentall
(1927) and Veihmeyer and Hendrickson (1927). Acco
when soil moisture is restricted to capillaries forme
a diameter of 0.001 mm. or smaller relatively little

to plants. He considers that particles coarser than clay are mainly responsible for retaining moisture available to plants, whereas clay aids in the conservation of this moisture by reducing its rate of movement. Livingston (1927), Keen (1927), Shull (1930), and others have pointed out that if little or no capillary water moves toward the roots then continual extension of the roots into new regions of the soil is essential to the absorption of an adequate supply of water."

"The volume of water made available daily by root growth was calculated for winter rye using the data published by Dittmer (1937). It was assumed that the roots contacted all soil particles in a cylinder 2 mm. in diameter and that 3.1 miles of roots were added daily. This amount of root extension would make available about 1.6 liters of water daily in a sandy loam soil at field capacity and about 2.9 liters in a heavy clay soil. It appears that at least under some conditions root extension might supply all the water required by a plant."

58. INFLUENCE OF DRY SOIL ON ROOT EXTENSION

A. H. Hendrickson and F. J. Veihmeyer. Plant Physiol. 6(3): 567-576. 1931.

Sunflower and beans were grown in moist soil contained in a waxed wire basket outside of which was soil at the wilting point. Results indicated that roots will not grow into soil which contains less moisture than the permanent wilting percentage.

59. ABSORPTION OF WATER THROUGH SUBERIZED ROOTS OF TREES

Paul J. Kramer. Plant Physiol. 21(1): 37-41. 1946.
(Reprinted by permission.)

With a potometer Kramer measured absorption through the suberized portions of shortleaf pine and dogwood roots. In dry soil in June shortleaf pine took 3.37 cubic millimeters of water per square centimeter of root per hour during the day; at night absorption was 1.32 cubic millimeters. In wet soil in August absorption was 1.73 and 0.92. Dogwood roots were considerably more permeable than pine.

When soil is too cold or dry for root elongation, "...a
rough suberized roots may be of major importance. The nu
nall suberized roots in the soil under a forest stand is large
possess considerable surface and absorb an appreciable am
ater [Coile, 1937]. Since the soil is usually moist during th
nen root growth is slowest, conditions are such that most or
e water required by evergreen trees probably can be absorb
e mature, suberized roots. Even in the summer some wate
ss absorbed through them, particularly when a rain follows
ring which root elongation has ceased, and the roots have b
iberized to their tips. Several days would be required follov
in for root growth to be resumed, but absorption through th
rtions of the roots doubtless begins immediately."

). MOVEMENT OF WATER VAPOR IN SOILS

Edward L. Breazeale, W. T. McGeorge, and J. F. Brea
Soil Sci. 71(3): 181-185. 1951. The Williams and Wilkir
(Reprinted by permission.)

Ninety percent saturated vapor was passed through a s
ining a growing tomato plant. Initially the soil was at field
hen soil dried and the plant wilted, soil samples were taken
il with calculated wilting percentage of 7.55, moisture cont
ilting under vapor flow ranged from 2.85 to 2.93. For anotl
ith a wilting percentage of 11.27, comparable values were 5
50. When the experiment was repeated with dry air, the m
ntent of soil samples at wilting point was very close to the
rcentage.

"There was no definite sign of wilt for 30 days where a
relative humidity of 90 per cent was passed through the soil
r the dry air, wilt appeared in 17 days."

"The preceding experiments show that a highly vegetat
mato, can survive for an extended period in soils at moistu
ell below the calculated wilting percentage with vapor as the
water."

61. PLANT ASSOCIATION AND SURVIVAL, AND THE BUILD-UP OF MOISTURE IN SEMI-ARID SOILS

J. F. Breazeale and F. J. Crider. Univ. Ariz. Agr. Expt. Sta. Tech. Bul. 53. 1934. (Reprinted by permission.)

"The pull or suction force of a plant for water amounts to about 7 or 8 atmospheres, that is, about 105 to 120 pounds per square inch, while the pull or adhesive force of the soil for water varies from zero in a saturated soil up to 25,000 or more atmospheres in a nearly dry soil. In an ordinary air-dry soil the pull is approximately 1,000 atmospheres. Therefore, when a plant is placed in a saturated soil, where the soil pull is practically zero, the balance of power is in favor of the plant, so water will move freely from the soil to the plant. As the percentage of water in the soil is reduced, the thickness of the moisture film is decreased and the pull of the soil for water increases. The pull of the plant for water remains fairly constant so at the wilting percentage the pull of the plant and the pull of the soil are balanced exactly; that is, each pulls about 8 atmospheres, and no water can be taken from the soil. If further loss of water from the soil takes place by direct evaporation from its surface, the pull of the soil will be increased rapidly. Under such conditions the water will move from the plant to the soil. The plant must either supply this water from that already stored up in its tissues, as in the case of the cactus, or it must absorb water from some other part of the soil which is above the wilting point, transport this water and exude it into the drier portions, or the plant will eventually wilt and die.

"It has been shown that a plant which has a tap root growing down into a moist subsoil, with its lateral roots in a surface soil which has been reduced to the wilting point, is able to draw a certain amount of water from the subsoil and to exude this water into the surface soil and keep the soil which is in direct contact with the feeding roots at near the wilting percentage. This phenomenon may enable a plant to tide over long periods of water stress."

"Most of the work which has been done hitherto upon the water requirements and wilting coefficients of plants in pot cultures indicates that a plant root cannot elongate into a soil which is below the wilting percentage. Under the conditions of these experiments this is unquestionably true, but in nature a plant does not always grow under the conditions which are imposed upon it in pot cultures. In humid regions parts of the roots of a plant are always in a subsoil where there is available moisture, while the surface roots may be in a soil which during a drought

is below the wilting percentage. It is the observat
that, under such conditions, a root can elongate in
wilting percentage and even absorb a limited amou
from such a soil.

"In the case of the mesquite or palo verde
arid regions, such as the Southwest, the condition
The tap roots may be in a relatively dry soil while
surface may be growing in a soil which is wet by o
depth of only a few inches. It is evident from expe
is taken up by these surface roots, transported to
exuded into the dry soil around the growing tips.
may grow into a soil mass which is below the wilti
the growing tip, owing to a constant exudation of m
come into film-contact with a soil below the wiltin

62. STUDIES IN TRANSPIRATION OF CONIFERC

G. A. Pearson. Ecology 5(4): 340-347, 192‹
(Reprinted by permission.)

Western yellow pine, Douglas-fir, bristlec
Engelmann spruce seedlings were planted in pots ‹
clay loam. During the growing season the pots we
Weighings were usually made every 3 or 4 days.

"...Observations were made in regard to t
lings toward drastic reduction of soil moisture. C
watering was discontinued in a series of pots whic
had been maintained at a constant moisture conten
approximately 9.5 per cent above wilting coefficie
question.... The rate of transpiration began to fall
at the end of a month it was in most cases less tha
mal. In the meantime the soil moisture had fallen
from two to five tenths of one per cent above the w
During the next two months, transpiration was alm
Two yellow pine pots were watered on November 1
ture content to 20 per cent. Immediately the tran:
to the level maintained before the water supply wa

"The capacity of plants for extracting wate
pends upon their osmotic power and upon the abilit
roots in search of moisture. In the seedling stage

into the deeper soil layers is often a necessary condition for survival. Tests of the wilting coefficient indicate no large or consistent differences between the species with respect to pulling power upon the water of the soil. More exhaustive tests may show consistent differences, but it is almost certain that they would not be sufficiently large to give any species an appreciable margin over the others."

"When we turn our attention to rooting habits, we need not look far for outstanding characteristics which distinguish one species from another. When confined to a limited soil mass, as in the case of potted seedlings, the compact, fibrous root systems of spruce and fir place these species at an advantage. Western yellow pine is able to adapt itself well to such conditions, but not so the bristlecone pine. In their natural habitat the pines are particularly aggressive as to rooting habits. Within three months after germination, the taproot of western yellow pine is down from 15 to 25 centimeters. In the same period a spruce seedling will have penetrated scarcely one-third of this depth. Douglas fir is intermediate between the pines and spruces. Lateral extension depends much upon the proximity of other trees or of other deep-rooted plants. The firs and spruces usually grow in dense stands and have shallow, compact root systems. The pines grow in more open stands and have deeper and more extensive root systems. Of all the factors which figure in survival on a dry site, depth of root penetration during the seedling stage is undoubtedly the most important. Rate of transpiration is equally important; but, as far as the results of this study may be regarded as an indication, the difference between various species is not sufficient to constitute a deciding factor in drought resistance."

63. WILTING AND SOIL MOISTURE DEPLETION BY TREE SEEDLINGS AND GRASS

R. D. Lane and A. L. McComb. Jour. Forestry 46(5): 344-349. 1948. (Reprinted by permission.)

Black locust, green ash, brome grass, and tomato seedlings were grown in pots of loam and sandy loam in a greenhouse. Beginning at field capacity, daily water losses were determined by weighing and drying until the plants wilted permanently.

Pots containing brome grass had lower daily soil-moisture contents on both soils than other plants.

"A comparison of the total weights of water absorb
weights of roots shows that tomato with the smallest root w
ed the least water, while brome grass with largest root wei;
the most water. The trees were intermediate with respect
weight and water absorbed. These data indicate that the we
absorbed were related to the dry weights of roots produced

"Comparing grass with trees, leaf area, and soil w.
appear to be negatively correlated. Locust developed the la
area and brome grass the smallest. Ash was intermediate.
of brome grass in this list is the reverse of its position in t
upon root weight; that is, brome grass had the smallest leaf
sorbed the most water."

"Brome grass reduced soil moisture to the lowest
tree species reduced it to an intermediate level, and tomato
percent of soil moisture. Over-all statistical analysis of th
that the differences among soil moisture percentages at per
were highly significant. Individual comparisons of the wilti
different plants in sandy loam showed that only with tomato
ence significant. On loam, however, there was a significant
between the wilting percents of any two of the indicator plan
among the wilting percents were greater on loam than on sa

	Mean soil-moisture percentage at pe	
	Sandy loam	Loa
Tomato	7.3	10.
Ash	5.3	7.
Locust	5.1	6.
Grass	4.7	5.

"The dry weight of roots may provide a rough mea;
ramification and of moisture-absorbing area in young plant;
think that the differences in the wilting percents obtained wi
indicator plants are due mainly to differences in the extent
cation."

64. A WATER COST OF RUNOFF CONTROL

A. R. Croft. Jour. Soil and Water Conserv. 5(1): 13-15.
1950.

Croft measured soil moisture on aspen, herbaceous, and bare
plots. Evapo-transpiration deficits at the end of the growing season
were 11, 8, and 3 inches respectively. "The reason water is saved
by the removal of aspen is found in the distribution of its roots with
respect to roots of herbaceous plants. The tree roots penetrate the
soil to a depth of six feet or more. This is two to three feet deeper
than herbaceous roots extend. Thus, three or four inches more water
is available to aspen roots than to herbaceous roots. This additional
available water probably accounts for the fact that the trees grow
vigorously for about a month longer in the fall than herbaceous vegetation. The latter begins drying as early as August 1, presumably
because it has used all the moisture available to its roots."

Croft doubts that thinning would increase water supplies, for
".. . additional water made available to fewer plants would merely result
in lengthening their growing season. Eventually, however, the soil
would dry out as much as with the unthinned stand."

"Water saving by altering plant cover probably can be achieved
only when deep-rooted species are removed, or are replaced by more
shallow-rooted plants."

65. AN OBJECTIVE LOOK AT THE VEGETATION-STREAM FLOW
RELATIONSHIP

Richard S. Sartz. Jour. Forestry 49(12): 871-875. 1951.
(Reprinted by permission.)

"With an unlimited supply of water available, some plants must
certainly have the inherent ability to transpire more water than others.
Except for stream-bordering and swamp vegetation, however, this
variation is not important because transpiration rates are usually limited
by the soil moisture available soon after precipitation has stopped."

"The amount of water that is lost through transpiration depends
on two factors associated with the soil. One of these is the depth of the
soil mantle. The storage opportunity of any soil profile is primarily

limited by its depth. The depth that plant roots penetrate may
be limited by the soil depth. Deeper root penetration means th
water is used because the root system has access to a larger r
This has an indirect effect on stream flow: during the period w
moisture deficits are being replenished, more of the precipitat
utilized in satisfying these deficits, and less contributes to stre

"The other factor is whether or not the plant roots exter
the water table. Where this condition exists, the transpiration
the growing season can greatly influence the volume of flow. T
of course, depends on what proportion of the watershed is affec

66. WATER AND TIMBER MANAGEMENT

Marvin D. Hoover. Jour. Soil and Water Conserv. 7(2): 7
1952.

"It is necessary to distinguish between evaporation and
ration because evaporation removes moisture mainly from the
soil, while transpiration withdraws water uniformly from the e
zone of plants. Differences in rooting habits of plants are imp
determining the water used in transpiration. On the Piedmont (
Carolina, for instance, pine trees take water to a depth of six f
broom sedge, with more shallow roots, draws only to a depth c
feet. The more water available to tree roots, the more they w
pire."

67. THE DEPENDENCE OF TRANSPIRATION ON WEATHER
    CONDITIONS

H. L. Penman. Jour. Soil Sci. 1:74-89. 1949.

Penman reviews equations for predicting evaporation f
water surfaces, and previous determinations of the ratios of tr
from turf to evaporation from open water. He points out that
are for conditions in which the grass always has plenty of wate
In many years, however, particularly in southern England, nat
suffers from lack of water, transpiration and growth-rates are
the problem of estimating the net deficit of water built up in the

remaining fundamentally meteorological, has biological and soils
aspects added to it."

"Although many factors may alter it, one can ascribe to herbage
a more or less definite root range in the soil, of the order of a foot
or so. In this depth of soil there will be, at field capacity, a certain
amount of water readily available for the plant upon which it can draw
as easily as the turf referred to in the experiments outlined in para-
graph 1. As an order of magnitude, the available water might be equivalent to 3 in. of rain, i.e., under the stated conditions the grass would
be able to transpire 3 in. of water without needing to draw on supplies
below the root zone, and without water-supply limiting the rate of
transpiration. This, in effect, is an acceptance of the Veihmeyer [1942]
concept of available water postulating a very narrow range of moisture
content, or more correctly soil suction, on one side of which water is
readily available, and on the other side of which the water is not available. The depth of the root zone, and the available water in it, might
depend upon the composition of the herbage, its manurial treatment,
the nature of the soil, the nature of the crop management, and the
rainfall in the early part of the growing season. For the last two, for
instance, one would expect frequent cutting to reduce the root range,
and would expect a dry spring to increase it. Specification of the
'root reservoir' of water will be almost entirely guess-work, but the
guess should not be inconsistent with the natural or imposed condition
of the soil and crop.

"The plant roots may be regarded as exerting a drying potential.
While they are using up the available water around them there will, of
course, be some water movement upward from below, but it is convenient
to make the slightly artificial assumption that the roots first use up the
neighbouring water before the movement from below attains any appreciable magnitude. The drying after the exhaustion of the root reservoir
should be the same as if the drying potential were applied to bare soil
initially at field capacity, and its progress will then be represented by
the drying curve obtained in laboratory experiments [Penman, 1941]
which compared the evaporation-rate from bare soil with that from open
water exposed to the same conditions. As this curve was the same for
a sandy soil (Woburn) and a clay (Rothamsted), it will be assumed that
it can be safely used for a range of soils."

The author constructs a drying curve with potential transpiration
in inches of water as the abscissa, actual as the ordinate; at the origin
(point 0) the soil is at field capacity. The root reservoir constant (in this
case 3 inches of available water in the upper foot of soil) is plotted as

point A. From point A a line of unit slope is drawn to the origin.
"From A, in the opposite direction we draw the moisture-depletion
curve for bare soil. The curve has a slope of unity up to nearly 1 in.
and then bends sharply to become, for practical purposes, a straight
line of slope of about 1 in 12. When the soil moisture is represented
by point 0, i. e., at the origin, the soil is at field capacity. As the
integrated drying potential increases, the actual drying remains equal
to it up to point A (i.e., the root reservoir has been transpired) and
slightly beyond it; when it exceeds the root reservoir by more than
about 1 in. the actual transpiration-rate decreases rapidly, i.e., the
soil-moisture deficit increases very slowly."

The curve was used to keep a running record of soil-moisture
deficit by months and to estimate the month when soil-moisture drainage
would begin.

"In the first trials a root constant of 3 in. was used in all
years and was reasonably successful, but it was noted that an improvement could be effected by increasing the root constant in years with
a dry spring, and decreasing it in years with a wet spring. Qualitatively,
this is acceptable. indeed, one might impose it a priori as a necessary
condition to be satisfied, but quantitatively there is no justification
other than expediency for the form adopted:

$$C = 5.0 - 0.6 \Sigma R$$

where C is the root constant in inches and Σ R is the sum of the April
and May rainfalls."

ROOTS

68. ROOT PENETRATION IN RELATION TO SOIL AERATION

R. E. Stephenson. Proc. 27th Ann. Report of 50th Ann. Meeting
Oregon State Hort. Soc. pp. 19-34. 1935. (Reprinted by permission.)

"Good orchard soils (examined by Schuster and Stephenson) contain air spaces which vary in size from those that are barely visible to
the eye (20 microns are just visible) (a micron is 0.001 millimeter) to
those as large as a lead pencil. A few spaces, such as old root channels or animal burrows, are larger than this. Russian investigators
found that the oxygen content of the soil air did not approach that of the
outside atmosphere until the soil particles were 500 microns diameter,
and the pore spaces correspondingly large. Air in smaller pores contained about one-fourth as much oxygen as the outside air."

"Examination of some thirty soils revealed few or no visible air
spaces where tree growth was unsatisfactory. Further examination revealed that few or no roots enter those horizons of any soil lacking air
space. Schuster's studies on root penetration in orchard soils has
shown a distinct correlation between lack of aeration and the absence
of tree roots. Roots passing through unaerated horizons are conduct-
ing rather than feeding roots which absorb moisture."

"When plants wilt, only the very small pores are filled with
water, and these because of their smallness cling to their water with
such force that plant roots are unable to pull it away. The pull exerted
by root cells trying to remove water from the soil over the wilting
range is estimated at 4 to 25 atmospheres. This is equal to a force of
60 to 375 pounds on a square inch of surface, trying to pull away the
water.

"These forces when converted to capillary diameters (calculations by M. R. Lewis) indicate pores from 0.11 to 0.60 microns diameter. Only particles or aggregates of particles of collodial size (1.0
to 2.0 microns down) afford such small pore spaces. A Sites clay with
50 per cent pore space, showed nearly 90 per cent of its pores full of
water at the wilting point. Then 90 per cent of the pores are of an order
of magnitude of 0.60 microns diameter or less. This is in the sixth
foot, where the total clay is 62 per cent and the two micron clay is 40
per cent.

"The moisture content of soils corresponding to the moisture
equivalent is about equal to the field capacity. Thomas gives 10 microns
as the diameter of pores (assuming pore diameter equals the particle
diameter) which are entirely filled with water at this moisture constant.

Ten micron pores correspond roughly to silt size particles. Smaller particles and their open spaces hold capillary water and, therefore, exclude the air. A 10 micron pore corresponds roughly to an attractive force equal to one fourth of an atmosphere, or less than four pounds on a square inch. Such water is readily removed by the roots of plants, provided the soil is aerated and penetrable.

"Soils high in clay and fine silt may be entirely water-logged. The Sites soil spoken of above contains more than 70 per cent of the clay and fine silt in the fifth and sixth feet. The pore spaces are all of an order of magnitude of 10 microns or less. The soil is entirely water logged at field moisture capacity. This condition has been described as water-logging without a water table."

"We have tentatively placed soils with less than 5 or 6 per cent of air space (non-capillary pores) in addition to the space occupied by water (capillary pores) in the class of undesirable soils. Such soils may have high moisture capacity, but room for root development and penetration is too limited. Soils with 10 or 12 per cent or more of air space over and above space occupied by water at field capacity are desirable, and favorable to deep penetration and strong root development. Soils between these two groups are somewhat marginal."

"The following table gives an idea of how the size of soil particles affects the rate of movement of moisture in soils.

Size of Soil Grain	Penetration Rate of Water	Relative Rate of Water Penetration
1 micron (collodial clay)	1 inch in 41 days	1
10 microns (fine silt)	1 inch in 10 hrs.	100
100 microns (fine sand)	1 inch in 6 min. 1 ft. in 1-1/5 hrs.	10,000
1000 microns (coarse sand)	1 inch in 3-3/5 sec. 1 ft. in 43 seconds	1,000,000

"The data (calculations by M. R. Lewis) are based upon Schicte formula assuming uniform size and spherical particles with most open packing. This formula was checked by King of Wisconsin on field soil. The rate above assumes a constant head (supply of free water), a downward movement, and a free outlet. Upward movement against gravity would be normally many times slower. In addition to indicating the impossibility of upward movement of water through fine pores, the data serve to indicate how a few large pore spaces, between either soil grains or aggregates of grains, aid materially in drainage and aeration of soils.

"More than a month for one inch of water to move through an inch of tight clay! The trees might be dead in less time. These tight subsoils in the fourth, fifth, and sixth feet depths, have wilting points as high as 40 per cent, when plants are grown in the greenhouse with roots in contact with the soil. How impossible for the moisture to reach roots two or three feet above!

"By contrast, Weaver states that young, actively growing roots in a permeable soil may grow at rates of half an inch to two inches a day. The roots go after the water when they can get through the soil. They can get through only when there are open, aerated pore spaces.

"Tight soils offer mechanical resistance to root penetration. No data on the size of tree roots is available. Weaver reports that some vegetables and grasses have roots as small as one-tenth-milli-meter (100 microns) diameter. The growing root tip with its root cap is probably larger than this. There are few or no open spaces between the soil particles in the Sites soil at the fifth and sixth feet depths larg-er than about 10 microns. Ninety per cent of the pores are less than this. A root 100 microns diameter must have a hard time penetrating an opening ten microns or less in diameter. (Try crawling into a gopher hole). Russell stresses the mechanical resistance which clays offer to root penetrations.

"Even root hairs, the smallest part of root systems, are 5 to 10 microns diameter, or larger than many of the open spaces in heavy clay soils. There are few roots, and few or no root hairs on the roots, in tight subsoils. The tree or other plant is unable to provide an ab-sorbing system (the small young roots and the root hairs are the ab-sorbing system) in heavy unaerated soil.

"The few roots that are found in heavy soils appear to have found entrance through worm holes, insect burrows, cracks, and old root channels. We have found mats of absorbing roots on the cleavage surfaces of cracks in the rock six and eight feet down. After the root finds a hole through a heavy horizon, and comes into open space or more permeable layer, profuse branching may occur. Most of the roots in heavy soils, however, are confined to the shallow depth of the surface, that is more permeable and better aerated. Often the first and second feet are noticeably full of holes, cracks, and cavities made by various means. These permit root development."

"Shallow soils are a handicap. Soil is only as deep as roots can penetrate and obtain air and moisture. Soils may be 20 feet to rock but if roots utilize only three feet that is the practical, or usable soil depth.

69. SOIL FACTORS IN RELATION TO ROOT GROWTH

W. Stephen Rogers. Trans. 3rd Inter. Cong. Soil Sci. 1:249-253. 1935.

A study of the root systems of apple trees 10 to 11 years old showed that: "In all cases the roots spread farther than the branches. In the sand, the root spread was two to three times the branch spread. In the loam and clay about 1.6 times the branch spread... The stem-root ratio [by weight] varied in the different soils. On loam it was about 2.2, on clay 2.1, and on sand about 0.9. Hence it required more than twice as much root to support a given amount of top on the poor soil as on the loam or clay [Rogers and Vyvyan, 1934]."

"The type of root is largely modified by the soil. In the poor sand, the roots are very long, thin and straight. In the clay, the roots were shorter and stouter, tapering and rapidly branching, and twisting in all directions. In the loam they were intermediate in character. In spite of this modifying influence of the soil, the roots of the different rootstocks retained their own distinctive characters."

"... In irrigated orchard... the soil dried almost equally at 1 foot and 3 feet since the roots of the trees and cover crop extended throughout the soil to this depth."

"... As the soil temperature rose, the root growth, as shown in both number and length of roots, increased also. A fall in temperature was followed by a fall in root growth. When the soil became drier, roo growth decreased although the temperature was rising. When irrigatioi was applied, the root growth rose again sharply.... It appears that root growth has varied directly with soil temperature, with sufficient moisture as a limiting factor."

"The new roots are white, with numerous tiny root hairs. After a period of two to four weeks these hairs shrivel up, and the cortex be-

comes suberized--finally sloughing off through action of soil insects
and organisms. The root is thus left loose in its hole until secondary
thickening begins. Many of the finer roots rot away entirely, leaving
numerous channels which must be of great value in aerating and draining the soil. Meanwhile new root growth continues to comb the soil
through and through. "

70. THE GROWTH OF FOREST TREE ROOTS

W. B. McDougall. Amer. Jour. Bot. 3(7): 384-392. 1916.

At the University of Illinois the author observed root growth of
silver maple, basswood, hickory, and white oak through a square foot
of glass set horizontally over roots beneath humus, and through a similar square at the end of a trench two feet deep. The glass was covered
with felt to keep out light. Readings at midday in summer injured or
killed roots by exposure to air; observations made at sunrise with two
years of record showed that growth started in early spring and continued as long as soil was wet until November or December.

McDougall believed that periodicity of growth is not due to internal causes. "The results recorded in the present paper show conclusively that the resting period of the roots studied are not fixed and
hereditary since, in 1914, although most of the roots under observation
had a summer rest, some of the hickory roots did not have; and in 1915
there was no summer rest period in any of the roots studied, unless it
occurred after September 1, which would be most unlikely...In 1914
there was very little rainfall from early spring until the end of August.
The soil thus became progressively drier and reached a minimum of
water content toward the end of August. The rate of root-growth also
gradually decreased and ceased entirely in most cases sometime in
July, to begin again only after the heavy rains of August 28. In other
words, the summer period of rest was only during the period of drought.
In 1915 there was no period of drought and, naturally, no rest period.
The hickory roots which did not have a rest period during the summer
of 1914 were some of the most deeply located roots upon which observations were made, and, naturally, the soil was not so thoroughly dry at
that depth as nearer the surface. It is probable that observations on
still deeper roots would show all roots located where adequate moisture was available growing throughout the period of drought. "

"It seems reasonable to conclude, then, that the summer rest
period, when it occurs, is due not to any inherent tendency toward

periodicity but to a lowering of the water supply. As to the winter rest period, the results show a close relation to temperature. But temperature to a certain extent controls the water supply, since a lowering of the temperature renders absorption increasingly difficult and thus reduces the amount of physiological water. In this case, therefore, the rest period is due indirectly to temperature but more directly to a decrease in the available water supply."

71. SEASONAL GROWTH OF GRASS ROOTS

Irene H. Stuckey. Amer. Jour. Bot. 28: 486-491. 1941. (Reprinte by permission.)

Grasses were grown in well-drained loam soil at the Rhode Island Agricultural Experiment Station. "For some of the species the whole root system was regenerated annually, with active production of new root growth beginning in October, continuing slowly through the winter and increasing rapidly after the spring thaw in March with its maximum in April. After the middle of June, few, if any, new roots were formed and there was no appreciable growth of existing roots until October. Most of the old roots disintegrated shortly after the new ones developed. These species included timothy, timothy S-50, meadow fescue, rough-stalked meadow grass, perennial rye grass, probably colonial bent, and redtops."

"With other species the development of roots during the first year was essentially the same as that described above, but only a smal. percentage of the roots disintegrated, and after the first spring only a few new roots developed. Most of the new roots developed during the second year were at the nodes of new rhizomes. The species with 'perennial' roots are Kentucky bluegrass, Canada bluegrass, crested wheat grass, and orchard grass."

72. WINTER ROOT GROWTH OF PLANTS

F. J. Crider. Sci. 68: 403-404. 1928.

At Boyce Thompson Southwestern Arboretum, Superior, Arizon. Crider observed roots of plants growing in large boxes provided with plate-glass fronts.

He found that roots of certain plants generally thought to be dormant in winter make definite continuous growth at this season. This was true of both deciduous and evergreen species. Notable examples were <u>Prunus persica</u>, <u>Prunus armeniaca</u>, <u>Covillea tridentata</u>, <u>Simmondsia californica</u>, <u>Cupressus arizonica</u>, and <u>Opuntia laevis</u>. Rate of root elongation per day varied from 9 millimeters in November as the maximum to 0.5 millimeter in February as the minimum. Growth was evidently affected by change in seasonal temperature of the soil, but there was no direct or close correlation between daily growth and soil temperatures. Average daily root elongation of peach (<u>P</u>. <u>persica</u>) for the winter period November 4, 1927, to March 31, 1928, was 2.10 millimeters. Average daily growth for November was 5.55 millimeters, for December 2.01 millimeters, for January 1.65 millimeters, for February 0.90 millimeters, and for March 1.16 millimeters.

Other plants under the same environmental conditions made no root growth in winter: <u>Citrus aurantium</u>, <u>Vitis vinifera</u>, <u>Prosopis velutina</u>, <u>Parkinsonia torreyana</u>. The period of their root inactivity began about the first of December and lasted until the latter part of March.

73. DISTRIBUTION OF ROOTS OF CERTAIN TREE SPECIES IN TWO CONNECTICUT SOILS

G. I. Garin. Conn. Agr. Expt. Sta. Bul. 454. Jan. 1942. (Reprinted by permission.)

Garin studied the root distribution of white pine, red pine, Norway spruce, white ash, and red oak on loamy sand and fine sandy loam.

"Root development of a forest plantation can be pictured as passing through four stages. The first stage is that of free root growth, when roots have space in which to develop without coming near the territory occupied by those of other trees. The second is that of root invasion, when the expanding root systems begin to intermingle and invade areas adjacent to other trees. This stage is reached at a very early age in the forest plantation. The third period, that of root competition, begins when root capacity is reached. In some soils this may be much sooner than in others. On poor dry soils this period, in most cases, precedes the closing of the tree crowns above the ground. Observations show that, in poor soils, roots of trees of about the same height and the same age spread more widely and occupy a much larger volume of soil than those in richer soils. On rich soils the stage of root competition

ay follow the closing of the crowns. The third period would pre
roughout the greater part of the life of the stand. A fourth stag
at of release from root competition, begins when mature trees
die and release a sufficiently large area from root competition
e new reproduction can become established. Trees at this stag
it have the vigor to replace to the point of 'root capacity' the are
:lease by dead trees, before new reproduction becomes establis

l. DEVELOPMENT AND ACTIVITIES OF ROOTS OF CROP PL

John E. Weaver, Frank C. Jean, and John W. Crist. Carne
Inst. Wash. Pub. 316. 1922. (Reprinted by permission.)

"The importance of root extent and distribution in a study
il-moisture is patent. These should determine not only what d
sample should be used, but also·the maximum depth to which s
es should be obtained. The time, method, and amount of the ap
ition of water for irrigation studied in the light of root developm
rnish a rich and varied field for investigating problems of the g
:ientific and economic importance. Conversely, the proper dra
swamps and boglands for pastures, meadows, afforestation, or
iltivated crops, should be determined with reference to root rel
f. Howard, 1916, 1918; Osvald, 1919.)"

"It seems not improbable that some of our best yielding c·
ay be able to outstrip others largely because of their greater ef
securing a larger and more constant supply of water and nutrie
hy certain artificial mixtures of grasses and other herbs may tl
pastures and meadows, while·others do less well, must depenc
rge degree upon competition of root systems. This is the case
itive grassland, where it is usual for 200 to 300 individuals or g
individual plants to grow in a single square meter, due to less(
impetition resulting from absorption at different soil-levels and
aximum above-ground activities at different times of the growir
in. "

"The investigations here recorded were carried out durin;
·owing-seasons of 1919 to 1921. Stations were selected at Peru ·
incoln, Nebraska, at Phillipsburg, Kansas, and at Burlington, '
ido. These stations have a mean annual precipitation of about 3
i, and 17 inches respectively. The differences in climate are c
:pressed in the type of natural vegetation. The true prairies at
iln give way southeastward along the Missouri near Peru to the

- 106 -

climax prairie, which is potentially chaparral or woodland, the grasses having possession only because of such disturbances as grazing, fire, mowing, etc. At Burlington, in eastern Colorado, a typical expression of the short-grass plains is found, while in north-central Kansas at Phillipsburg short-grasses intermingle with the taller ones and constitute mixed prairie (Clements, 1920; Weaver, 1920). Crops were grown at the several stations under measured environmental conditions for the purpose of determining not only the nature of the root system, but especially also its distribution and extent at various stages of growth. The work was conducted under field-crop conditions and methods of tillage in order that the results might faithfully portray the root relations of crops as grown under usual farm practice. Moreover, extensive experiments have been conducted both in the greenhouse and under field conditions to determine the active working-level of the roots of cereals and other crop plants as regards the absorption of water and nutrients at various stages in their growth."

At Peru, Nebraska, the surface 1 to 1.5 feet of soil was a dark-colored silt loam. This overlaid a deep mellow loam.

"All the cereals, including corn, possessed a root system in which there was a definite group of more or less horizontal, spreading roots lying within the first 1 to 1.3 feet of soil, and a second group of deeply penetrating roots extending into the subsoil to depths of 6 or 7 feet."

"The Early Ohio potato differed from the other plants in that the same group of roots which at the outset formed the shallow portion of the system subsequently became the deeper portion by turning more or less abruptly to the vertical position and growing downward."

"The more superficial roots reached their maximum development first. In most cases this occurred about the time the top had reached an intermediate stage of growth; the deeper roots developed coordinately with the top and thus balanced water absorption and transpiration."

"Oats reduced the soil moisture to a greater degree than any of the other small cereals. Corn in its later stages of growth was an extravagant user of water. The potato showed the greatest variation in the number and extent of its roots."

At Lincoln, Nebraska, the soil was a silt loam, much more compact than that at Peru. Crops were not so deeply rooted as those at Peru.

At Phillipsburg, soil was a silt loam of the Colby series. R
depth exceeded that at Lincoln. With high water loss in surface soil
June and July, crops developed root systems that went far into the
deeper moist soils.

At Burlington the soil was a fine sandy loam with hardpan at
to 2. 5 feet which was not penetrated by the roots.

"The differences found in the lesser height-growth, smaller
yield, and less extensive underground parts, going from the more
mesophytic eastern stations to those of greater aridity westward, c
relate directly in nearly every way with the growth of native vegetat
whether trees, weeds, or species of the native grass land are consi
The native vegetation growing through a long period of years integra
the climatic conditions during its growth. Thus it is not only an expression of the present conditions, as is true largely of rapidly mat
ing crops, but is to a large extent a record of conditions that have o
tained during a period of many years. "

"This increasingly greater root depth as one goes eastward f
the short-grass plains into regions where the deep subsoil is consta
moist is in agreement with determinations made on the root depth o
cereals at 14 stations during 1919. Using root extent in the true pr.
as unity, the relative depth in mixed-prairie and short-grass plains
as follows: Working depth of rye, 100: 92: 69; oats, 100: 95: 79; wi
wheat, 100: 93: 61. Maximum depth of rye, 100: 90: 65; oats, 100:
77; winter wheat 100: 80: 51.

"... In nearly all cases where the roots of crop plants were
excavated, the total development below the cultivated soil-layer was
great and usually much greater than that in the surface soil. Amon
native plants, the bulk of the root system in the great majority of ca
lies below the surface foot, and the same holds true for many crop
plants, including especially the fall-planted cereals. The dependen
of plants upon the deeper-seated portion of their root systems is we
illustrated in times of drought, where the vegetation remains unwil
and crops do fairly well even after the water in the surface 6 inches
soil has been nearly or entirely exhausted. "

75. THE SUBSOIL

Eric Winters and Roy W. Simonson. Advances in Agronomy 3:1-92. Academic Press, Inc. New York. 1951. (Reprinted by permission.)

"It is common to speak of deep-rooted and shallow-rooted plants, although the two are not distinct groups. Deep-rooted plants could more appropriately be designated as those which can develop deep root systems under favorable soil and moisture conditions. For example, alfalfa, sweet clover, and apple trees (Malus spp.), are generally considered to be deep-rooted plants. Working in Kansas, Grandfield and Metzger (1936) found that alfalfa removed moisture from permeable soils to depths of 25 ft., whereas Myers (1936) noted the roots of two-year-old sweet clover at depths of 13 ft. Sweet (1929) reported apple tree roots well below 10 ft. in certain soils of southern Missouri, and similar observations were made by Browning and Sudds (1942) in West Virginia. Generally speaking, deep penetration occurred in soils which were relatively open and permeable throughout the profile. By way of contrast, the authors have observed alfalfa and sweet clover roots which failed to penetrate claypans and fragipans at depths of 18 to 30 in. Roots extended down to the pans but not into them. Sweet (1929) reports that the roots of apple trees were confined to that part of the soil above the hardpan or claypan. Browning and Sudds (1942) noted that the root systems were shallow in soil types with heavy subsoils. Thus, plants which are ordinarily deep-rooted may be restricted in root penetration by the character of the subsoil."

"Although shallow-rooted plants usually have the major part of the root system in the upper soil horizons, this varies with the character of the soil profile. For example, corn roots are almost completely restricted to the A horizon by the claypan in Putnam silt loam in northern Missouri, whereas they extend to depths of 5 or 6 ft. in the friable Marshall and Monona soils of western Iowa. This variation illustrates the flexible rooting habits of some plants. Because of that flexibility, the adaptations of shallow-rooted plants to soils is influenced less by root penetration relationships of subsoil layers than is the adaptation of deep-rooted plants."

"The differences in the ability of the roots of plants to penetrate certain kinds of B horizons is worth noting. Taking corn and alfalfa as examples, the root systems commonly extend into and through the B horizons of zonal soils. It is true, however, that alfalfa and other perennial plants are better able than corn to penetrate B horizons that are high in clay, provided the soils are well drained and well aerated. This difference in root penetration is probably related to the length of

the growing season and the length of life of the plants rather than
differences in rate of penetration. Corn grows during a relative
short season and its period of root elongation is limited, probabl
ing completed by tasseling time. Alfalfa grows from early sprin
late fall, a much longer period. Furthermore, alfalfa root pene
can continue beyond the one season as the plants grow in subsequ
years. "

"...Simple mechanical difficulties in root growth can be
pected as a general rule in claypans. Coupled with restricted ae
these mechanical difficulties provide adequate reason for the lim
penetration by most plant roots. "

"The difficulties of root penetration in fragipans and hard
are due to one or more of the following: high bulk density or vol
weight, low pore space, restricted aeration, or low fertility stat

"Deep rooting can seldom be expected in soils with well-c
oped fragipan or hardpan layers as parts of their profiles. Field
servations indicate that roots seldom penetrate such layers. Co
quently, it does not seem probable that subsoils consisting of fra
or hardpan layers can be improved by the growing of deep-rootec
plants. It may be possible in exceptional instances. Chevalier (
reports that a laterite crust disappeared in forty years under a g
stand of forest. Comparable experiences with fragipans and mos
hardpan layers, however, seems to be lacking. "

"The gradual deterioration of structure in soils such as F
clay under cultivation suggests that trees were able to improve p
properties of soils, especially in the deeper horizons (Bradfield,
Lutz and Griswold (1939) state that tree roots may influence soil
phology by pushing soil material aside as they grow, by leaving c
nels when they decay, and by the mixing of soil materials when t
blown over by the wind. Their studies of Podzols and Brown Poc
soils in profile pits in New Hampshire indicated significant distu
and mixing of horizons by windthrow of individual trees in ordin
forest stands. Page and Willard (1946) suggest that the gradual de
disappearance of coarse tree roots originally present in soils un
forest may be reflected in gradual deterioration of physical prop
of soils under cultivation. Yeager (1935) found that trees growin
Chernozem and Humic-Gley soils of eastern North Dakota had ab
97 per cent of their roots, on the average, in the uppermost 4 fe
Inasmuch as trees are not the indigenous vegetation on Chernoze
growing conditions are less favorable for them than they are on p

soils in more humid regions. Even so, some tree roots penetrated to depths of 10 feet and many to depths of 4 feet. More direct information on the influences of tree roots on soil properties may be taken from experience with shifting cultivation in the Belgian Congo (Kellogg and Davol, 1949). Rotation systems are followed in which crops are planted in corridors or strips that alternate with the native forest. Optimum rotations provide for cultivation of a corridor from four to eight years, after which it is kept in forest about as long. This practice is essential to the maintenance of satisfactory physical properties of the soils. Chevalier (1949) also emphasizes the importance of 'bush fallow' in maintaining soils in good physical condition in French Guinea. Problems of structure maintenance in the cultivation of Latosols seem to differ greatly from those common to podzolic soils. For the most part, granulation is encouraged and sought for in podzolic soils, whereas it must be controlled and sometimes prevented in Latosols. These various observations indicate that tree roots may serve in several ways in the improvement of subsoil conditions or in the maintenance of satisfactory conditions for plant growth. As with data on the effects of other types of roots, however, precise information as to the kinds and magnitude of effects are still lacking. "

76. EFFECTS OF DIFFERENT INTENSITIES OF GRAZING ON DEPTH
    AND QUANTITY OF ROOTS OF GRASSES

    J. E. Weaver. Jour. Range Managt. 3 (2): 100-113. 1950.

         Quantity of roots was determined in three pasture types near Lincoln, Nebraska, by washing roots free from soil monoliths. Monoliths of Carrington silt loam were taken in a high-grade pasture mostly in little bluestem but with about 15 percent big bluestem, in a mid-grade pasture with about half bluegrass and little bluestem that had been weakened by grazing, and in a low-grade pasture mainly in bluegrass and blue grama.

         "[There were] ... remarkable changes in root depths in the three grades of pasture. As the native mid and tall grasses weakened and died and were replaced by low-growing bluegrass and blue grama, both depth of soil occupied by roots and the amount of root material decreased greatly. Oven-dry weights of roots and percentage decrease at each soil level are shown in Table 1.

Table 1

Dry weight in grams of underground plant material at the s
soil depths in three grades of pasture and percentage decre
from the high-grade type

Depth	High-grade	Mid-grade	Decrease	Low-grade	Dec
ft.	gm.	gm.	%	gm.	%
0 -0.5	65.50	52.20	20	28.14	57
.5 - 1	9.10	8.55	6	4.13	55
1 - 2	6.93	6.52	6	2.66	62
2 - 3	2.39	.95	60	.14	94
3 - 4	.40	.17	58	- - -	100
Total	84.32	68.39	19	35.07	58

"The chief cause of the greater weight of materials
face 0-0.5 foot compared with that of the second layer is th
abundance of stem-bases and rootstocks near the soil surfa
often equal the weight of the roots. They, of course, are lij
worn-down bluestems, and especially in bluegrass and blue
But the roots at all depths decreased greatly and especially
third and fourth foot....Loss of roots in the deeper soil or d
the roots in this part results in a much restricted volume c
which water and nutrients may be absorbed."

"The degeneration of the bluestem grasses under d
tensities of grazing was further examined. This was done l
of representative individual bunches. One bunch of little bl
selected from a portion of the pasture where grazing for se
had been very light or none. The last year's stubble was thi
tinuous, and 8 inches tall. The bluestem was in a very vigc
dition. A second bunch was selected in the mid-grade area
it had been grazed closely for at least two years. There w:
little debris left from preceding years. The individual tuft
the bunch were abundant but more or less separated by bar
the crown was somewhat open. The third clump was from i
a low-grade area where little bluestem still persisted but '
weakened condition. The bunch was very open, much bare
exposed because of this and the lack of a good mulch. The
of grass were short and much stunted. Many fragments of
were present. A monolith of soil 12 inches wide and 4 feet
was taken directly below each bunch."

"Decrease in the density of the rootmass at all levels from the high-grade to the low-grade pasture is clearly evident. The roots were almost 5 feet deep in the first sample, about 4 feet in the second, but they extended to only about 3 feet in the third."

"The high weight of the first sample in the shallow soil is characteristic of this species and is due in a large measure to the weight of underground stem-bases and short rootstocks. Decrease in weight at all depths in the second sample is very great. The weight, compared with that of the first sample, decreases regularly with depth from 19 to 91 percent (Table 2). Total weight of the second sample was 55 percent less than the first. In the second sample the roots were not only fewer than in the first but also finer. Their diameter was only a half to a third as great. Some dead roots were found. In the third sample differences were even more marked. Branches were fewer, many roots were dead, and debris from decaying roots was abundant. There were no roots in the fourth foot.... It seems clear that root deterioration is from the tips upward toward the crown. This sequence has been noted several times and the actual process was observed in Sudan grass as a result of frequent clipping [Peralta 1935]."

Table 2

Dry weight of underground plant materials of little bluestem at the several depths in the three grades of pasture and percentage decrease from the high-grade type

Depth	High-grade	Mid-grade	Decrease	Low-grade	Decrease
ft.	gm.	gm.	%	gm.	%
0 - 0.5	44.60	18.99	57	10.17	77
.5 - 1	2.74	2.21	19	1.63	40
1 - 2	2.59	1.61	38	1.07	59
2 - 3	1.20	.64	47	.21	83
3 - 4	.75	.07	91	---	--
Total	51.88	23.52	55	13.08	75

"Similar studies were made on root deterioration of big bluestem, except that bluegrass invaded the area occupied by this species in the low-grade pasture. The process of deterioration of the root system was about the same as that of little bluestem.

"Here the loss in root materials below the first foot
reased with depth. Weight of the root system in mid-grade
as only half that in high-grade, and in low-grade pasture i
ne-fourth as great."

"Just as the forage yield may be increased by good p
1anagement, so too the root systems of the grasses will be
omewhat in proportion. A good top that produces much nut
>rage and a good root system that can withstand drought an
1uch food for early growth in spring go hand in hand. A de
f non-vigorous grasses is usually also one in which the roo
re absorbing water and nutrients only in the upper portion

INDEX OF AUTHORS

	Reference number
Loustalot, A. J.	32
Lunt, H. A.	51
Lyon, T. L., Buckman, H. O., and Brady, N. C.	1
McDougall, W. B.	70
Martin, E.	24
Mather, J. R.	47
Parshall, R. L.	17
Pearson, G. A.	62
Penman, H. L.	10, 11, 38, 6
Penman, H. L., and Schofield, R. K.	9
Rich, L. R.	41
Richards, L. A., and Wadleigh, C. H.	5
Rogers, W. S.	69
Rowe, P. B., and Colman, E. A.	43
Russell, Sir E. J.	2
Sartz, R. S.	65
Schneider, G. W., and Childers, N. F.	33
Schofield, R. K.	35
Shreve, F.	18
Spalding, V. M.	31
Stephenson, R. E.	68
Stone, E. L., Jr.	37
Stuckey, I. H.	71
Thornthwaite, C. W.	34, 46,
van Bavel, E. H. M., and Wilson, T. V.	49
Veihmeyer, F. J.	16
Veihmeyer, F. J., and Hendrickson, A. H.	4
Wadleigh, C. H.	56
Wadleigh, C. H., and Gauch, H. C.	50
Weaver, J. E.	76
Weaver, J. E., Jean, F. C., and Crist, J. W.	74
Weaver, J. E., and Mogensen, A.	29

	Reference number	Page
Whitfield, C. J.	25	42
Winters, E., and Simonson, R. W.	75	109
Woodruff, C. M.	15	30
Zingg, A. W.	44	69

'D CLIMATIC FACTORS RELATED TO
E GROWTH OF LONGLEAF PINE

D. C. McClurkin

'HERN FOREST EXPERIMENT STATION

st Service, U.S. Department of Agriculture

SOIL AND CLIMATIC FACTORS RELATED TO THE GROWTH OF LONGLEAF PINE [1]

D. C. McClurkin

The purpose of this investigation was to find a means of predicting site quality for stands of longleaf pine (Pinus palustris Mill.) in terms of the soil and climatic factors that may be related to tree growth. In areas where the land is cutover or abandoned, the conventional means of estimating site quality--the trees themselves--is not present. Therefore, it is desirable to have a method that is not dependent on the presence of trees, but rather is based on permanent mappable features of the area such as the climate and the physical properties of the soil.

The study showed that in the Gulf Coastal Plain region longleaf site quality--in terms of the height of dominant trees at 50 years of age--can be predicted from two factors. The first of these is the amount of rainfall that the site receives during the first six months of the year. The second is the depth of soil to the least permeable horizon.

Since both soil depth and amount of rainfall can readily be determined, the study findings can be applied by practical land managers who wish to evaluate their sites. Readers who are interested only in this phase of the study may wish to turn directly to the section on field application, pages 9-12.

---

[1] Condensed from a dissertation presented in partial fulfillment of the requirements for the Ph. D. degree at Duke University, Durham, North Carolina. This study was conducted under the direction of T. S. Coile and F. X. Schumacher of the School of Forestry, Duke University.

CLIMATE AND SOILS

The band of longleaf pine from Mississippi to Texas is approximately fifty miles wide. The climate of this narrow strip is relatively uniform, being temperate to subtropical and subhumid (6) 2/. Rainfall for the first six months of the year increases from a low of 25 inches in Texas to a high of 33 inches in southern Mississippi and southeastern Louisiana. The growing season averages 250 days.

During Cretaceous times, all this area was a shallow sea which retreated near the end of the era. The sediments of the Coastal Plain, which were brought down by streamflow or left behind by the retreating sea, are slightly consolidated sands, clays, and marls, and dip gently seaward.

In those parts of the Coastal Plain where longleaf pine is found, the most prominent soil series are in the Norfolk-Ruston and Caddo-Beauregard catenas (5). All series in these catenas belong to the red and yellow podzolic soils. These soils, which are products of podzolization and laterization, are characteristic of warm-temperate, humid regions. Their internal drainage ranges from good to poor.

METHODOLOGY

During the past quarter century, numerous investigations have been made in the United States to describe or predict site quality of land for tree growth in terms of soil characteristics. These investigations have followed two general lines: one approach has been the study of chemical properties of the soil; the other approach has been the study of physical properties of the soil profile or particular horizons. Thus far, it has been in the realm of physical properties that the most usable correlations between site quality and soil have been found. Consequently, the soil phase of this study was, in the main, restricted to physical characteristics. All the literature pertinent to the subject has been brought together in a thorough and comprehensive review by Coile (1).

2/ Underscored numbers in parentheses refer to Literature Cited, page 12.

For this study a circular sample plot, 0.2-acre in size, was taken in each of 143 separate stands of longleaf pine from Mississippi to the western limit of the species' range in east Texas (fig. 1). Stands which were selected for study showed evidence that the dominants were of uniform age, had grown under conditions of good stocking, and had never been suppressed (fig. 2). No stand was included unless it was at least twenty years old. Trees younger than this vary considerably in height growth and probably do not fully reflect site and competition conditions. Stands of similar age and soil type were not taken closer than one mile from each other.

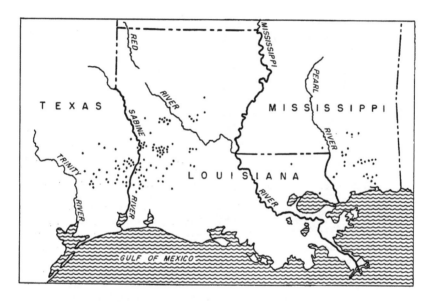

Figure 1.--Location of sample plots.

For eight dominant trees on each plot, the diameters at breast height were measured to the nearest one-tenth inch, and the ages at breast height were determined with an increment borer. On the first 84 plots, two of the eight trees were also bored at one foot above ground, to obtain age at stump height. The total height of each of the eight trees was measured to the nearest foot with an Abney level.

Soils data were obtained from four post holes dug on each plot in approximately cardinal directions from the plot center. The holes

were sunk to a depth of forty inches or to the C horizon if it occurred sooner. A composite sample of each horizon was collected for laboratory analysis. The thickness, internal drainage, texture, and consistence of each horizon were recorded on the plot tally sheet. The last two values were estimated by manipulation in the hand. If mottling occurred, its depth was recorded. Four soil auger holes also were bored to the least permeable horizon, giving additional checks on the depths of the different horizons.

No attempt was made to classify the soils according to family and series, but only plots on which the soils were homogeneous were sampled.

Figure 2. --A well-stocked even-aged longleaf stand typical of many sampled.

Four topographic features were recorded: position on slope, percent slope, class of surface drainage (a reflection of position on slope and percent slope), and aspect. For areas with no slope, the aspect was recorded as zero.

moisture equivalent values than the same soil without organic matter. This is a justifiable procedure when one considers that the quantity of organic matter in Gulf Coastal Plain soils varies with the age of the stand. Indeed, organic matter may be lacking on recently abandoned or barren land--the very places where it is most useful to be able to predict site in terms of permanent soil features.

Imbibitional water value (the numerical difference between moisture equivalent and xylene equivalent) was determined for the least permeable horizon. The imbibitional value is a measure of the kind and amount of clay present in the soil.

ANALYSIS OF DATA

The method of least squares was used for all statistical determinations. The initial step in the statistical analysis was to express the height in feet of the dominant stand in its logarithmic equivalent. Equating the logarithm of height to a constant plus the reciprocal of age resolves the sigmoid-shaped growth curve of longleaf pine into a more manageable linear expression (3). Once the general form of the equation is decided upon, the regression can be written as follows:

$$Y = b_o + b_1x_1 + b_2x_2 + b_3x_3 + \ldots + b_nx_n$$

This system of analysis, together with the solution of the variables and their respective variances, is described by Schumacher and Chapman (4).

Since longleaf pine lives an unpredictable period of years in the grass stage, it was decided to relegate measurements of age to stump height rather than introduce an average age from seed to breast height. Age at stump height was obtained by adding two years to the age at breast height; it was computed from the 84 plots in which age at stump height and at breast height were determined.

Longleaf pine, in contrast to other southern pines, often appears in rather open, savannah-like stands (fig. 3). Under these conditions, it was thought, tree heights might be significantly lower than in comparable well-stocked stands. To test this hypothesis for these data, a separate regression was set up in which stand density per acre was expressed in terms of tree size, age, and numbers, and in combinations of these factors. The resulting equation was used to calculate the stand density of each of the 143 plots.

Figure 3.--Longleaf pine often grows in open, savannah-like stand
In this study, however, stand density did not significantly influenc
height growth.

Since the range of plots extends from eastern Mississippi into Texas, it was necessary to consider possible environmental effects that might be correlated with height growth and longitude. Rather than employ the actual degrees of longitude as a variable, it was decided to use rainfall, which decreases in all directions from a center near east-central Louisiana. The rainfall variable was further refined by using the first six months' rainfall, since presumably this is the most critical moisture period in terms of height growth for southern pine.

Ralston's earlier study of site index of longleaf pine in the Atlantic Coastal Plain showed that topographic position is a significant factor in determining height growth (2). Typically, this factor is used to divide the data into four classes of surface drainage: excessive, good, imperfect, and poor. Of the 143 plots measured in the present study, however, 133 were classed as having good drainage, and therefore only that one drainage class was recognized in the analysis.

In Ralston's work, depth to mottling also was a significant variable. In the data presented here, the scarcity of imperfectly or poorly drained sites virtually precluded any considerations of depths to mottling, since mottling rarely occurs in well-drained profiles.

RESULTS

The statistical solution of the data showed the amount of January-June precipitation to be more important than any other variable. Of the soil factors measured, depth to the least permeable horizon was the most significant.

The preliminary regression was solved by removing first the constant and then age. All subsequent variables were removed in the order that would account for the greatest remaining sum of squares each time. The results of this regression showed the effect of rainfall to be so great that the soil variables showed no significant effect on height growth. Therefore a second regression was solved in which rainfall was not eliminated until last. Even so, the effect of rainfall completely overshadowed the effect of the significant soil variables.

Two more regressions were solved, one excluding rainfall and the other using the joint variation in rainfall and soil depth. The latter of these showed the most practical results and was used for calculating table 1 (page 11).

- 7 -

In the final regression, the following variables were significant:
1. Reciprocal of age
2. Depth to least permeable horizon times rainfall
3. Depth to least permeable horizon.

When solved by least squares for its appropriate coefficients, the equation for tree height derived from the final regression reads as follows:

Log of tree height
$$\text{(in feet)} = 1.9995 - \frac{6.492}{\text{Age}} + \text{DLP}[0.0002636(R) - 0.006734]$$

where: DLP = Depth to least permeable horizon in inches
R = First six months' rainfall in inches.

The error of estimate is 8.26 percent. After the age is equated to 50 years, the equation is given in terms of site index by the following expression (graphically illustrated by fig. 4):

Log of site index = $1.8697 + 0.0002636(R)(DLP) - 0.006734(DLP)$

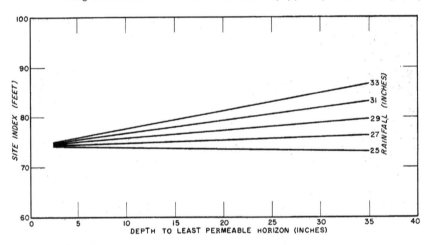

Figure 4. --Site index plotted over depth to least permeable horizon, by rainfall classes.

Stand density, either with or without rainfall, did not prove significant in any of the regression schemes. This would indicate that there was no significant variation of height growth within the range of stocking measured.

Moisture equivalent variables of the surface soil proved to be significant at the five-percent level when rainfall variables were omitted. In the presence of rainfall variables, all soil moisture variables and their interactions became nonsignificant. This lack of correlation was due in part to the rather limited variation of these moisture values.

Depth to the least permeable horizon was not significant in the first two regressions. However, in the third regression in which rainfall was omitted, depth was significant at the five-percent confidence level. In the fourth regression, with rainfall interactions, depth became significant at the one-percent level.

Rainfall is significant regardless of the time of its removal from the regression. It is perplexing that increase in rainfall should be so strongly correlated with improvement in site, particularly when one considers that the average first six months' rainfall at the western limit of longleaf growth in the Gulf Coastal Plain is higher than that of the Atlantic Coastal Plain. Hence, average rainfall can scarcely be considered a limiting factor in this area. It is likewise highly improbable that a range of only eight inches of rainfall (in six months) could bring about any marked changes in soil properties capable of accounting for the range of site indices encountered. For these reasons, it is highly doubtful that any cause-and-effect relation between rainfall and site is being displayed here. Instead, rainfall would seem to be strongly correlated with some controlling climatic variable as yet not measured directly.

Rainfall times depth to the least permeable horizon was highly significant at five percent in the fourth regression, indicating an interaction between amount of moisture and growing space for roots.

FIELD APPLICATION OF RESULTS

The results of this study may readily be used to find the site index of any given area in the longleaf pine belt of the Gulf Coastal Plain. The procedure is as follows:

1. Determine the average amount of rainfall that the area receives during the first six months of the year. For all practical purposes, this can be done merely by locating the area on the map containing the rainfall contour lines (fig. 5).

2. Bore enough holes in the soil on the area to determine the average depth to the least permeable soil horizon.

3. Read the site index from table 1.

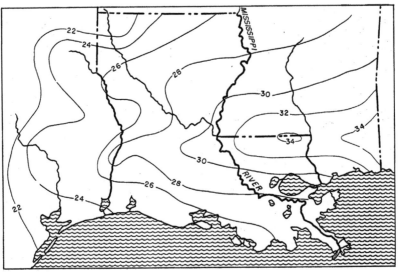

Figure 5.--Average rainfall in inches, January through June (6).

For example, on an area where the depth to the least permeable layer is about 22 inches, and where January-June rainfall averages from 28 to 30 inches, the site index is 77. That is, dominant longleaf pines can be expected to reach 77 feet in height by the time they are 50 years of age. It must be remembered, however, that the heights predicted by table 1 apply only to even-aged, unsuppressed stands. The figures do not hold for stands that develop under an existing forest canopy.

Table 1. --Site index of longleaf [1]

Depth to least permeable horizon (inches)	Rainfall, January through June				
	24 - 26 inches	26 - 28 inches	28 - 30 inches	30 - 32 inches	32 - 34 inches
	- - - - - - - Site index - - - - - - -				
0-5	74	74	74	75	75
6-10	74	74	75	76	77
11-15	73	74	76	77	78
16-20	73	75	76	78	80
21-25	73	75	77	80	82
26-30	72	75	78	81	84
31-35	72	75	79	82	85
36+	72	75	80	83	87

[1] Average total height in feet of dominant trees at age 50 years.

The main problem in applying these instructions is to take enough depth measurements on a site to truly know the average depth to the least permeable horizon. There may also be some difficulty, at times, in identifying the least permeable horizon.

If the area in question is fairly flat, or if it is of uniform slope, four or five holes per acre might well suffice, especially if the depth measurements do not vary more than a few inches from hole to hole. If the topography is rolling, one would expect marked changes in subsoil depth between the bottom and the top of a slope. On long, sustained slopes where wide areas are comparable, different productive capacities can be separated without much difficulty. In sharply rolling topography such as occurs in the upper Coastal Plain, the average of the measurements on lower, middle, and upper slopes will have to suffice.

In areas where the least permeable horizon is a heavy clay, it is easily distinguished from the more permeable horizons that overlie it. Often, however, the least permeable horizon found under longleaf stands is not markedly different in texture from the surrounding horizons. In such cases slight changes in color may help to tell one ho-

rizon from the next, and if the soil is rubbed between thumb
finger the least permeable horizon will feel finer in textur
other horizons.

LITERATURE CITED

(1) Coile, T. S.
     1952. Soil and the growth of forests. Advances in
          my 4: 329-398.

(2) Ralston, C. W.
     1949. Soil factors related to the growth of longleaf
          the Atlantic Coastal Plain. Ph. D. dissertat
          Univ. Library, Durham, N. C.

(3) Schumacher, F. X.
     1939. A new growth curve and its application to timl
          studies. Jour. Forestry 37: 819-820.

(4) _____ and Chapman, R. A.
     1948. Sampling methods in forestry and range mar
          Duke Univ. School Forestry Bul. 7 (revised

(5) U. S. Department of Agriculture.
     1938. Soils and men. Agr. Yearbook 1938, 123

(6) _____
     1941. Climate and man. Agr. Yearbook 1941, 1;

OCCASIONAL PAPER 133 June 1954

PREVENTING AND CONTROLLING
WATER-CONDUCTING ROT IN BUILDINGS

A F. Verrall

SOUTHERN FOREST EXPERIMENT STATION

Philip A Briegleb, Director

Forest Service, U.S. Department of Agriculture

PREVENTING AND CONTROLLING
WATER-CONDUCTING ROT IN BUILDINGS

A. F. Verrall
Southern Forest Experiment Station

Most decay in buildings is caused by a number of fungi that attack wood previously wetted by rain seepage, leaky plumbing, and condensation. This decay does not extend appreciably beyond the area wetted. Occasionally, however, a water-conducting fungus attacks. These fungi, like other decayers, start at some point where the wood is moist, but if conditions are right they can moisten and advance in wood that would otherwise be too dry to decay. This is the basis for the misnomer "dry rot" for the work of these fungi. Fortunately, the water conductors are of infrequent occurrence. When they do occur, damage may be extensive and necessitate costly replacements.

During the past 25 years, the Division of Forest Pathology has studied more than 35 cases of water-conducting rot in the southern States. The present analysis of these cases shows the main structural features and conditions associated with attack and how control can best be attained. It is hoped that this analysis will allay fears caused by some alarmist publications in the past and also disclose to the building owner less drastic and less costly control measures than have been commonly recommended.

DESCRIPTIONS OF WATER-CONDUCTING FUNGI

There are two important water-conducting fungi: Merulius lacrymans Fr. and Poria incrassata (B. and C.) Burt. The former is one of the most common building decay fungi in Europe (4). [1] It is occasionally found in the northern United States, but is of little or no importance in the South.

---

[1] Underscored numbers in parentheses refer to Literature Cited, p. 14.

Poria incrassata, hereafter referred to as Poria, is by far the most destructive water-conducting fungus in the United States. It is most prevalent in the warmer, more humid coastal regions of the South, East, and West but has been found in most of the States. Baxter (1) suggests that its occurrence in the North may be due to the importation of infected lumber from the South or from the West Coast. Even in the southern and West Coast regions Poria is not common.

Poria has been found mainly in buildings. This may mean that in buildings the fungus finds the protected locations best suited to its development. Since its fruiting structure is fragile and would soon disappear under most outdoor conditions, the fungus might be noticed less than would species with durable fruiting bodies. Hence the restriction in occurrence may be more apparent than real. The fruiting body, vegetative mycelium, rhizomorphs, and cultures have been described (5, 6).

When a water conductor occurs in a building, the extent and rapidity of attack may be spectacular (6, 9). Poria causes a brown cubical rot, usually starting in the substructure--the sills, first-floor joists, nailing strips, wall plates, etc. (fig. 1). Later it may work up into the walls and even attack the second floor. By this time all structural and trim items in the area of the original attack may be damaged. Oftentimes owners are unaware of the attack until a floor fails, doors settle, sunken areas appear in baseboards, or a yellowish-white fungus felt appears under a rug or in a cupboard. When the wall or floor is opened for repairs, extensive yellowish-white mycelial mats (fig. 2) are usually found between sheathing and building paper or between two wood surfaces as finish and subfloors.

Figure 1.--Flooring badly decayed by Poria.

Figure 2. --Mycelial growth of Poria on the back of a wood panel.

When extensive damage occurs, rhizomorphs (water-conducting strands) from a small fraction of an inch to 2 inches in diameter can usually be found (fig. 3). They extend from the soil or other constant water source up into the wood structure. It is these rhizomorphs and well-developed mycelial fans that permit Poria to conduct water to wood normally too dry to decay. When decay is more restricted, definite rhizomorphs may be lacking but the mycelial mats show a tendency to form strands. Laboratory studies[2] have shown that Poria can conduct water at least 5 feet up wood columns. Observations in buildings indicate that this fungus can conduct water at least 20 feet horizontally and 12 feet vertically.

_____

[2] Scheffer, T. C. Importance of conducted water and "auto-humid-ification" in decay by the "dry-rot" fungus Poria incrassata. U. S. Dept. Agr. Div. Forest Path. Unpublished report. Nov. 1941.

All fungi liberate water as
one of the end products of wood
decay. The brown rotters, which
include the water conductors, are
particularly proficient in this respect (4). This "autohumidifica-
tion" undoubtedly aids in the development of water-conducting
fungi. However, there seems to
be no basis for the fear that Poria
when once well established in a
building, can exist with no other
source of water than autohumid-
ification. Studies by Scheffer 2/
suggest that the critical relative
humidity for the progress of Poria
in freely exposed wood not exceeding 2 inches in thickness, and
with the fungus isolated from
sources of water other than auto-
humidification, would be somewhat above 95 percent. He points
out that such humidity would have

Figure 3.--Water-conducting
rhizomorphs growing on the ends
of piled lumber.

to be continuous, a condition
seldom encountered in buildings.
Thus it appears that, under
most conditions in buildings,
the fungus must depend on conducted water for continued development.
There may be some exceptions to this generalization.

Two other physiological factors in the growth of Poria are important in its occurrence and control. Among wood rotters, it is one of
the most sensitive to drying (11). In wood with about 8 percent moisture
it died in 24 hours; in wood with 13 to 21 percent moisture it died in 12
days or less. This is in contrast to other common building decay fungi
like Lenzites saepiaria, which survive several years in air-dry wood
(10). The other important factor is Poria's ability to decay heartwood
of durable species. Both laboratory studies (5) and field observations
show that Poria will seriously attack the heartwood of baldcypress,
western redcedar, and redwood as well as less durable common building
woods such as pine and Douglas-fir. In contrast, it shows no unusual
resistance to such common wood preservatives as creosote, zinc
chloride, or pentachlorophenol (2, 7). It does, however, show some
resistance to copper fungicides (7).

SURVEY OF PORIA CASES IN THE SOUTH

Since 1928, observations have been made [3/] on cases of water-conducting rot as they were found or reported. Notes were taken on structural features and other factors that might explain why the attack occurred. In many cases, the buildings were watched for several years after control recommendations had been carried out and repairs had been made. The cases in buildings in the South are summarized in table 1.

Sometimes complete information was lacking, but it appears that most attacks occur in relatively new houses or in older houses in which recent structural changes have been made. The most logical interpretation is that Poria often is introduced into a building in infected lumber. One known exception to this is the Dothan, Alabama, case (table 1). Even though structural changes had been made 2 years before decay was noticed, they were not involved in the decay. Instead, the decay was centered on an old outside wall against which coal had been piled. Because Poria mycelial mats have been observed on coal in other cases, it is assumed that the fungus was introduced into the Dothan house on coal from an infected lumber yard.

All the evidence points to little dissemination of this fungus by air movement of spores. Were this not true, it would be more prevalent in houses with features known to favor attack but which have stood for years without attack. The available information points to a need for better protection of lumber and building supplies in lumber storage sheds as an important factor in prevention. The buyer is fairly sure of getting lumber free of live Poria if he insists on the lumber being dry, i.e. below 20 percent moisture content (and preferably below 18 percent) for framing and sheathing, and below 12 percent for all siding and trim (11).

Source of water supporting decay

The worst decay observed was associated with sites with continuously wet soil at some place around the foundation. The wet soil was due to poor drainage, leaky plumbing, lack of provision for carrying roof runoff away from the foundation, continual watering of flower beds, or poorly designed concrete porches that allow excessive seepage into the dirt fill under the slab. In general, the more water present the more

---

3/ Some of the observations were made by C. Hartley, R. M. Lindgren, T. C. Scheffer, and C. A. Richards.

Table 1.--Summary of factors associated with water-conducting rot in buildings in the South

Location	Type of building	Constructed	Altered [2]	Decay first noticed	Inspected	Wet soil	Leaky plumbing	Rain leaks	Green lumber	Leaky downspouts	Joists or sills	Dirt-filled porches	Foundation forms	Siding, trim, sheathing	Lattice work	Wood foundation	Wood on groundline	Concrete slab	Inadequate ventilation	Miscellaneous
ALABAMA																				
Dothan [3]	House	1920	1944	1946	1948	P	P				P*									Decay centered where coal was piled against siding.
Mobile	House	1880?	Yes	1940	1940		P	P			P*	P*	P	P		P	P	P		Dirt-filled porch added at unknown date.
Selma [1]	House	1858	1930	1937	1940	P														Concrete porch added in 1930. Soil dry except at porch.
FLORIDA																				
Jacksonville	House	1926	No	1931	1931															
Lake Wales	House	1920?	Yes	1931	1931						P*					P*	P*	P	Decay in unventilated wooden locker added to basement.
Lake Wales	House	1920?	No	1931	1931						P*					P*	P	Partition wall plate on wet concrete basement floor.
Orlando	House	1925?	No	1931	1931											P*			
Penny Farms	Houses	1927	No	1930	1931											P*		Several houses involved.
Pensacola	House		1930	1930	1931	P													Repaired several times.
Port St. Joe [3]	5 houses	1935?	No	1946		P													
Tampa	House	1928?	No	1931	1931														
Tampa	House	1928?	No	1931	1931														
GEORGIA																				
Bainbridge [2]	House	1931?	No	1941	1941						P*								Rain probably seeped into dirt fill.
LOUISIANA																				
Baton Rouge	House			1940		P	P				P*	P*	P*						Joist on pile of mortar left from removing chimney.
New Orleans [3]	House	1880?	Yes	1942		P	P				P*							P*	Untreated sleepers in groundline concrete slab.
New Orleans	House	1850?	Yes	1941		P													Sill to new basement room laid on soil.
New Orleans [3]	House	1900?	Yes	1937		P					P*	P*						P*	Rich, wet, flowerbed soil against siding. Untreated nailing strips on soil in basement.
New Orleans	House	1927?		1937	P	P													
Southport [3]	House		Yes	1940	1941														
MISSISSIPPI																				
Gulfport [3]	House	1926	No	1944	1944	P	P				P*	P*		P*					Wood sheathing to ground between stucco and brick piers.
Meridian [3]	House	1890?	Yes	1940	1940	P	P												Dirt-filled porches added at unknown date.
S. CAROLINA																				
Marion [3]	House	1893	1947	1950	1951	P		P*			P	P*							Termite-damage repairs in 1947. Suspect use of infected wood.
TEXAS																				
Denison	House	1910?	1949	1950	1951	P					P	P	P					P	Decay in new sapwood addition made in 1949.
Terrell [2]	House		1939	1940															Wood skirting touching soil.
LOUISIANA																				
New Orleans	Store		1940	1940	P	P					P*	P*	P*						
New Orleans [2]	Duplex		Yes	1931?	1931	P					P								
MISSISSIPPI																				
Gulfport	Motel	1935?	No	1938	1940	P										P*			Wood debris piled up to joists.
Hattiesburg	School	1926	No	1929	1929	P													
N. CAROLINA																				
Cherry Point [3]	Apt.	1944	No	1946	1948	P						P*						P*	Excessive floor washing.
FLORIDA to LOUISIANA	7 lumber sheds					P		P			P*								Decay often worked up into stored lumber.

-6-

extensive the damage. In some cases there was a possibility that the main source of water was not the soil but water in wood wetted directly by rain seepage or leaky plumbing. It is assumed that after the fungus is established in such wood, it can progress without ground contacts as long as the leaks remain.

A less common source of water appeared to be operative in the case of the apartment building at Cherry Point, North Carolina. There were no contacts with the soil or moist concrete, and no condensation was evident. Apparently the fungus gained entrance in infected green framing lumber. Because of the tight foundation closure, the water in the green lumber plus metabolic water from the decay may have dried so slowly that decay to the point of wood failure occurred without other additions to the water in the wood. Replacements with dry wood remained sound.

In the South no cases have so far been found in which condensation was an important source of water that supported decay. Such instances have been found in the Washington, D. C., area and--provided ground connections are also broken--are controllable by the use of soil covers (3).

Lack of substructure ventilation may increase the severity of attack but is not a prerequisite of heavy damage. Some of the severest cases observed were in houses on open brick piers. There are usually protected places, such as wall interiors. or the contact areas of flooring with joists or between floor layers, in which the high humidities needed for fungus development are maintained during active decay. Ventilation is probably most important in the borderline cases in which water is derived mainly from green lumber or from condensation.

Structural details associated with Poria attacks

With the exception of the Cherry Point and Dothan cases, all attacks appeared to start at contacts of untreated wood and moist soil or groundline concrete slabs. These contacts, in the order of prevalence, are discussed below.

(1) Dirt-filled porches with an untreated sill or header in contact with the dirt fill or separated from it only by asphalt-impregnated sheathing paper (fig. 4A). The dirt-filled porch has led to so much decay and termite trouble that extreme caution should be exercised in including this type of construction in a building. It is much safer to use a self-supporting slab with the subslab area open to the crawl

Figure 4. --Some common practices leading to attack by water-conducting fungi. A, The dirt fill for this porch is separated from the untreated sill only by sheathing paper. B, Non-waterproofed groundline concrete slab. C, Untreated wood in contact with soil. D, If these forms are left after the stoop slab is poured, they will make a wooden contact between the soil and the sill of the house.

space or cellar. When the slab is a step or two lower than the house floor the fill may be entirely below the wood, in which case there is little danger. Attack may also be prevented by the insertion of a continuous non-corrosive metal shield. Such a shield should extend down from under the siding, over the outside and underside of the sill to the crawl space, and there should bend downward at 45° for 2 inches. There is also a possibility that attack of both decay and termites can be prevented by treating the fill with fungicides and insecticides before the slab is poured. Tests on the effectiveness of soil fungicides have been started. The easiest way, however, is to avoid the dirt fill.

When attack involves a dirt fill the usual correction consists of removing a section of the foundation and excavating the fill away from the sill. The excavation should be sufficient to permit crawling along the sill for inspection. The concrete slab and foundation exposed by the excavation should be scraped clean and painted or sprayed with creosote or with 5 percent pentachlorophenol in a petroleum oil. In one case a 12-inch strip of the porch slab was removed next to the sill and enough fill removed to insert a new pressure-treated sill. The excavation was filled with creosote-treated cinders and the slab was repaired. This, however, did not leave as finished-appearing a job as the usual excavation from the underside.

(2) Joists or sills touching the soil. This group included cases in which sills or joists touched piles of earth thrown up in laying plumbing lines or digging furnace pits; non-leveling of natural ground humps; sills or nailing strips of untreated wood laid directly on the soil, particularly in basements or unexcavated store floors; and a pile of mortar left during removal of a fireplace. Indirect contacts included were a pile of cut-off lumber ends left under the building and touching a joist, and contacts of joists and stumps left under the house.

Many of these cases never would have occurred had a little care been exercised during construction. Where it is necessary to have a floor on the ground, a concrete slab is best (but see point 3 below for precautions about wood coverings). If wood of any kind must be placed directly on the ground it should be treated with such suitable preservatives as creosote or pentachlorophenol, with solution retentions of at least 6 pounds per cubic foot.

(3) Untreated wood on moist groundline concrete slabs (fig. 4B). Even when water-conducting rot does not occur, this condition usually leads to floor buckling and damp living quarters. Poria can become established in wall plates and nailing strips and can secure enough water from the soil through the slab to cause extensive decay. When severe water-conducting rot is associated with wooden floors and walls on a damp concrete slab, there is no sure way of control except to remove the floor, replace the wall plates with pressure-treated wood, and add a moisture-proof membrane to the slab, being sure that it extends under the wall plates. New nailing strips and subflooring should be pressure-treated. This is the most expensive and discouraging type of Poria attack to correct. Many of the houses attacked in this way have been torn down completely and replaced with other types of construction. When the attack is not too severe there is a possibility that control could be secured by removing wooden floors and replacing badly damaged wood in the walls, using treated plates. Then the unattacked plates could be given some protection by flowing 5 percent pentachlorophenol over and under them. A non-wood floor would then be used. There is no record of such a treatment but it should stand a good chance of success.

Because of the difficulty and expense of correcting an attack in wooden houses laid on groundline slabs, care should be taken to insure that the construction is safe from the start. Concrete slabs for houses should be elevated so the top surface is 12 inches above grade, the slab poured on a layer of gravel, and a water-proof membrane used either below or on top of the slab. No grade stakes should be left in the slab. When these recommendations are followed, there usually is no trouble with decay. Nevertheless it is best also to use treated wall plates, nailing strips, and subflooring.

As a substitute for treated wall plates, the current tendency is to place wall plates on sheet metal flashing bent up over the sides of the plates. Sometimes asphalt roofing is used as flashing. Such metal or roofing flashing should not be used as a substitute for a complete water-proof membrane when slabs are to be covered with wood flooring.

An analogous condition exists in damp basements. It is best to use only treated wood or non-wood materials in contact with basement floors.

(4) Sheathing, siding, and trim in contact with the soil (fig. 4C). In most cases this was due to careless grading or the building up of flower beds. Rich, mulched flower beds seem to be an ideal place for the rooting of a well-developed rhizomorph system. As long as siding and trim is kept to the usually recommended minimum of 6 inches from the soil, there should be no danger of its being an entrance point for Poria. A number of houses were found, including one with heavy Poria attack, in which the sheathing was brought down to the soil to act as a support for metal lath and stucco. The wood between the stucco and the foundation piers could be removed only by removing the stucco.

(5) Wooden foundation. The house in Denison, Texas, was on Osage-orange foundation piers. These apparently were unattacked and were not replaced during repairs. However, in several other houses the fungus entered through cypress and pine foundation piers. Replacement with concrete or brick piers stopped further attack. In many lumber storage sheds, cypress or heart pine foundations have been used for lumber piles. This has led to frequent attack in stored lumber. The use of concrete footings extending 6 to 12 inches above grade has stopped the trouble.

(6) Wooden forms left on foundations, steps, and porch slabs (fig. 4D). Although only two cases were traceable to this condition, the common practice of leaving forms in hidden places is risky. The cost of

so small that it is almost inconceivable that people leave
ecay does not occur, termite attack may.

PREVENTION OF ATTACK

10ugh no random sampling of houses has been made, it is
m general observations that many of the features mentioned
nonly occur without <u>Poria</u> attack. The logical explanation is
is of limited occurrence and is distributed mainly in infected
1terial. For example, although it is a common practice to
s under concrete porch slabs or back of concrete steps, only
attack at this point were found.

s raises a real problem in control, i.e., one of convincing
a given construction feature can be dangerous even though it
ways lead to trouble. The added cost of avoiding dangerous
1ring original construction or even (in many cases) of re-
danger after construction is small. Hundreds of dollars for
1 often be saved by a few hours of extra labor during or after
1n.

main precautions for preventing attack in new construction

Use only uninfected lumber below 20 percent moisture
content, i.e., fully air-dried or kiln-dried.

Be sure the construction plan calls for no untreated wood
in contact with the soil. Framing generally should be 18
inches above grade and trim at least 6 inches. If wooden
steps, lattice between foundation piers, or any other wood
is likely to be on or near the ground, it should be adequately
treated with a preservative.

Be sure that changes in grading do not create wood-soil
contacts.

Remove all forms used in pouring concrete. Also remove
wood debris from the crawl space.

Provide adequate drainage to prevent water accumulations
near or under the house.

6. Do not build enclosed partitions, boxed steps, bins, rooms in a damp basement unless treated wood or n materials are used.

7. Beware of dirt-filled porches. See the previous sec cases associated with this.

8. If a groundline concrete slab is used, be sure it is r proof. See the previous section on cases associatec

9. If condensation occurs under a basementless house, soil cover (3).

In summary, prevention of water-conducting rot is ess the use of dry sound lumber in construction and the use of desi keep untreated wood away from the soil or damp concrete or m

WHAT TO DO WHEN PORIA ROT IS FOUND

The common recommendations for stopping an attack by have been drastic (6, 8, 9, 12). They usually call for removin decayed wood and all apparently sound wood within 2 or 3 feet decay, as well as the correction of structural faults. The req for the removal of apparently sound wood within 2 or 3 feet of infection turns the correction of even a minor attack into a ma costly operation.

The laboratory studies showing the sensitivity of Poria and observations in attacked buildings show that this drastic c unnecessary. The evidence is that not only the adjacent sound any infected wood with sufficient strength can be left if the sou outside water is removed. If the attack has been severe, repl of weak wood will be costly, but if the attack is discovered ear times no replacements are necessary. The most difficult cor occur with houses on groundline concrete slabs (for control m see point 3, page 9).

Correction always starts with a search for the source and its removal. This means breaking all the contacts betwee and the wood of the building. Foundations in the decayed part building must be inspected for rhizomorphs that connect the w the soil or moist concrete. Any rootlike structures should be from the foundation. In brick, concrete block, and stone wor rhizomorphs may penetrate voids or loose mortar and be invis

- 12 -

the outside. If the mortar is crumbly, it is safest either to remove a few top courses of brick and replace them with fresh mortar or to insert a continuous metal termite shield between the foundation and the sill. Foundation surfaces over which rhizomorphs have spread should be brushed with a clear preservative like 5 percent pentachlorophenol in a light petroleum solvent--or, if staining and odor are not objectionable, creosote will do. Other factors in moisture control may involve repair of leaks in plumbing, roofs, siding, or downspouts; provisions for better surface drainage; or the addition of a soil cover (3) if condensation occurs under a basementless house.

After these sources of water are removed, decayed wood usually dries out fast. If adequate foundation vent area is present(2 square feet per 150 linear feet of outside foundation plus 1/3 of 1 percent of the ground floor area), infected substructures dry out rapidly. Nevertheless, when walls and floors are opened to determine the extent of damage or to replace wood, leaving them open for a few days will hasten drying. Tight floor coverings like linoleum may impede drying of decayed wood under them, and probably should be lifted until drying is complete.

Poria has been found alive in the soil under buildings several years after decay had been stopped. Tests have been started to determine the value of fungicides in killing the water-conducting fungi in soil. If effective, soil fungicides might sometimes reduce the cost of control and be an added safeguard against recurrence.

Often it is simple for the expert to stop a Poria attack. Almost each case, however, is different,and all exigencies cannot be covered here. It is suggested that the home owner hire an expert repairman who will scrupulously follow the advice given here. In some cases encountered in this survey, all necessary corrections were not made during initial repairs. This almost always proved to be false economy, and led to a second or even a third major repair job. During the course of this survey no decay recurred in any case in which the owner followed the control recommendations completely.

LITERATURE CITED

1. Baxter, D. V. 1940. Some resupinate polypores from the region of the Great Lakes. XI. Papers Mich. Acad. Sci., Arts, and Letters 25: 145-170.

2. Carswell, T. S., and Hatfield, I. 1939. Pentachlorophenol for wood preservation. Ind. and Eng. Chem. 31: 1431-1435.

3. Diller, J. D. 1950. Reduction of decay hazard in basementless houses on wet sites. U.S. Dept. Agr., Forest Path. Spec. Release 30, 4 pp.

4. Findlay, W. P. K. 1953. Dry rot and other timber troubles. Hutchinson's Sci. and Tech. Pub. 267 pp. London.

5. Humphrey, C. J. 1923. Decay of lumber and building timbers due to Poria incrassata (B and C) Burt. Mycologia 15: 258-277.

6. _____ and Miles, L. E. 1925 (reprinted 1929). Dry-rot in buildings and stored materials and how to combat it. Ala. Polytech. Inst. Cir. 78, 24 pp.

7. Leutritz, J., Jr. 1946. Wood-soil contact culture technique for laboratory study of wood-destroying fungi, wood decay and wood preservation. Bell System Tech. Jour. 25: 102-135.

8. Milbrath, D. G. 1934. Wood decay in buildings. Calif. Dept. Agr. Monthly Bul. 23: 95-102.

9. Richards, C. A. 1933. Decay in buildings. Proc. Amer. Wood-Preservers' Assoc. 29: 389-397.

10. Scheffer, T. C., and Chidester, M. S. 1948. Survival of decay and blue-stain fungi in air-dry wood. South. Lumberman 177(2225): 110-112.

11. _____ 1943. Significance of air-dry wood in controlling rot caused by Poria incrassata. U.S. Dept. Agr., Forest Path. Spec. Release 17, 6 pp.

12. Weber, G. F. 1930 (revised 1938). Dry rot of lumber in storage and in buildings. Univ. Fla. Agr. Expt. Sta. Press Bul. 424, 2 pp.

NEW TREE-MEASUREMENT CONCEPTS:
HEIGHT ACCUMULATION, GIANT TREE, TAPER AND SHAPE

L. R. Grosenbaugh

SOUTHERN FOREST EXPERIMENT STATION
Philip A Briegleb, Director
Forest Service, U.S. Department of Agriculture

NEW TREE-MEASUREMENT CONCEPTS:
HEIGHT ACCUMULATION, GIANT TREE, TAPER AND SHAPE

L. R. Grosenbaugh
Southern Forest Experiment Station

An entirely new concept of tree measurement was announced by the author in 1948 (11). Since the original theory and applications have subsequently been broadened considerably, it seems advisable to publish the entire development in readily usable form, along with other material helpful in tree measurement.

In its essence, the new concept consists of selecting tree diameters above d.b.h. in diminishing arithmetic progression (equivalent to saying that outside-bark diameter is treated as the independent variable); then tree height to each such diameter is estimated, recorded, and accumulated. This reverses the classical tree measurement concept which regarded height as the evenly-spaced independent variable, and which estimated and recorded diameter at regular intervals of height.

The greatest advantage of the new concept lies in the ease with which individual trees can be broken down into various classes of material and recombined with similar portions of other trees, using punched cards, business machines, or both. Sections varying in length and grade (from different trees) can be sorted by grade-diameter groups within which lengths will be additive. The ability of modern business machines to accumulate has been largely wasted in conventional procedures where height is treated as the independent variable. Their capability can be much more fully exploited in the new procedure using diameter as the independent variable.

Another advantage lies in the convenience of being able to use height, the sum of heights, and the sum of height-accumulations to calculate volume and surface of individual trees (or of classes containing portions of several trees). Final tabulations can be readily converted into surface, cubic volume, or volume by a number of log rules; no volume tables are required. Little extra work is involved in tabulating volumes to several different merchantable tops, or in segregating volume by grades.

There are already in existence dendrometers which are well adapted to locating diameters in diminishing progression along the bole of a standing tree--the Biltmore pachymeter (4), the Clark dendrometer (6), the Bitterlich relascope (1) and spiegel-relascope (2), and the Bruce slope-rotated wedge-prism (3), to mention a few. However, anyone currently content to use arbitrary volume tables not based on valid sampling of the given tree population, or to estimate butt-log form-class and accept arbitrary upper-log taper assumptions can use the new technique with the crudest instruments and still not exceed the error currently ignored or tolerated in commonly used volume-table techniques.

This paper not only describes the theory and application of the new concept but also makes available for the first time a giant-tree table, and discusses tree taper and shape.

HEIGHT ACCUMULATION--A NEW CONCEPT

If a peeled 16-foot log with a small-end d.i.b. (diameter inside bark) of 8 inches, and a large-end d.i.b. of 10 inches is stood upright on its large end, and if heights (H) are measured as the number of 4-foot sections occurring beneath diameters in the complete progressive series

2 inches, 4 inches, 6 inches, 8 inches, and 10 inches

the record will appear as follows:

d.i.b. (inches)	H Number
2	4
4	4
6	4
8	4
10	0

It will be noticed that all diameters in the series which are smaller than the small end of the log have the same height as the small end.

Now if these heights are differenced (upper minus lower) and if each difference in height is expressed as (L), the calculation will appear as follows:

d.i.b. (inches)	L Number	H
2	0	4
4	0	4
6	0	4
8	4	4
10	0	0

Finally, if heights are accumulated so that each entry indicates the progressive total (H') of all heights recorded beneath it, and L, H, and H' are then summed, the calculation will appear as follows:

d.i.b. (inches)	L	H Number	H'
2	0	4	16
4	0	4	12
6	0	4	8
8	4	4 = ΣL	4
10	0	0	0
	4 = ΣL	16 = ΣH	40 = ΣH'

Now these three sums can be converted to un
surface) by referring to Appendix A. Where taper-
where height is expressed in 4-foot units, and whe

(i.e., where it is not necessary to correct for ba
diameter measurements), the board-foot volume by t
can be calculated as: $2\Sigma H' - 5\Sigma H + 4\Sigma L = 80 -$
This is identical with the volume obtained by the
rule for 16-foot logs: $(d.i.b. - 4)^2 = 16$. The t
is discussed in Appendix E.

Of course, the height-accumulation techniqu
over the conventional method in the simple case ab
technique has real utility in more complicated tim
measurement situations.

This technique can be applied when d.o.b. (
is measured but volume inside bark is desired. Fc
tration, it will be postulated that bark constitut
d.o.b. (i.e., that mean $\frac{d.i.b.}{d.o.b.}$ ratio is .90), that
measured above a 1/2-foot stump whose d.o.b. is as
greater than d.b.h., that it has been decided to a
2 inches, and that lengths, heights, and height-ac
expressed as number of 4-foot sections.

The heights to regularly diminishing diamet
with five 16-foot logs (or twenty 4-foot sections)
chantability might be estimated as in the boxed pc
tabulation:

	d.o.b. (inches)	L	H
		Number of	4-foot sectic
	2	0	20
	4	0	20
	6	0	20
	8	0	20
merch. top	10	4	20 = Σ L
	12	2	16
	14	5	14
	16	6	9
	18	2	3
d.b.h.	20	1	1
		20 = Σ L:	143 = Σ H:

Note that there __must__ be an H and an H' for
step in a complete progression between (but exclud
even though no merchantable material is found outs
the merchantable top d.o.b. of 10-inches. Merchar
repeated for each progressive diameter less than t

- 4 -

however. Actually, if the complete H column (commencing with the H corresponding to d.b.h. and including repeated heights outside the box) is entered in a 2-register bookkeeping machine which can transfer subtotals from first to second register without clearing, then Σ H and Σ H' can be automatically and simultaneously computed, and Σ L will be the last entry in the H column. Some people may prefer to invert the column of diameters so that the corresponding H and H' columns progress downward as on the bookkeeping machine tape, but this is merely a matter of personal preference, as long as the arithmetic accumulation proceeds in a direction which is up the tree.

Thus, Σ H' = 603, Σ H = 143, and Σ L = 20 for a 20-inch five-log tree described in terms of H as 20: 1: 3: 9: 14: 16: 20, with 4 more 20's needed to complete the diminishing diameter progression. Appendix A gives coefficients A, B, C appropriate for these accumulations where unit-height is 4 feet, where taper-step is 2 inches, and where mean $\frac{d.i.b.}{d.o.b.}$ ratio is .90. The appropriate A, B, C coefficients convert the accumulations into surface or cubic volume (inside or outside bark) and into board-foot volume according to International 1/4-inch log rule, Scribner formula log rule, or Doyle log rule. This is demonstrated below for surface, cubic volume, and International 1/4-inch volume:

Surface inside bark = 1.88(143) + .942(20) = 288 sq. ft. i.b.

Cubic vol. inside bark = .141(603) + .0236(20) = 85.5 cu. ft. i.b.

Int. 1/4-inch vol. = 1.29(603) - 1.34(143) - .284(20) = 580 bd. ft.

Although the simple tally and bookkeeping-machine technique just outlined is well adapted to many situations, a more useful method involves punch-card breakdown of the tree into sections classified by grade (or utility) and d.o.b. Lengths of such sections can then be machine-sorted and accumulated by grade-diameter classes, and need not be converted to volume till the last step. If mark-sensing is employed, more than one mark-sensed card may be needed per tree, depending on how much information is desired. It is assumed that plot information will be recorded on a separate mark-sensed card. In addition to plot number, species, defect class, and priority class, only d.b.h. and the graded length of each taper-step need be recorded in the field on the mark-sensed tree-card. For the tree used in the preceding example, the latter information would be tallied as d.b.h. and successive L's with their grades A, B, C, D, etc., thus:

20: 1A: 2A: 6B: 5B: 2C: 4C

In actual practice, of course, grades would be coded numerically.

This would require only 14 of the 27 columns on a mark-sensed card. Occasionally it might be more convenient for the estimator to record H instead of L, but ordinarily this would waste mark-sensed columns and would require differencing later, if a breakdown by grades

is desired. The 27-column mark-sensed cards will later be
punched, and in some cases may suffice for volume calculat:
junction with a 3-register bookkeeping machine. Generally,
will be used to reproduce a card for each tree section, so
easily be sorted and accumulated, after which $\Sigma L, \Sigma H$, and
directly obtained by the progressive totalling process on :
machine such as the IBM 402 or 403.

To illustrate quite simply how the mechanics of the
recombining procedure might work, suppose that (in addition
inch tree described above) there were also an 18-inch tree,
and graded lengths (L) recorded as 18: 1A: 3B: 4B: 4C: 3D:
cards would be punched for each length in each tree, with a
master card in each summarized d.o.b.-grade group to facil:
sive totalling. A series of machine sorts and tabulations
each d.o.b. class) would break down the trees into their el
would recombine and summarize them thus:

d.o.b. of sections

	20	18	16	14	12	10	8	6	4	2
Tree #1	1A	2A	6B	5B	2C	4C	0	0	0	0
Tree #2		1A	3B	4B	4C	3D	0	0	0	0
Total	1	3	9	9	6	7	0	0	0	0
A	1	3	0	0	0	0	0	0	0	0
B			9	9	0	0	0	0	0	0
C					6	4	0	0	0	0
D						3	0	0	0	0
Total	1	3	9	9	6	7	0	0	0	0

The individual tree sums and progressive sums are n(
to illustrate that they agree with the later partition by ɛ

The $\Sigma H'$, ΣH, and ΣL for each grade can be converte
using the A, B, C coefficients in Appendix A, if an approp:
ratio is chosen. Although for most purposes this ratio is
estimated as the sample-based ratio $\sqrt{\dfrac{\text{volume i.b.}}{\text{volume o.b.}}}$ with li:
tion between tabled values of A, B, C, occasionally it may
able to sample both d.i.b. and d.o.b. at regular intervals
to derive two ratios: $\dfrac{\Sigma(\text{d.i.b.})^2}{\Sigma(\text{d.o.b.})^2}$ and $\dfrac{\Sigma(\text{d.i.b.})}{\Sigma(\text{d.o.b.})}$. These

applied to outside-bark volume and surface formulae respectively, and the resultant formulae should be multiplied by the log rule formulae given in Appendix E to derive A, B, C coefficients more appropriate than those obtained by linear interpolation from Appendix A. Where volume or surface including bark is desired, the coefficients given for $\frac{d.i.b.}{d.o.b.}$ ratio = 1.00 are appropriate without any adjustment.

If it is desired to break down volume or surface summaries by d.o.b. class, first number each d.o.b. class in the complete series consecutively from the smallest to the largest (beginning with number 1), then accumulate the ordinals progressively from the beginning, then multiply each length (L) by its corresponding ordinal (I), or its ordinal accumulation (II), thus:

d.o.b. (inches)	(I) Numbered consecutively	(II) Accumulated progressively	L	(I)(L) H	(II)(L) H'
2	1	1	0		
4	2	3	0		
6	3	6	0		
8	4	10	0		
10	5	15	7	35	105
12	6	21	6	36	126
14	7	28	9	63	252
16	8	36	9	72	324
18	9	45	3	27	135
20	10	55	1	10	55
			$35 = \Sigma L$	$243 = \Sigma H$	$997 = \Sigma H'$

Each L, when multiplied by its corresponding entry in column I, will give the appropriate H for that d.o.b., and when multiplied by its corresponding entry in column II will give the appropriate H'. Of course, cumulative volumes of material larger than any specified d.o.b. could be obtained easily by merely assigning zeros in the L column for all smaller d.o.b.'s, and then getting ΣL, ΣH, $\Sigma H'$ in the usual fashion. Volumes or surfaces in any desired diameter class or group of diameter classes are then obtained by multiplying H', H, L or their desired sums by the A, B, C coefficients from Appendix A in the usual fashion.

One shortcut might be mentioned which frequently may furnish an acceptable approximation. Where the average $\frac{d.i.b.}{d.o.b.}$ ratio is exactly .932 (or where $\frac{volume\ inside\ bark}{volume\ outside\ bark}$ ratio is .869), if taper-step of 2 inches and unit-height of 4 feet have been adopted, the board-foot volume by the International 1/4-inch log rule can be easily calculated, thus:

$$\text{Board feet (Int. 1/4")} = (1.38)\left(\Sigma H' - \Sigma H - \frac{\Sigma L}{5}\right)$$

$$(1.38)\left(603 - 143 - \frac{20}{5}\right) = (1.38)(456) = 629 \text{ b}$$

There are three general situations in which sligh
ariations of the height-accumulation technique may be f

(1) Where it is desired to grade different porti
ample tree, the entire job should be done on punched ca
ith the field tally made directly on mark-sensed cards.
e graded and described in terms of d.b.h. and a series
rather than H's), thus:

$$18: 1A: 3B: 4B: 4C: 3D$$

eparate cards should be reproduced for each d.o.b. star
18 inches, and after desired sorts are made, an IBM 40
ng machine will accumulate ΣL, ΣH, $\Sigma H'$ for each clas
s desired, provided that $L = 0$ is supplied for any d.o.
he complete progression.

(2) Where it is not desired to grade portions of
n entire tree can be punched on an individual card, wit
ecorded under its appropriate d.o.b. field starting wit
tarting with 18 inches for tree #2 in the tabulation on

1: 4: 8: 12: 15: (with 15: 15: 15: 15 completing th

or 1: 3: 4: 4: 3: (with 0: 0: 0: 0 completing the L

ach class for which a separate volume is desired will h
ummarized for each d.o.b. field occurring in it. The H
including repeat heights or zeros needed to complete th
or each class can then be entered in a 2- (or 3-) regis
achine able to transfer subtotals from register to regi
learing. Examples of suitable 2-register bookkeeping m
he H's can be entered are the Burroughs Sensimatic F-50
ash Register 30210 (1-6). Examples of suitable multipl
nes in which the L's can be entered are Burroughs Sensi
ational Cash Register 30412 (17) or 3100, any of which
he 3 registers needed.

(3) Where it is not desired to use punched cards
or each sample tree will consist of its d.b.h. followed
eights, thus:

$$18: 1: 4: 8: 12: 15$$

ith a taper-step (T) = 2 inches and d.b.h. of 18 inches
$\frac{.b.h.}{T}$ = 9 entries, so 1, 4, 8, 12, and 15, 15, 15, 15,

entered in a 2-register bookkeeping machine to obtain ΣH and $\Sigma H'$ for the individual tree; ΣL would, of course, be equal to the last entry (15).

In all three situations, it will usually be found desirable to employ the volume sampling techniques described above merely to get volume/basal area ratios which are applicable to a larger sample of basal area which has been stratified by species and height. A brief discussion of this system may be found on page 7 of Shortcuts for Cruisers and Scalers (12).

All the examples above have employed a taper-step of 2 inches and unit-height of 4 feet because such interval and unit are most convenient for United States measure, with stump height occurring 1 unit below breast height. However, Appendix A also gives A, B, C coefficients for taper-step of 1 inch and unit-height of 1 foot. Such interval and unit are recommended only with very accurate dendrometers or tree diagrams where unusually precise measures are desired; 4 rather than 1 unit-heights will occur between d.b.h. and stump-height. Also, Appendix A gives A, B, C coefficients for the most convenient interval and unit in the metric system: taper-step of 5 centimeters and unit-height of 1 meter. Assuming 1 unit-height below European breast-height of 1.3 meters would imply a stump height of about 30 centimeters. Of course, the retention of decimal fractions of a meter in height measurements would not change the coefficients. No board-foot volume coefficients were provided for metric intervals, since such units are not popular in countries using the metric system.

A, B, C coefficients for intervals and units other than those tabulated may be derived from the basic formulae given in Appendix E.

Existing tables of volume (cubic-foot or b
factory in that none of them gives volumes for ev
length and diameter (measured in fractional inche
in practice. To overcome this drawback without i
excessively, an accumulative table is needed. Th
values in Appendix B ab initio, retaining 6 to 9
the final rounding to 5 digits. All columns have
tion formulae involving the sums of numbers in ar
the sums of their squares, and the sums of their
International log rule with 1/4-inch kerf are bas
published by H. H. Chapman (7), i.e., board feet
for a 4-foot section, with taper allowance of 1/2
To minimize accumulative and rounding error, howe
this to a form in which taper was an implicit joi
and diameter (12).

In effect, Appendix B is a huge upright pe
ing on a 50-inch base and tapering upward for 400
1/8-inch per foot. To use it, enter at the desir
mark the surface or volume found opposite this di
column a distance equal to the desired log length
volume found there, and subtract the marked value

As an example, the board-foot volume of a
diameter and 81 feet long would be:

$$
\begin{array}{r}
8396 \\
- 1447 \\
\hline
6949
\end{array}
$$ board feet (Int. 1/4-

Cubic volume of this log would be

$$
\begin{array}{r}
1004.1 \\
- 171.6 \\
\hline
832.5
\end{array}
$$ cubic feet

Surface of this log would be:

$$
\begin{array}{r}
1085.9 \\
- 167.4 \\
\hline
918.5
\end{array}
$$ square feet

Of course it is desirable to measure more
in long logs or trees, but occasionally that is a
to obtain, and a taper-assumption of 1/8-inch per
amiss.

A particularly handy use of the giant-tree
d.i.b. at the top of the first 16-foot log (or ot

measured or estimated. The extra column of tabular length 0 - 80 feet
at the end of Appendix B can be cut out and used as a sliding scale with
the table. If it is laid over the regular length column (positioned
between the d.i.b. column and the surface column) with 16 feet on the
sliding scale opposite the measured d.i.b., the volume between any specified heights above stump on the tree may be computed by subtracting the
volumes which appear opposite these heights. As an example, if d.i.b.
at the top of the first 16-foot log is 17-1/8 inches, the diameter,
height, and cubic volume readings will look as follows after the sliding
length (or height) scale is positioned:

d.i.b.	Sliding height scale	Cubic volume
Inches	Feet	Cubic feet
	etc.	etc.
	19	1 749.7
	18	1 748.2
17	17	1 746.6
Set 16 feet at 17-1/8 inches ──→ ←──	16	1 745.0
	15	1 743.4
	14	1 741.8
	13	1 740.1
	12	1 738.4
	11	1 736.7
	10	1 735.0
18	9	1 733.2
	8	1 731.4
	7	1 729.6
	6	1 727.8
	5	1 726.0
	4	1 724.1
	3	1 722.2
	2	1 720.2
19	1	1 718.3
	0	1 716.3

The volume in the 0 - 8 foot portion of the tree would be:

$$\begin{array}{r} 1\ 731.4 \\ -\ 1\ 716.3 \\ \hline 15.1 \text{ cubic feet} \end{array}$$

while the volume in the 9 - 18 foot portion of the tree would be:

$$\begin{array}{r} 1\ 748.2 \\ -\ 1\ 731.4 \\ \hline 16.8 \text{ cubic feet} \end{array}$$

Higher portions not shown could be handled similarly. Grades or use-
classes could be assigned in the field to sections between various tree
heights, and the actual differencing could be done later on bookkeeping
machines or punched-card machines.

When used as above, the giant-tree table has most of the
tages of form-class volume tables with arbitrary upper-log tape
only one table is needed, and that table is in a form facilitat
breakdown of the tree into portions of varying lengths for each
lengths both cubic and board foot volumes can be easily compute
cubic volume, surface, and length of portions of a tree are kno
can be converted to International 1/4-inch, Scribner, or Doyle
by coefficients in Appendix E.

It is obvious that results will not be as accurate as th
determined earlier by the new height-accumulation method, becau
giant-tree table assumes taper of 1/8 inch per foot between dia
measurements, whereas the height-accumulation method utilizes a
taper. Appendix B is very handy, however, in that it is equall
to get surface or volume for a log or tree of any length, by en
with a single diameter (to the nearlest 1/8 inch) at any specif
point on the log or tree (small end, large end, middle, or any
else). A 1/8-inch (rather than a 1/10-inch) diameter interval
implicit in International 1/4-inch log rule assumptions where l
are desired in whole feet.

TREE TAPER AND SHAPE

When viewed from the butt, the taper of a tree can be defined as its loss in diameter divided by the length affected. The shape of different portions of a tree can be visualized as a function of taper. Where taper tends to increase, tree shape resembles a paraboloid; where it tends to remain constant, tree shape resembles a conoid; where it tends to decrease, tree shape resembles a neiloid. The corresponding profiles of the surfaces of these tree shapes viewed from outside the tree would be convex, linear, and concave.

In general, trees tend to be neiloidal till butt swell disappears, then conoidal, and finally paraboloidal in the upper portions. This fact is helpful where d.b.h., merchantable height, and merchantable top (d.o.b.) have been measured, and where it is desired to distribute taper between d.b.h. and top d.o.b. by eye for use in the height-accumulation technique. Suppose a tree 18 inches in d.b.h. has 16 four-foot sections from a 1/2-foot stump to an 8-inch merchantable top d.o.b. It can usually be arbitrarily assumed that there is a two-inch diameter increase and one four-foot section below d.b.h. This would leave 15 sections above d.b.h. to be distributed among $\frac{18-8}{2}$ = 5 two-inch taper-steps. An absolutely conoidal taper (except for the neiloidal butt swell) would be recorded as:

5 taper-steps above d.b.h.
containing 15 four-foot sections

$$1 : 3 : 3 : 3 : 3 : 3$$

However, lacking any intermediate measures, the eye may still detect a convex (or paraboloidal) tendency near the top, and a better estimate of composite tree shape would be:

5 taper-steps above d.b.h.
containing 15 four-foot sections

$$1 : 4 : 4 : 3 : 2 : 2$$

When only d.b.h. and merchantable height are estimated, and no information is available on top d.o.b. or shape above d.b.h., a crude conoidal assumption has often been used in volume-table work (especially with hardwoods) with reasonably satisfactory results. The International log rule and the giant-tree table discussed earlier are examples of such an assumption (2 inches of taper per 16 feet). If a taper assumption of slightly less than 2 inches per 16 feet were deemed adequate for the hypothetical tree above, the estimate of tree shape for use in the height-accumulation technique would be:

use four units of length per taper-step,
until total 15 units is exhausted.

$$1 : 4 : 4 : 4 : 3$$

ate size is contained in the butt log. However, this butt-log fc
class is usually either estimated or assumed, as is shape of the
log and taper above the first log. If d.o.b. of the butt log is
measured in several places, and if $\frac{d.i.b.}{d.o.b.}$ ratios for a species gr
not vary excessively, the height-accumulation technique will perm
more accurate description of both the lower and upper portions of
than will the use of form-class volume tables which assume arbitr
butt-log shape and upper-log tapers. In addition, the height-acc
technique is much better adapted to grading portions of the tree
business-machine compilation.

If the effort involved in getting objective height-accumul
measurements is deemed excessive, the height-accumulation techniq
can be adapted to the same sort of eye-estimates and assumptions
are involved in the usual use of butt-log form-class volume table
Below is an illustration showing how height-accumulation informat
about tree butts implicitly includes eye-estimated butt-log form-
information, plus additional shape and taper information. For cc
venience, a taper-step of 2 inches, a unit-height of 4 feet, and a
$\frac{d.i.b.}{d.o.b.}$ ratio of .90 have been employed; a different $\frac{d.i.b.}{d.o.b.}$ relat
ship would, of course, lead to different form-class values.

Lengths (number of 4-ft. sections per 2 inches of taper)	Taper from d.b.h. to d.o.b. at top of first 16-ft. log		Implied butt-log form-class of various d.b.h.'s			
	- - Inches - -		10 ins. d.b.h.	20 ins. d.b.h.	30 ins. d.b.h.	40 ins. d.b.h.
			- - - - - Percent - - - - -			
1:7	.9		82	86	87	88
1:6	1.0	(1)	81	86	87	88
1:5	1.2		79	85	86	87
1:4	1.5		76	83	86	87
1:3	2.0	(2)	72	81	84	86
1:2:5	2.4		68	79	83	85
1:2:4	2.5		68	79	83	84
1:2:3 or 1:1:6	2.7		66	78	82	84
1:1:5	2.8		65	77	82	84
1:2:2 or 1:1:4	3.0	(3)	63	76	81	83
1:1:3	3.3		60	75	80	82
1:2:1 or 1:1:2	4.0	(4)	54	72	78	81
1:1:1:5	4.4		50	70	77	80
1:1:1:4	4.5		50	70	76	80
1:1:1:3	4.7			69	76	80
1:1:1:2	5.0	(5)		68	75	79
1:1:1:1	6.0	(6)		63	72	76
1:0:1:1:2	7.0	(7)		58	69	74
1:0:1:1:1	8.0	(8)		54	66	72
1:0:0:1:1:2	9.0	(9)		50	63	70
1:0:0:1:1:1	10.0	(10)			60	68

Any forester who prides himself on being able to estimate butt-log form-class can readily adapt his talent to the height-accumulation techniques; he can also check his eye-estimates readily with one of the dendrometers discussed earlier. Pole calipers or tapes of various kinds (8)(9) are already in existence, and are well adapted to checking tapers on the most important portion of a tree--its butt log. Probably the most convenient hand-held dendrometer for use on upper as well as lower portions of the tree would be a slope-rotated wedge-prism (3) on which height or slope can be read. A horizontal target of adjustable width attached to d.b.h. will allow the observer to position himself or adjust his instrument so that the split-image of the target just fails to overlap. He can then slowly raise his line of sight up the tree trunk, and the height where split-image of the trunk just fails to overlap will be where d.o.b. equals the desired width for which the target has been adjusted.

It will be found convenient to have the hypsometer scale graduated to read height in number of four-foot sections with a 40-foot base-line assumed, and to have the rotating wedge-prism manufactured or adjusted to effect a maximum deviation of 143.25 minutes (equivalent to about 4.167 prism-diopters or enough to exactly juxtapose the direct and the split image of a 20-inch horizontal target at a distance of 40 feet).

In using such an instrument with a tree 20 incl
flat terrain, the observer would successively occupy :
28, 24, 20 feet, etc., distance from tree center and
at each point the height at which the split image of
pulled apart, or separated. In each case, this figur
read from the hypsometer scale multiplied by an appro
such as 1.0, .9, .8, .7, .6, .5, etc. (computed as

$$\frac{\text{actual distance to d.b.h.}}{\text{base for which hypsometer was graduated}}\Big).$$

The reader can easily infer for himself modifi
technique necessary where trees are more or less than
d.b.h., where the terrain slopes, or where trees lean.

Technicians studying the effect of erroneous a
taper or shape may be interested in Appendix D, which
izing how scaled volume estimates employing erroneous
tions may be improved by shortening the interval betw
diameters.

Literature Cited

(1) Bitterlich, W.
  1949. Das relaskop. Allgemeine Forst- und Holzwirtschaftliche
    Zeitung 60 (5/6): 41-42.

(2) _____.
  1952. Das spiegel-relaskop. Oesterreichs Forst- und
    Holzwirtschaft 7(1): 3-7.

(3) Bruce, David
  1955. A new way to look at trees. Jour. Forestry 53:
    (in press).

(4) Burton, R. G.
  1906. The Biltmore pachymeter. Forestry Quarterly 4:8-9.

(5) Clark, J. F.
  1906. The measurement of sawlogs. Forestry Quarterly 4: 79-
    91.

(6) _____
  1913. A new dendrometer or timber scale. Forestry Quarterly
    11: 467-469.

(7) Chapman, H. C.
  1921. Forest mensuration (page 493). John Wiley and Sons,
    Inc., New York City. 553 pages.

(8) Ferree, M. J.
  1946. The pole caliper. Jour. Forestry 44: 594-595.

(9) Godman, R. M.
  1949. The pole diameter tape. Jour. Forestry 47: 568-569.

(10) Grosenbaugh, L. R.
  1948. RS-SS, MENSURATION, Tree studies, volume calculations.
    Mimeographed memorandum report dated August 25, 1948,
    with Appendices A through F. U. S. Forest Service,
    South. Forest Expt. Sta. 24 pages.

(11) _____
  1948. Forest parameters and their statistical estimation.
    Proceedings of the Auburn Conference on Statistics
    Applied to Research in the Social Sciences, Plant
    Sciences, and Animal Sciences. Sept. 7-9, 1948.
    Statistical Laboratory, Alabama Polytechnic Institute,
    Auburn, Alabama. Abstract of paper on page 46.

(12) _____
  1952. Shortcuts for cruisers and scalers. U. S. Forest Service,
    South. Forest Expt. Sta. Occas. Paper 126. 24 pages.

(BLANK)

Height-accumulation coefficients converting $A\Sigma H' + B\Sigma H + C\Sigma L$ to surface or volume under various assumptions as to taper-step, unit-height, and mean ratio: d.i.b./d.o.b.

Given:
Taper-step = 2 inches
Unit-height = 4 feet

Surface coefficients for square feet

Mean ratio: d.i.b. d.o.b.	A	B	C
1.00	0	2.09	1.047
.95	0	1.99	.995
.90	0	1.88	.942
.85	0	1.78	.890

Volume coefficients for cubic feet

Mean ratio: d.i.b. d.o.b.	A	B	C
1.00	.175	0	.0291
.95	.158	0	.0263
.90	.141	0	.0236
.85	.126	0	.0210

Volume coefficients for board feet (International 1/4")

Mean ratio: d.i.b. d.o.b.	A	B	C
1.00	1.59	-1.48	-.308
.95	1.44	-1.41	-.296
.90	1.29	-1.34	-.284
.85	1.15	-1.26	-.270

Volume coefficients for board feet (Scribner)

Mean ratio: d.i.b. d.o.b.	A	B	C
1.00	1.58	-1.78	-1.06
.95	1.43	-1.69	-1.06
.90	1.28	-1.61	-1.04
.85	1.14	-1.52	-1.02

Volume coefficients for board feet (Doyle)

Mean ratio: d.i.b. d.o.b.	A	B	C
1.00	2.00	-5.00	4.00
.95	1.50	-4.75	4.09
.90	1.42	-4.50	4.19
.85	1.44	-4.25	4.28

Given:
Taper-step = 5 centimeters
Unit-height = 1 meter

Surface coefficients for square meters

Mean ratio: d.i.b. d.o.b.	A	B	C
1.00	0	.157	.0785
.95	0	.149	.0746
.90	0	.141	.0707
.85	0	.134	.0668

Volume coefficients for cubic meters

Mean ratio: d.i.b. d.o.b.	A	B	C
1.00	.00393	0	.000664
.95	.00354	0	.000591
.90	.00318	0	.000530
.85	.00284	0	.000473

Metric equivalents:

1 centimeter = .393700 inches
1 meter = 3.28083 feet
1 sq. meter = 10.7639 sq. feet
1 cu. meter = 35.3145 cu. feet

Given:
Taper-step = 1 inch
Unit-height = 1 foot

Surface coefficients for square feet

Mean ratio: d.i.b. d.o.b.	A	B	C
1.00	0	.262	.131
.95	0	.249	.124
.90	0	.236	.118
.85	0	.223	.111

Volume coefficients for cubic feet

Mean ratio: d.i.b. d.o.b.	A	B	C
1.00	.01091	0	.00182
.95	.00984	0	.00164
.90	.00884	0	.00147
.85	.00788	0	.00131

Volume coefficients for board feet (International 1/4")

Mean ratio: d.i.b. d.o.b.	A	B	C
1.00	.0995	-.185	-.0339
.95	.0898	-.176	-.0309
.90	.0806	-.167	-.0278
.85	.0719	-.158	-.0246

Volume coefficients for board feet (Scribner)

Mean ratio: d.i.b. d.o.b.	A	B	C
1.00	.0988	-.223	-.207
.95	.0892	-.212	-.203
.90	.0800	-.201	-.199
.85	.0714	-.190	-.195

Volume coefficients for board feet (Doyle)

Mean ratio: d.i.b. d.o.b.	A	B	C
1.00	.125	-.625	.125
.95	.113	-.594	.126
.90	.101	-.562	.128
.85	.090	-.531	.129

N. B. Explanation of terms ("taper-step," "unit-height," "mean ratio d.i.b./d.o.b.") and symbols ($\Sigma H'$, ΣH, ΣL) is given on page 4.

Other information on use of tables can be found on pages 6 and 9.

Giant-tree table: Diameter inside bark (taper: 1/8 inch per foot) with cumulative length, surface, and volume

D.I.B. (inches)	Length	Surface	Cubic volume	Int.1/4" volume
	Feet	Square feet	Cubic feet	Board feet
0	400	2 618.0	1 818.1	
1/8	399	2 618.0	1 818.1	
2/8	398	2 618.0	1 818.1	
3/8	397	2 617.9	1 818.1	
4/8	396	2 617.8	1 818.0	
5/8	395	2 617.6	1 818.0	
6/8	394	2 617.4	1 818.0	
7/8	393	2 617.2	1 818.0	
1	392	2 617.0	1 818.0	
	391	2 616.7	1 818.0	
	390	2 616.4	1 818.0	
	389	2 616.0	1 818.0	
	388	2 615.7	1 818.0	
	387	2 615.3	1 818.0	
	386	2 614.8	1 818.0	
	385	2 614.3	1 818.0	
2	384	2 613.8	1 817.9	
	383	2 613.3	1 817.9	
	382	2 612.7	1 817.9	
	381	2 612.1	1 817.9	
	380	2 611.5	1 817.8	
	379	2 610.8	1 817.8	
	378	2 610.1	1 817.7	
	377	2 609.4	1 817.7	
3	376	2 608.6	1 817.7	--
	375	2 607.8	1 817.6	--
	374	2 607.0	1 817.6	14 752
	373	2 606.1	1 817.5	14 752
	372	2 605.2	1 817.4	14 752
	371	2 604.3	1 817.4	14 752
	370	2 603.3	1 817.3	14 752
	369	2 602.3	1 817.2	14 752
4	368	2 601.3	1 817.1	14 751
	367	2 600.2	1 817.0	14 751
	366	2 599.1	1 816.9	14 751
	365	2 598.0	1 816.8	14 751
	364	2 596.8	1 816.7	14 751
	363	2 595.6	1 816.6	14 751
	362	2 594.4	1 816.5	14 750
	361	2 593.1	1 816.4	14 750
5	360	2 591.8	1 816.2	14 750
	359	2 590.5	1 816.1	14 749
	358	2 589.2	1 815.9	14 749
	357	2 587.8	1 815.8	14 748
	356	2 586.4	1 815.6	14 748
	355	2 584.9	1 815.5	14 747
	354	2 583.4	1 815.3	14 747
	353	2 581.9	1 815.1	14 746
6	352	2 580.3	1 814.9	14 745
	351	2 578.7	1 814.7	14 745
	350	2 577.1	1 814.5	14 744
	349	2 575.5	1 814.3	14 743
	348	2 573.8	1 814.1	14 742
	347	2 572.1	1 813.8	14 741
	346	2 570.3	1 813.6	14 740
	345	2 568.5	1 813.3	14 739
7	344	2 566.7	1 813.1	14 738
	343	2 564.9	1 812.8	14 737
	342	2 563.0	1 812.5	14 735
	341	2 561.1	1 812.2	14 734
	340	2 559.1	1 811.9	14 733
	339	2 557.1	1 811.6	14 731
	338	2 555.1	1 811.3	14 730
	337	2 553.1	1 810.9	14 728
8	336	2 551.0	1 810.6	14 726
	335	2 548.9	1 810.2	14 724
	334	2 546.8	1 809.9	14 723
	333	2 544.6	1 809.5	14 721
	332	2 542.4	1 809.1	14 719
	331	2 540.1	1 808.7	14 717
	330	2 537.9	1 808.3	14 714
	329	2 535.5	1 807.9	14 712
9	328	2 533.2	1 807.4	14 710
	327	2 530.8	1 807.0	14 707
	326	2 528.4	1 806.5	14 705
	325	2 526.0	1 806.1	14 702
	324	2 523.5	1 805.6	14 699
	323	2 521.0	1 805.1	14 697
	322	2 518.5	1 804.6	14 694
	321	2 515.9	1 804.0	14 691
10	320	2 513.3	1 803.5	14 688
10				
1/8	319	2 510.7	1 803.0	14 684
2/8	318	2 508.0	1 802.4	14 681
3/8	317	2 505.3	1 801.8	14 678
4/8	316	2 502.6	1 801.2	14 674
5/8	315	2 499.8	1 800.6	14 670
6/8	314	2 497.0	1 800.0	14 667
7/8	313	2 494.2	1 799.3	14 663
11	312	2 491.3	1 798.7	14 659
	311	2 488.4	1 798.0	14 655
	310	2 485.5	1 797.3	14 651
	309	2 482.5	1 796.6	14 646
	308	2 479.5	1 795.9	14 642
	307	2 476.5	1 795.2	14 637
	306	2 473.4	1 794.5	14 633
	305	2 470.4	1 793.7	14 628
12	304	2 467.2	1 792.9	14 623
	303	2 464.1	1 792.1	14 618
	302	2 460.9	1 791.3	14 613
	301	2 457.7	1 790.5	14 607
	300	2 454.4	1 789.6	14 602
	299	2 451.1	1 788.8	14 596
	298	2 447.8	1 787.9	14 591
	297	2 444.4	1 787.0	14 585
13	296	2 441.1	1 786.1	14 579
	295	2 437.6	1 785.2	14 573
	294	2 434.2	1 784.2	14 567
	293	2 430.7	1 783.3	14 560
	292	2 427.2	1 782.3	14 554
	291	2 423.6	1 781.3	14 547
	290	2 420.0	1 780.2	14 540
	289	2 416.4	1 779.2	14 533
14	288	2 412.8	1 778.1	14 526
	287	2 409.1	1 777.1	14 519
	286	2 405.4	1 776.0	14 511
	285	2 401.6	1 774.8	14 504
	284	2 397.9	1 773.7	14 496
	283	2 394.0	1 772.6	14 488
	282	2 390.2	1 771.4	14 480
	281	2 386.3	1 770.2	14 472
15	280	2 382.4	1 769.0	14 463
	279	2 378.5	1 767.7	14 455
	278	2 374.5	1 766.5	14 446
	277	2 370.5	1 765.2	14 437
	276	2 366.4	1 763.9	14 428
	275	2 362.4	1 762.6	14 419
	274	2 358.3	1 761.2	14 410
	273	2 354.1	1 759.9	14 400
16	272	2 349.9	1 758.5	14 390
	271	2 345.7	1 757.1	14 381
	270	2 341.5	1 755.6	14 370
	269	2 337.2	1 754.2	14 360
	268	2 332.9	1 752.7	14 350
	267	2 328.6	1 751.2	14 339
	266	2 324.2	1 749.7	14 328
	265	2 319.8	1 748.2	14 317
17	264	2 315.4	1 746.6	14 306
	263	2 310.9	1 745.0	14 295
	262	2 306.4	1 743.4	14 283
	261	2 301.9	1 741.8	14 271
	260	2 297.3	1 740.1	14 260
	259	2 292.7	1 738.4	14 247
	258	2 288.1	1 736.7	14 235
	257	2 283.4	1 735.0	14 223
18	256	2 278.7	1 733.2	14 210
	255	2 274.0	1 731.4	14 197
	254	2 269.2	1 729.6	14 184
	253	2 264.4	1 727.8	14 170
	252	2 259.6	1 726.0	14 157
	251	2 254.8	1 724.1	14 143
	250	2 249.9	1 722.2	14 129
	249	2 244.9	1 720.2	14 115
19	248	2 240.0	1 718.3	14 101
	247	2 235.0	1 716.3	14 086
	246	2 230.0	1 714.3	14 071
	245	2 224.9	1 712.3	14 056
	244	2 219.8	1 710.2	14 041
	243	2 214.7	1 708.1	14 026
	242	2 209.6	1 706.0	14 010
	241	2 204.4	1 703.9	13 994
20	240	2 199.1	1 701.7	13 978

Giant-tree table: Diameter inside bark (taper: 1/8 inch per foot) with cumulative length, surface, and volume (cont'd)

D.I.B. (inches)	Length	Surface	Cubic volume	Int.1/4" volume	D.I.B. (inches)	Length	Surface	Cubic volume	Int.1/4" volume
	Feet	Square feet	Cubic feet	Board feet		Feet	Square feet	Cubic feet	Board feet
20					30				
1/8	239	2 193.9	1 699.5	13 961	1/8	159	1 667.7	1 420.4	11 785
2/8	238	2 188.6	1 697.3	13 945	2/8	158	1 659.8	1 415.5	11 745
3/8	237	2 183.3	1 695.0	13 928	3/8	157	1 651.8	1 410.4	11 705
4/8	236	2 177.9	1 692.7	13 911	4/8	156	1 643.9	1 405.4	11 664
5/8	235	2 172.6	1 690.4	13 894	5/8	155	1 635.9	1 400.3	11 623
6/8	234	2 167.1	1 688.1	13 876	6/8	154	1 627.8	1 395.2	11 582
7/8	233	2 161.7	1 685.7	13 859	7/8	153	1 619.8	1 390.0	11 541
21	232	2 156.2	1 683.4	13 841	31	152	1 611.7	1 384.8	11 499
	231	2 150.7	1 680.9	13 822		151	1 603.5	1 379.5	11 456
	230	2 145.1	1 678.5	13 804		150	1 595.4	1 374.2	11 414
	229	2 139.6	1 676.0	13 785		149	1 587.2	1 368.8	11 371
	228	2 134.0	1 673.5	13 766		148	1 578.9	1 363.5	11 327
	227	2 128.3	1 671.0	13 747		147	1 570.7	1 358.0	11 284
	226	2 122.6	1 668.4	13 728		146	1 562.4	1 352.5	11 239
	225	2 116.9	1 665.8	13 708		145	1 554.0	1 347.0	11 195
22	224	2 111.2	1 663.2	13 688	32	144	1 545.7	1 341.5	11 150
	223	2 105.4	1 660.5	13 668		143	1 537.3	1 335.9	11 105
	222	2 099.6	1 657.8	13 648		142	1 528.9	1 330.2	11 059
	221	2 093.8	1 655.1	13 627		141	1 520.4	1 324.5	11 013
	220	2 087.9	1 652.4	13 606		140	1 511.9	1 318.8	10 967
	219	2 082.0	1 649.6	13 585		139	1 503.4	1 313.0	10 920
	218	2 076.0	1 646.8	13 563		138	1 494.8	1 307.2	10 873
	217	2 070.1	1 644.0	13 542		137	1 486.2	1 301.3	10 825
	216	2 064.1	1 641.1	13 520		136	1 477.6	1 295.4	10 777
23	215	2 058.0	1 638.2	13 497	33	135	1 469.0	1 289.4	10 729
	214	2 051.9	1 635.3	13 475		134	1 460.3	1 283.4	10 680
	213	2 045.8	1 632.3	13 452		133	1 451.6	1 277.3	10 631
	212	2 039.7	1 629.3	13 429		132	1 442.8	1 271.2	10 582
	211	2 033.5	1 626.3	13 406		131	1 434.0	1 265.1	10 532
	210	2 027.3	1 623.2	13 382		130	1 425.2	1 258.9	10 482
	209	2 021.1	1 620.1	13 358		129	1 416.3	1 252.7	10 431
24	208	2 014.8	1 617.0	13 334	34	128	1 407.5	1 246.4	10 380
	207	2 008.5	1 613.8	13 310		127	1 398.5	1 240.1	10 329
	206	2 002.2	1 610.6	13 285		126	1 389.6	1 233.7	10 277
	205	1 995.8	1 607.4	13 260		125	1 380.6	1 227.3	10 224
	204	1 989.4	1 604.2	13 235		124	1 371.6	1 220.8	10 172
	203	1 983.0	1 600.9	13 209		123	1 362.5	1 214.3	10 119
	202	1 976.5	1 597.5	13 184		122	1 353.5	1 207.7	10 065
	201	1 970.1	1 594.2	13 158		121	1 344.3	1 201.1	10 011
25	200	1 963.5	1 590.8	13 131	35	120	1 335.2	1 194.5	9 957
	199	1 957.0	1 587.4	13 105		119	1 326.0	1 187.8	9 902
	198	1 950.4	1 583.9	13 078		118	1 316.8	1 181.0	9 847
	197	1 943.7	1 580.4	13 050		117	1 307.6	1 174.2	9 792
	196	1 937.1	1 576.9	13 023		116	1 298.3	1 167.4	9 736
	195	1 930.4	1 573.3	12 995		115	1 289.0	1 160.5	9 679
	194	1 923.7	1 569.7	12 967		114	1 279.6	1 153.5	9 622
	193	1 916.9	1 566.1	12 939		113	1 270.3	1 146.5	9 565
26	192	1 910.1	1 562.4	12 910	36	112	1 260.8	1 139.5	9 508
	191	1 903.3	1 558.7	12 881		111	1 251.4	1 132.4	9 450
	190	1 896.4	1 555.0	12 852		110	1 241.9	1 125.2	9 391
	189	1 889.5	1 551.2	12 822		109	1 232.4	1 118.0	9 332
	188	1 882.6	1 547.4	12 792		108	1 222.9	1 110.8	9 273
	187	1 875.7	1 543.5	12 762		107	1 213.3	1 103.5	9 213
	186	1 868.7	1 539.7	12 731		106	1 203.7	1 096.2	9 153
	185	1 861.7	1 535.7	12 700		105	1 194.1	1 088.8	9 092
27	184	1 854.6	1 531.8	12 669	37	104	1 184.4	1 081.3	9 031
	183	1 847.5	1 527.8	12 638		103	1 174.7	1 073.8	8 970
	182	1 840.4	1 523.7	12 606		102	1 165.0	1 066.3	8 908
	181	1 833.3	1 519.7	12 574		101	1 155.2	1 058.7	8 845
	180	1 826.1	1 515.6	12 541		100	1 145.4	1 051.1	8 782
	179	1 818.9	1 511.4	12 509		99	1 135.6	1 043.4	8 719
	178	1 811.6	1 507.2	12 476		98	1 125.7	1 035.6	8 655
	177	1 804.3	1 503.0	12 442		97	1 115.8	1 027.8	8 591
28	176	1 797.0	1 498.8	12 409	38	96	1 105.9	1 020.0	8 526
	175	1 789.7	1 494.5	12 375		95	1 095.9	1 012.1	8 461
	174	1 782.3	1 490.1	12 340		94	1 085.9	1 004.1	8 396
	173	1 774.9	1 485.8	12 306		93	1 075.9	996.1	8 330
	172	1 767.4	1 481.4	12 271		92	1 065.8	988.1	8 263
	171	1 760.0	1 476.9	12 235		91	1 055.7	979.9	8 197
	170	1 752.4	1 472.4	12 200		90	1 045.6	971.8	8 129
	169	1 744.9	1 467.9	12 164		89	1 035.4	963.6	8 061
29	168	1 737.3	1 463.3	12 127	39	88	1 025.2	955.3	7 993
	167	1 729.7	1 458.7	12 091		87	1 015.0	947.0	7 924
	166	1 722.1	1 454.1	12 053		86	1 004.7	938.6	7 855
	165	1 714.4	1 449.4	12 016		85	994.4	930.2	7 786
	164	1 706.7	1 444.7	11 978		84	984.1	921.7	7 715
	163	1 699.0	1 439.9	11 940		83	973.8	913.1	7 645
	162	1 691.2	1 435.1	11 902		82	963.4	904.6	7 574
	161	1 683.4	1 430.2	11 863		81	952.9	895.9	7 502
30	160	1 675.5	1 425.4	11 824	40	80	942.5	887.2	7 430

(BLANK)

Giant-tree table: Diameter inside bark (taper: 1/8 inch per foot)
with cumulative length, surface and volume (cont'd)

D.I.B (inches)	Length	Surface	Cubic volume	Int.1/4" volume
	Feet	Square feet	Cubic feet	Board feet
40				
1/8	79	932.0	878.5	7 358
2/8	78	921.5	869.6	7 285
3/8	77	910.9	860.8	7 211
4/8	76	900.3	851.9	7 137
5/8	75	889.7	842.9	7 063
6/8	74	879.1	833.9	6 988
7/8	73	868.4	824.8	6 913
41	72	857.7	815.6	6 837
	71	846.9	806.4	6 761
	70	836.1	797.2	6 684
	69	825.3	787.9	6 607
	68	814.5	778.5	6 529
	67	803.6	769.1	6 450
	66	792.7	759.6	6 372
	65	781.7	750.1	6 292
42	64	770.7	740.5	6 213
	63	759.7	730.8	6 132
	62	748.7	721.1	6 052
	61	737.6	711.4	5 970
	60	726.5	701.5	5 888
	59	715.4	691.7	5 806
	58	704.2	681.7	5 723
	57	693.0	671.7	5 640
43	56	681.7	661.7	5 556
	55	670.5	651.6	5 472
	54	659.2	641.4	5 387
	53	647.8	631.2	5 302
	52	636.4	620.9	5 216
	51	625.0	610.5	5 129
	50	613.6	600.1	5 043
	49	602.1	589.6	4 955
44	48	590.6	579.1	4 867
	47	579.1	568.5	4 779
	46	567.5	557.9	4 690
	45	555.9	547.2	4 600
	44	544.3	536.4	4 510
	43	532.6	525.6	4 419
	42	520.9	514.7	4 328
	41	509.2	503.7	4 237
45	40	497.4	492.7	4 144
	39	485.6	481.6	4 052
	38	473.8	470.5	3 958
	37	461.9	459.3	3 865
	36	450.0	448.0	3 770
	35	438.1	436.7	3 675
	34	426.1	425.3	3 580
	33	414.2	413.9	3 484
46	32	402.1	402.4	3 387
	31	390.1	390.8	3 290
	30	378.0	379.1	3 193
	29	365.9	367.5	3 095
	28	353.7	355.7	2 996
	27	341.5	343.9	2 897
	26	329.3	332.0	2 797
	25	317.0	320.0	2 696
47	24	304.7	308.0	2 595
	23	292.4	295.9	2 494
	22	280.1	283.8	2 392
	21	267.7	271.6	2 289
	20	255.3	259.3	2 186
	19	242.8	247.0	2 082
	18	230.3	234.6	1 978
	17	217.8	222.1	1 873
48	16	205.3	209.6	1 767
	15	192.7	197.0	1 661
	14	180.1	184.3	1 554
	13	167.4	171.6	1 447
	12	154.7	158.8	1 339
	11	142.0	145.9	1 231
	10	129.3	133.0	1 122
	9	116.5	120.0	1 012
49	8	103.7	106.9	902
	7	90.8	93.8	792
	6	78.0	80.6	680
	5	65.0	67.3	568
	4	52.1	54.0	456
	3	39.1	40.6	343
	2	26.1	27.1	229
	1	13.1	13.6	115
50	0	.0	.0	0

Feet: 80 79 78 77 76 75 74 73 72 71 70 69 68 67 66 65 64 63 62 61 60 59 58 57 56 55 54 53 52 51 50 49 48 47 46 45 44 43 42 41 40 39 38 37 36 35 34 33 32 31 30 29 28 27 26 25 24 23 22 21 20 19 18 17 16 15 14 13 12 11 10 9 8 7 6 5 4 3 2 1 0

CUT OUT STRIP ALONG DOTTED LINES TO OBTAIN A SLIDING SCALE WHICH CAN BE POSITIONED IN BODY OF TABLE AS DESIRED

(BLANK)

Relative Mid-diameters and Volumes of Various Solids of Revolution
For Given Ratios of Smallest Diameter (d) to Largest Diameter (D)

Ratio $\frac{d}{D}$ (Percent)	Ratio of $\frac{\text{Middle Diameter}}{D}$				Ratio of $\frac{\text{Frustum Volume}}{\text{Cylinder Volume}}$			
	Para-boloid	Conoid	Neiloid	Sub-neiloid*	Para-boloid	Conoid	Neiloid	Sub-neiloid*
	%	%	%	%	%	%	%	%
0	70.71	50.00	35.36	35.36	50.00	33.33	25.00	25.00
2.5	70.73	51.25	39.98	37.92	50.03	34.19	27.34	26.27
5	70.80	52.50	42.79	40.35	50.12	35.08	28.92	27.56
7.5	70.91	53.75	45.19	42.66	50.28	36.02	30.37	28.89
10	71.06	55.00	47.38	44.86	50.50	37.00	31.79	30.25
12.5	71.26	56.25	49.40	46.98	50.78	38.02	33.20	31.64
15	71.50	57.50	51.33	49.02	51.12	39.08	34.62	33.06
17.5	71.79	58.75	53.18	51.00	51.53	40.19	36.04	34.52
20	72.11	60.00	54.96	52.92	52.00	41.33	37.47	36.00
22.5	72.48	61.25	56.69	54.78	52.53	42.52	38.93	37.52
25	72.88	62.50	58.37	56.60	53.12	43.75	40.42	39.06
27.5	73.34	63.75	60.01	58.37	53.78	45.02	41.93	40.64
30	73.82	65.00	61.61	60.10	54.50	46.33	43.47	42.25
32.5	74.35	66.25	63.19	61.81	55.28	47.69	45.04	43.89
35	74.92	67.50	64.73	63.47	56.12	49.08	46.64	45.56
37.5	75.52	68.75	66.25	65.10	57.03	50.52	48.27	47.27
40	76.16	70.00	67.75	66.71	58.00	52.00	49.94	49.00
42.5	76.83	71.25	69.24	68.29	59.03	53.52	51.64	50.77
45	77.54	72.50	70.70	69.84	60.12	55.08	53.36	52.56
47.5	78.28	73.75	72.15	71.37	61.28	56.69	55.13	54.39
50	79.06	75.00	73.58	72.89	62.50	58.33	56.92	56.25
52.5	79.86	76.25	74.99	74.38	63.78	60.02	58.74	58.14
55	80.70	77.50	76.39	75.85	65.12	61.75	60.61	60.06
57.5	81.57	78.75	77.77	77.30	66.53	63.52	62.50	62.02
60	82.46	80.00	79.15	78.74	68.00	65.33	64.43	64.00
62.5	83.39	81.25	80.52	80.16	69.53	67.19	66.40	66.02
65	84.34	82.50	81.88	81.57	71.12	69.08	68.40	68.06
67.5	85.31	83.75	83.22	82.96	72.78	71.02	70.43	70.14
70	86.31	85.00	84.55	84.33	74.50	73.00	72.50	72.25
72.5	87.34	86.25	85.88	85.70	76.28	75.02	74.60	74.39
75	88.39	87.50	87.20	87.05	78.12	77.08	76.73	76.56
77.5	89.46	88.75	88.50	88.39	80.03	79.19	78.90	78.77
80	90.55	90.00	89.81	89.72	82.00	81.33	81.11	81.00
82.5	91.67	91.25	91.11	91.04	84.03	83.52	83.35	83.27
85	92.80	92.50	92.40	92.35	86.12	85.75	85.62	85.56
87.5	93.96	93.75	93.68	93.65	88.28	88.02	87.93	87.89
90	95.13	95.00	94.95	94.93	90.50	90.33	90.27	90.25
92.5	96.32	96.25	96.23	96.21	92.78	92.69	92.65	92.64
95	97.53	97.50	97.50	97.48	95.12	95.08	95.07	95.06
97.5	98.76	98.75	98.75	98.75	97.53	97.52	97.52	97.52
100	100.00	100.00	100.00	100.00	100.00	100.00	100.00	100.00

*A subneiloid is a fictitious solid of revolution whose volume equals that of a
cylinder with diameter equal to $\frac{d + D}{2}$, and whose mid-diameter can be postulated
on the assumption that Newton's prismatoidal formula holds true. Its volume is
an easily calculated close approximation to that of a neiloid.

Bias Due to Assumed Taper or Shape

Bias in volume and surface caused by erroneous assumptions as to taper from a single measured small-end diameter can be expressed as functions of taper:

Let d = diameter in inches at small end of portion of tree

       T = actual taper in inches between large and small end of portion

       $\widehat{T}$ = assumed taper in inches between large and small end of portion

       L = length in feet of portion

       K = actual shape divisor (2 if paraboloid, 3 if conoid, 4 if neiloid or subneiloid, which last term is defined in the footnote to Appendix C; of course, empirical values which are between or beyond the integers listed may be more appropriate)

       $\widehat{K}$ = assumed shape divisor

$$\text{Volume bias (cubic feet)} = (.005454\ L)(\widehat{T} - T)\left(d + \frac{\widehat{T} + T}{K}\right)$$

$$\text{Surface bias (square feet)} = (.1309\ L)(\widehat{T} - T)\ \text{for a conoid}$$

If a dendrometer is used with the height-accumulation technique, $\widehat{T} - T = 0$, since actual taper is measured, and no bias is created. However, any method or volume table which assumes rather than measures taper may be subject to considerable bias.

Bias in volume and surface caused by erroneous assumptions as to shape (or distribution of taper) between two measured diameters is much smaller in magnitude, as can be seen below. If the smaller diameter is at least 70 percent of the larger diameter, bias due to erroneous shape assumption is inconsequential, as can be inferred from Appendix C.

$$\text{Volume bias (cubic feet)} = (.005454\ LT^2)\left(\frac{1}{\widehat{K}} - \frac{1}{K}\right)$$

$$\text{Surface bias (square feet)} = \text{negligible variation between conoid and other shapes}$$

It is quite apparent that the conoidal shape assumption will give rise to very little bias in practical application of the height-accumulation method, but that any method or volume table which assumes both taper and shape from a single measured diameter may be subject to huge bias.

The formulae from which previously given functions were derived are:

The formulae from which previously given functions were derived are:

$$\text{Volume (cubic feet)} = (.005454\ L)\left(d + \frac{T}{2}\right)^2 + S$$

where S is $\frac{T^2}{4}$ for paraboloid, $\frac{T^2}{12}$ for conoid, and 0 for subneiloid.

$$\text{Surface (square feet)} = (.2618\ L)\left(d + \frac{T}{2}\right) \text{ for conoid.}$$

The increase in precision attributable to breaking up long logs into several shorter portions each having a single measured diameter is well known in log scaling. It restricts log-rule taper and shape assumptions to shorter lengths and this prevents bias from building up when multiplied by length. The author has derived a function express-ing the scaled volume of a long conoidal log as a joint function of log-rule assumptions and scaling interval, together with log diameter at small end, log length, and actual log taper.

Let long-log parameters be denoted as follows (for a conoidal log):

    d = diameter in inches at small end

    L = total length in feet

    T = actual constant <u>rate</u> of taper, in inches per foot

Let scaling interval be denoted as follows:

    K = scaling interval in feet (a constant)

Let log rule assumptions be denoted as follows:

    $\textcircled{T}$ = fictitious constant rate of taper assumed by log rule (in inches per foot: 1/8 for International, 0 for Scribner and Doyle)

    M = minimum length to which taper assumed by log rule will apply

        (4 ft. in case of International, irrelevant in case of Scribner and Doyle)

    N = d - 1.6136   for Int. 1/4   (milling frustum diameter)

        d - 1.255   for Scribner

        d - 4   for Doyle

    C = .0498 L   for Int. 1/4 in. (kerf, length, and scale factors)

        .0494 L   for Scribner

        .0625 L   for Doyle

    V = $C(N^2 - 2.60)$ for Int. 1/4 in. (volume of minimum length)

        $C(N^2 - 7.01)$ for Scribner

        $C(N^2)$ for Doyle

Let certain joint functions be denoted as follows:

    P = (L - K)(T)

    $\textcircled{Q}$ = (K - M)($\textcircled{T}$)

    R = LT

    $\textcircled{S}$ = K $\textcircled{T}$

Then the log-rule board foot volume of a long log with small end diameter (d), length (L), actual rate of taper (T), and with diameters taken at constant interval (K) will be:

$$V + (C)\left(\frac{(P + ⓠ)(P + ⓠ + R + 6N) - ⓢ(K - M)(T - ⓣ)}{6}\right)$$

If, instead of the constant interval K above, diameters had been measured at a longer constant interval K', and if the board foot volume obtained by using interval K were subtracted from that obtained by using interval K', the difference in board foot volume would be:

$$(C)(K - K')(T - ⓣ)\left(\frac{P + ⓠ + P' + ⓠ' + R + 6N + ⓣ(K + K' - M)}{6}\right)$$

Primed symbols denote quantities containing K', and circled symbols denote quantities containing taper assumed by log rule (zero in case of Scribner and Doyle).

As an example, consider a 20-inch log of 80-feet length tapering at a constant rate of 1 inch per 16 feet. If actual diameters were measured at 16-foot intervals, the basic formula above would compute board foot volume as 1,778 board feet by International 1/4-inch log rule and 1,680 board feet by Scribner formula log rule. If, however, the scaling interval were lengthened and actual diameters were only measured at 20-foot intervals instead of at 16-foot intervals, the International 1/4-inch volume of the same log would scale 22 board feet higher than previously, while the Scribner formula volume of the same log would scale 21 feet lower than previously.

The above calculations can be accomplished or verified by tabular methods, but the two previous formulae give some insight into the mechanism through which erroneous taper assumptions in log rules and interval between actual scaling diameters affect log scale.

APPENDIX E

Theory of Height Accumulation

The bole of a tree can be considered as the sum of several frusta of solids of revolution, each generated by a different function having the generalized form $D^2 = BL^{K-1}$, where:

B = an appropriate constant.

D = d.o.b. (diameter outside bark) at one end of frustum.

T = taper-step or difference in diameter between small and large ends of frustum.

H = height above stump at which D is measured.

L = difference in height between small and large ends of frustum.

K = shape divisor (2 if paraboloid, 3 if conoid, 4 if neiloid or subneiloid, which last term is defined in the footnote to Appendix C).

If d.o.b. of the bole of a tree be measured at two points, and if the height above stump of those two points also be measured, then the author has shown (10) that the volume of any frustum cut by planes normal to the tree axis and passing through the two points can be expressed as:

$$\frac{\pi T^2}{4}\left[L\left(\frac{D}{T}\right)^2 + L\left(\frac{D}{T}\right) + \frac{L}{K}\right]$$

The surface of any such conoidal frustum can be expressed as:

$$\pi T\left[L\left(\frac{D}{T}\right) + \frac{L}{2}\right]\left[\sqrt{1 + \left(\frac{T}{2L}\right)^2}\right]$$

where D T, and L are all measured in the same scale units (i.e., all in feet, or all in inches, or all in meters, etc.).

- 30 -

is apparent that if taper-step (T) is kept constant, and if stump
r is chosen so that it is a multiple of T, and if $X = \frac{D}{T}$, then X's
cur in diminishing arithmetic progression as successive D's ascend
e.

The entire tree volume above stump then can be calculated as:

$$\frac{\pi T^2}{4}\left[\Sigma X^2 L + \Sigma XL + \frac{\Sigma L}{K}\right]$$

entire tree surface can be calculated as:

$$\pi T\left[\Sigma XL + \frac{\Sigma L}{2}\right]\left[\sqrt{1 + \left(\frac{T}{2L}\right)^2}\right]$$

Since X occurs in an arithmetic progression diminishing up the
le till it reaches 1, appropriate combinations of the sum of
pward progressive totals ($\Sigma L'$) and the sum of second upward
sive totals ($\Sigma L''$) can be substituted for ΣXL and $\Sigma X^2 L$. The
lume can then be calculated as:

$$\frac{\pi T^2}{2}\left[\Sigma L'' + \frac{\Sigma L}{2L}\right]$$

tree surface as:

$$\pi T\left[\Sigma L' + \frac{\Sigma L}{2}\right]\left[\sqrt{1 + \left(\frac{T}{2L}\right)^2}\right]$$

It is obvious that the radical constituting the last factor
ill be negligible (1.00001356) where tree taper averaging about
s in 16 feet is involved. It is also obvious that if H = height
D above stump, then $\Sigma H = \Sigma L'$, and $\Sigma H' = \Sigma L''$. Thus, assuming
age taper of 2 inches per 16 feet (merely to allow evaluation of
ical), the above expressions can be written as:

$$\text{Tree volume} = \frac{\pi T^2}{2}\left[\Sigma H' + \frac{\Sigma L}{2K}\right]$$

$$\text{Tree surface} = \pi T\left[\Sigma H + \frac{\Sigma L}{2}\right]\left[1.00001356\right]$$

accumulation theory stemmed from the author's derivation of these
e in 1948 (10). As long as a reasonable taper-step (T) is
i, the conoidal assumption (with K = 3) will give rise to negli-
rror in either formula, as can be seen from Appendices C and D.

- 31 -

J. F. Clark (who devised the International 1
nally conceived of his board-foot fule as a functio
face, and cubic volume (5). However, derivation of
the case of the International 1/4-inch log rule, th
log rule, and the Doyle log rule required explicit
formulae not published till much later (12). On pa
cited reference (Shortcuts for Cruisers and Scalers
explicit functions leading to the deduction that bo
the case of 16-foot logs can be expressed as:

Board-foot volume (International 1/4-inch kerf) =
  9.1236 (cu.ft.vol.) - .70846 (sq.ft. surface) + .

Board-foot volume (Scribner log rule) =
  9.057 (cu.ft. vol.) - .852 (sq.ft. surface) - .11

Board-foot volume (Doyle log rule) =
  11.459 (cu.ft. vol.) - 2.387 (sq.ft. surface) + 1

Although there is no change in the International co
length changes, there is change in the Scribner and
For a 12-foot log, the Scribner .852 would become .
become .158; the Doyle 2.387 would become 2.269, an
1.399.

The various board-foot volume coefficients A, B, C
are based on coupling these formulae for 16-foot lo
basic height-accumulation formulae derived earlier.
A, B, C coefficients have been made in Appendix A f
where 12-foot logs are commonly cut and Scribner or
used. However, such coefficients may be easily cal
formulae given above.

SOME
FIELD, LABORATORY, AND OFFICE PROCEDURES
FOR SOIL-MOISTURE MEASUREMENT

SOUTHERN FOREST EXPERIMENT STATION
Philip A Briegleb, Director
Forest Service, U. S. Department of Agriculture

ACKNOWLEDGMENT

This publication is based on work accomplished by
the Vicksburg Infiltration Project and other research
centers of the Southern Forest Experiment Station,
U. S. Forest Service, for the Waterways Experiment Station, Corps of Engineers, U. S. Army.

CONTENTS

BORATORY, AND OFFICE PROCEDURES
IL-MOISTURE MEASUREMENT

easing extent, soil-moisture records are con-
luations of plant-soil-water relations governing
orest and agricultural crops. The use of water
getation growing in different types of soil, sea-
ties for soil moisture under various kinds of
·oot occupancy and growth--in these and a host
il moisture is one of the measured variables.

noisture is a relatively simple task, but it does
xperience with the instruments and techniques.
er, like those in an earlier publication, [1] rep-
of experience of this kind.

the eight articles deal with bulk density, the
t; the second, how to evaluate it. Both are re-
expression of soil-moisture content. The third
rences in moisture-tension data secured from
The next three papers describe equipment to
impling and reading electrical resistance units.
laboratory and field calibration of fiberglas
th roots, their measurement and concentration
ited to the evaluation of soil-moisture records
is.

---

rement with the fiberglas instrument. Southern
s. Paper 128, 48 pp. 1953.

One of the important phases of the
Vicksburg Infiltration Project has been th
Defined as the ratio of the weight of oven·
occupied in the field, bulk density is expr
meter. Its use is principally for converti
percent by weight to depth or area-inches

Between March 1951 and Decembe
bulk density determinations were made or
and loess near Vicksburg. Of these, app
three methods. No attempt was made to
cedure with another, for the overall obje‹
pediently, enough values at the various s‍
of soil-moisture records in inches depth.
obtaining the values, certain procedures
and some improved equipment was develc

Description of Prc

The three procedures evaluated aі
determinations were designated "block, "
picnometer. "

Block procedure

This procedure involves measure:
of soil 12 inches square by 3 inches in de

First, a 14- by 14-inch plot of so
on the ground as a sampling area. All oɪ
moved down to the mineral soil. A spike
close to the 14-inch square; its purpose i
which the depth of successive 3-inch sam

Figure 1. -- Taking a soil sample by the block method.

A shallow trench, approximately the width of a spade, is dug a-round the sampling area, thereby forming a soil column 14 inches square. A wooden frame 14 inches square (inside dimension) 1/ is gently forced down over the soil column until the top of the frame is just above the soil surface. The frame is leveled and the vertical distance from the top of the reference spike to the top of the frame is measured and recorded. The outline of the 12-inch sample square is then scratched on the surface of the 14-inch square.

The volume of the first sample must now be corrected for any un-evenness in the surface of the soil. This can be done by placing a straight-edge across the frame and measuring, at several points within the 12-inch square, the vertical distance between the straightedge and the soil surface. The average distance, multiplied by the area of the block, gives the proper correction.

Next, the soil is carefully cut away and removed from underneath the frame until the top of the frame can be lowered exactly 3 inches below the first measurement from the reference spike. The frame is again leveled, and the soil block trimmed to final dimensions, down to the top of the frame (fig. 1).

1/ The frame must not be warped or twisted. The model in figure 1 was made of nominal 2 by 4 lumber.

The soil sample is then removed flush with the top of the frame and placed in moisture-proof bags. Successive samples below the surface are obtained similarly (except that no correction is made for uneven surface), with the frame being forced down exactly 3 inches each time a sample is taken.

Representative portions are taken for moisture determinations from each block-sample. After the block of soil and the sample portions are weighed, bulk density is computed:

$$\text{Bulk density} = \frac{\text{Oven-dry weight of soil in grams}}{\text{Volume of soil block in cubic centimeters}}$$

San Dimas procedure

In this procedure a sampler [2] is used to obtain a core of soil 2.72 inches in diameter and 3 inches in depth, and with a volume of 288 cubic centimeters.

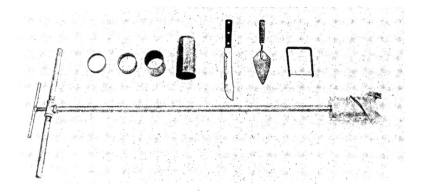

Figure 2.--The San Dimas sampler and equipment used in determining bulk density. (Photo by Waterways Experiment Station, Corps of Engineers.)

This sampler (fig. 2) has an outer cylinder equipped with spiral flanges with sharp cutting points, and an inner cylinder with 3 removable

---

[2] The sampler was designed at the San Dimas Experimental Forest, California Forest and Range Experiment Station, U. S. Forest Service. Information for making it was furnished by J. S. Horton.

brass sleeves. The largest sleeve is the size of the desired core; the two smaller sleeves fit above and below the large one, and serve as buffers to protect the soil core. Flanges are provided on the outer cylinder to cut the soil and bear it upward so that binding is diminished. When the handle attached to the outer cylinder is rotated, the inner cylinder is forced downward into the soil. Rotation of the inner cylinder is prevented by holding back on the handle of the rod enclosed in the main shaft. The design permits the inner cylinder to penetrate undisturbed soil.

After the sampler is removed from the sampling hole, the inner cylinder is pushed out of the outer cylinder and then detached from the rod. Next the sleeves containing the soil core are removed from the inner cylinder and the soil is trimmed flush with the ends of the largest sleeve. Then the core is pushed out of the sleeve into a 16-ounce soil can and taken to the laboratory, where it is oven-dried at 105° C. The head of the sampler is substantially longer than the desired core: about 5 inches of soil are removed in order to secure a 3-inch core. Where successive 3-inch samples are desired, as at the Vicksburg Infiltration Project, they must be taken alternately from 2 sample holes.

Calculation:

$$\text{Bulk density} = \frac{\text{Oven-dry weight in grams of soil}}{288}$$

Air-picnometer procedure

The air-picnometer was introduced by Russell (2)[3] to measure the volume of air-filled pores in soil. He pointed out that if the procedure is carried a step further and the weight of the soil obtained, then moisture content and bulk density for most soils can be estimated from nomographs.

The sample cores used in making the determinations at Vicksburg were taken from excavated pits by means of a Coile-type sampler (1) in conjunction with a cylindrical brass sample holder that fits closely into the air-picnometer tube. The sample holder is 2 inches in inside diameter and 1-3/8 inches high (fig. 3).

After the soil core is obtained, it is placed in the air-picnometer. Pressure is applied and the mercury scale reading recorded. Then the core (plus cylinder) is weighed, after which it may be discarded.

---

[3] Underscored numbers in parentheses refer to Literature Cited, page 11.

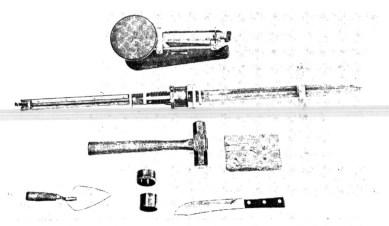

Figure 3. -- The air-picnometer and equipment for use in determining
bulk density. (Photo by Waterways Experiment Station, Corps of
Engineers.)

Calculations:

(a) Obtain air space from a previously constructed
calibration curve (2).

(b) On a nomograph [4] relating air space, moisture
content, and field weight of core (fig. 4), connect
with a straightedge the percent of air space and the
weight of the core (without cylinder), and read the
percent of moisture by weight.

(c) On a bulk-density nomograph (fig. 5), connect
weight of core sample and percent of water and
read the bulk density along the extended straightedge.

---

[4] Nomographs in figures 4 and 5 were developed by Dr. W. A. Raney,
Mississippi Agricultural Experiment Station, State College, Mississippi.
Calculations for these nomographs are based on a soil specific gravity
of 2.65; therefore the nomographs cannot be used to determine moisture
content and bulk density of soils with a high organic content.

ure 4.—Nomograph for estimating percent of moisture by weight
from weight of soil core and air space.

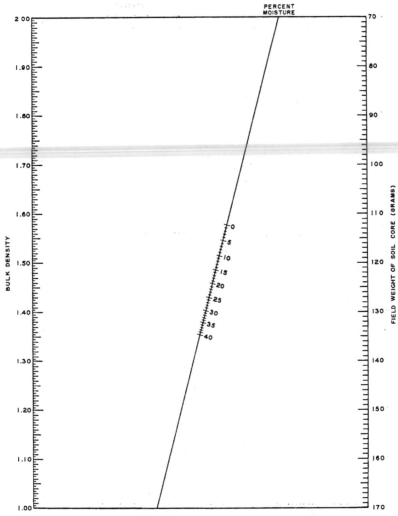

Figure 5.—Nomograph for estimating bulk density from weight of soil core and moisture content.

Comparison of Procedures

The procedures were evaluated by comparing mean bulk densities and by taking into consideration such factors as disturbance of sampling area (as the need to dig pits), time required to take the sample, personal skill required to make the determination, and utility of the procedure in dry soil.

The block method gave slightly lower bulk densities than the San Dimas and air-picnometer methods. For instance, in a comparison of 168 blocks and 149 San Dimas cores from four sites, the mean block value was 1.32 and the mean San Dimas value was 1.39, a difference significant at the one-percent level. In another comparison, block samples taken at various depths at one site were 0.04 lower than comparable air-picnometer values, a difference which was also significant at the one-percent level.

These results, however, are very likely biased. Block samples are feasible only when the soil is moist, and moist soil gives lower bulk density values than dry soil. Subsequently, in what was perhaps the most direct comparison, picnometer cores were taken out of 30 blocks of soil from 4 experimental sites, and the bulk density values obtained by the two procedures were treated statistically as paired samples. In this comparison, the mean differences were not significant at the five-percent level.

The San Dimas values tended to run higher than air-picnometer values, but the differences were small. In one case, the San Dimas bulk densities averaged 0.02 greater than the picnometer values, a non-significant difference at the five-percent level.

The other factors taken into consideration in evaluating the procedures were rated on a 1-2-3 system, with 1 representing the best and 3 the poorest:

Rating factor	Block	San Dimas	Air-picnometer
Disturbance of sampling area	3	1	3
Compression or disturbance of soil sample	1	3	2
Personal skill required	3	1	2
Utility in dry soil	3	3	1
Time required	3	2	1

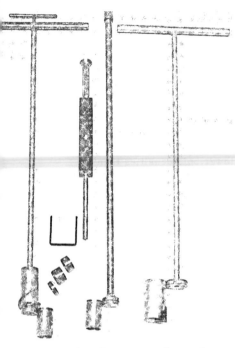

Considering all points or factors, the San Dimas and air-picnometer procedures appear about equally superior to the block procedure.

Sampling Equipment

The combined advantages of the San Dimas and air-picnometer procedures were realized by developing sampling devices similar to the San Dimas sampler but with a cylinder of the same diameter as the air-picnometer cylinder. Used with the air-picnometer, the sampler eliminates digging a pit and thus permits thorough sampling with little disturbance of an area.

Figure 6. --Two types of samplers and a hole-cleaner for obtaining soil cores for bulk-density determination. (Photo by Waterways Experiment Station, Corps of Engineers.)

Two samplers and a hole cleaner were made (fig. 6). One sampler is patterned after the San Dimas but is scaled down so that the inner cylinder (with its sleeves) fits the air-picnometer. The smaller size, and the fact that the pitch of the augering flanges is not so steep as on the original San Dimas model, makes it easier to get this sampler into the soil. The other sampler is driven into the soil with a King-tube type hammer. The inner cylinders of the two samplers are interchangeable. The "Little San Dimas" sampler is used in moist soil and the hammer-driven model in dry soil. The hammer-driven sampler has a small hole through the driving head, with an outlet near the head of the shaft, to permit escape of air. The outer cylinders and handles of both samplers are marked off in 3-inch increments, so that with either instrument successive 3-inch soil layers can be easily sampled from a single sampling hole.

The wire cheese cutter and the hole-cleaner (fig. 6) are accessories. The cutter is used to separate the buffer sleeves, and soil within the sleeves, from the sample. Sometimes it is preferable to slip the buffer sleeves off the ends of the core and trim the soil flush with the ends of the core cylinder with a sharp, straight-edged butcher knife.

- 10 -

The hole-cleaner has the same outside diameter as the outside edges of the cutting flanges on the Little San Dimas. When used with a hammer-driven sampler it serves as a hole-widener as well as a hole-cleaner. Sharp cutting knives welded to the bottom and side of the instrument remove the soil and scoop it into the detachable soil-collecting compartment. The cleaner is marked off in the same 3-inch increments as the samplers.

Field tests have shown these devices to be a rapid and convenient means of obtaining soil samples at various depths with very little disturbance of the experimental area. The core samples can be used without the air-picnometer for bulk density determinations. For soils containing over 5 percent organic matter it is recommended that the bulk density be determined by drying and weighing. For this computation, the oven-dry weight of the soil is divided by 71 cc, the volume of the core sleeve or sample.

<u>Literature Cited</u>

(1) Coile, T. S.
    1936. Soil samplers. Soil Sci. 42: 139-142.

(2) Russell, M. B.
    1949. A simplified air-picnometer for field use. Soil Sci. Soc. Amer. Proc. 14: 73-76.

AS IT AFFECTS SOIL-MOISTURE RECORDS

K. G. Reinhart

In many studies of the hydrologic cycle which include .
f soil moisture, it is necessary to convert soil-moistu
:ent of oven-dry weight to inches depth of water. This
out some of the relationships and difficulties involved,
ts procedures for making the conversion.

Moisture content in percent of the oven-dry weight of
close the actual amount of water present. However, n
lue by bulk density gives water volume per cubic unit o
:. Then multiplication by soil depth converts cubic me.
ar:

$$\left(\frac{\text{water volume}}{\text{soil volume}}\right) (\text{soil depth}) = \frac{\text{water volume}}{\text{soil area}} = \text{water d}$$

Bulk density, volume weight, and apparent specific gr
:ally synonymous. In this paper the term bulk density.
Defined as the ratio of the weight of oven-dry soil to tl
ipies in the field, it is generally expressed in grams pe
ieter, or $B = \dfrac{\text{oven-dry weight of soil in grams}}{\text{field volume of soil in cubic centimeters}}$

Bulk density is not easy to determine, even for small
cted by texture, organic content, and structure; these i
y remain fairly constant, at least within short periods (
so affected by changes in moisture content of the soil t
ıg and shrinking caused by addition and loss of water.
phenomenon is warranted because it results in change
y within short periods and thus complicates determinin;
lk density value applicable to any specified depth over (

Shrinking and Swelling of Soils

The shrinking and swelling of soil is accompanied by both vertical and horizontal changes. Shrinkage in a vertical plane may lower the soil surface; this lowering can be detected by appropriate methods (5). Vertical shrinkage may form voids below the soil surface, but the weight of the soil above tends to reduce the size of such voids and observation indicates that they have negligible volume. Shrinkage in a horizontal plane, however, causes cracks quite apparent at the soil surface.

Vertical shrinkage results in the soil occupying a smaller volume than it occupied before the shrinkage took place; the volume represented by the difference in elevations before and after shrinkage becomes a part of the atmosphere. Since the same mass of soil occupies a smaller volume, the bulk density must necessarily be increased.

Horizontal shrinkage, on the other hand, has no effect on average bulk density as long as void spaces are included in the soil volume. If the bulk density values are to be used in computation of area inches of water, they must represent the whole area, including that portion given over to cracks. If samples are taken only between cracks, bulk density values will be too high. Where soil cracks are large, conventional procedures for obtaining samples cannot be used and sampling becomes more difficult.

Because shrinkage affects bulk density and its determination, it also affects soil-moisture records. Its influence can be described according to the type of soil-moisture sampling: either gravimetric or by electrical-resistance instruments.

When soil-moisture records are taken by gravimetric means, the same depth increments of soil are removed at each sampling regardless of fluctuations in the elevation of the soil surface due to shrinking and swelling. This procedure is proper, for the sole objective is to obtain periodic moisture measurements for specified depths. The bulk density of each sample may be obtained for conversion of soil-moisture content from percent by weight to inches depth. Because of the time and effort required, this is rarely done. Another procedure sometimes advocated is to use bulk densities obtained from bulk density-moisture regressions. Horizontal shrinkage presents no problem in gravimetric sampling so long as bulk density values used for conversion are derived from samples which include the cracks in the soil where they exist.

The effects of shrinking and swelling are of greater import to the measurement of soil moisture when electrical-resistance units are used.

veight (M), bulk density (B), and thickness (D) of
he following equation:

$$d = \frac{(M)\,(B)\,(D)}{100}$$

;iven mass of soil varies only in vertical dimensi·
onstant, since $B = \dfrac{\text{constant oven-dry weight}}{(\text{constant soil area})(\text{variable s}}$

$\dfrac{(\text{constant weight})(D)}{(\text{constant area})(D)} = \dfrac{\text{constant weight}}{\text{constant area}}$. Hence

ilculated as $\dfrac{(M)\,(B)\,(D)}{100}$, regardless of fluctuatio

ınsequently, any pair of values for (B) and (D) can
y are both determined under the same moisture c
:, assume that the bulk density of a certain 3-incl
at a moisture content of 30 percent by weight (thi
of water in the 3-inch layer). Now suppose that ɛ
bulk density of 1.35 at 10 percent. As the bulk d
ɣ 135/130, the same mass of soil must now occup
much space. Since this change is entirely in a ve
ıe height of the layer has been reduced to (3) (130
rom 3 to 2.889 inches. The inches of water in th
noisture content is therefore

$$\frac{(10)\,(1.35)\,(2.889)}{100} = 0.390 \text{ inch}$$

seen that the same answer will be obtained if the
instead of 1.35 and 2.889, are used for bulk dens
respectively.

ssume that at 10 percent moisture the bulk densit
reased to 1.35 but that the depth of the soil layer
ıuld give too high a value; i.e., $\dfrac{(10)\,(1.35)\,(3)}{100} = 0$.

above reasoning is based on the premise that the
its installed at various depths in the soil retain th
ı respect to the adjacent soil despite any slight ra
the soil layers with shrinking and swelling.

The single value of bulk density used in conversion from percent to inches should correspond to the bulk density existing when the units are installed and the soil depths are measured. Both installation of units and bulk density sampling are most easily accomplished when the soil is moist, at field capacity or slightly below (3).

In field calibration of electrical-resistance units, gravimetric samples are obtained. As the thickness of the layer represented by each unit varies with moisture content, the depth of sampling should be adjusted if the amount of shrinking and swelling is great enough to have any appreciable effect. This question will be considered below.

Determining Amount of Shrinking and Swelling

Determination of the amount of shrinking and swelling is of interest with respect to both the calibration sampling with electrical-resistance units and, as already noted, the method using gravimetric samples alone.

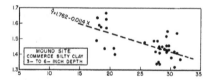

In the vicinity of Vicksburg, over 1,200 bulk density samples were obtained at different depths and over a range of moisture contents in loess silt loams and water-deposited clay. Linear regressions and correlation coefficients were computed to determine the nature and degree of relationship. The methods were somewhat similar to those reported by Bethlahmy (1).

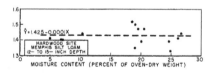

The regression coefficients obtained varied widely by location and depth, often without apparent reason. Typical regressions for three sites are given in figure 1. Of 90 regressions computed, 28 were statistically significant and 66 showed decreasing bulk density

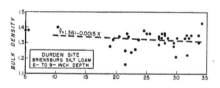

Figure 1.--Typical regressions of bulk density on moisture content.

with increase in moisture content. Signifi
(including the upper regression in figure 1
cause samples were taken (at low moistur∢
cracks and thus gave bulk density values tl
gressions might be interpreted as implyin₤
swelling of the soil. Such a conclusion wo
shown below.

   In the regressions, moisture conte
weight. Samples differing in bulk density,
of water per unit volume, will have differe
amount of water and weight of soil per uni₮
into computation of moisture percent by w∢
figure 2. Thus, if numerous samples are
usual variation in bulk density, and with tʰ
per unit volume, a regression of bulk dens
obtained even though no shrinking or swelʲ

   To investigate this effect further, ₁
moisture content in percent by volume wer
cent by weight, for the 0- to 3-inch and 12
The two regressions for each depth were c
samples. In the following equations, Y is
cent by weight, and Z is moisture percent

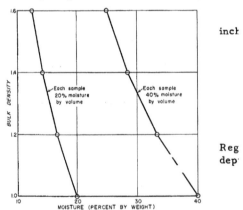

incʰ

Reg
depʳ

Figure 2. --Relationship of moisture
content by weight to bulk density,
when moisture percent by volume is
constant.

- 16 -

For the 0- to 3-inch depth, when percent-by-weight values were
ed, 62 percent of the variation was attributable to difference in
)isture content; using percent-by-volume values reduced this to 31
rcent. Comparable values for the 12- to 15-inch depth were 18 and 5
rcent.

These correlations show that a stronger influence of moisture
ntent on bulk density is indicated when moisture content is expressed
percent by weight rather than as percent by volume. The fact that
ight of the soil appears in the numerator of bulk density and in the
nominator of moisture percent by weight accounts for the stronger
rerse relation (or negative correlation) between these two.

The natural variation in bulk density within a given area results
varying amounts of pore space being available for the storage of
ter. If a constant value for the specific gravity of soil is assumed,
naximum or saturation moisture content in percent by weight can be
mputed for any given bulk density. The formula is:

$$\text{Maximum moisture content} = \frac{\text{specific gravity - bulk density}}{(\text{specific gravity})(\text{bulk density})} \times 100$$

$$= \frac{100}{\text{bulk density}} - \frac{100}{\text{specific gravity}}$$

us, a soil of bulk density 1.60, for example, will be saturated at 25
rcent moisture content by weight; a soil of bulk density 1.30 will be
turated at 39 percent. Intermediate bulk densities will be saturated
intermediate moisture contents because of the inverse relationship
tween bulk density and maximum moisture percent. This is illustrated
· two soils in figure 3. In this figure, the solid lines show the max-
um moisture content possible for any bulk density within the range en-
untered (a soil specific gravity of 2.65 is assumed) and the dashed
es represent the regression determined from field sampling. Simi-
·ities in slopes of the two lines are very noticeable.

The inverse relation between moisture content and bulk density
obvious at saturation (MC = $\frac{100}{B}$ - constant). At lower moisture con-
its it is less apparent. The question is largely one of relative distri-
tion of pore sizes in any two samples having a different bulk density.
the only difference lies in the amount of pore space of the sizes drain-
by gravity, there would be no difference in the moisture content (by
lume) in two samples subjected to the same tension at field capacity
below. However, if the sample having lower bulk density also has

- 17 -

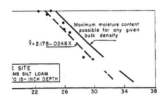

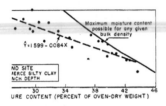

3. --Typical regressions of
:nsity on moisture content
wet conditions.

more pore space in the
classes, it will contain
at a given tension than
sample. Since the latt
seems more likely, an
ples obtained at any on
the same soil depth ten
same tension, a correl
tween bulk density and
content can be expecte(
part from any shrinkin,
that might be associate
moisture loss or moist

To provide info
this relationship with t
field capacity, an analy
made of tension-moistt
data. At each of sever
three sites, five replic
were obtained from the
Bulk density and moist

etermined for each core. Regression and correlation
n moisture content in percent by weight at 60 cm. tens
ty) and bulk density were computed for each depth (tabl
ne-third of the correlation coefficients and regression
tatistically significant. Thirty out of 31 of the regress
were negative, an indication that, on the average, the
r bulk density will have a higher moisture content in pe
at field capacity. Thus random field samples taken in

oefficients of regression of bulk density on moisture content in percent by weight at 60-cm. tension (f
and corresponding correlation coefficients

| Durden site | | | | Hardwood site | | Pine site | | |
| Pit No. 1 | | Pit No. 2 | | | | Pit No. 1 | | |
Correlation coefficient	Regression coefficient	Correlation coefficient	Regression coefficient	Correlation coefficient	Regression coefficient	Correlation coefficient	Regression coefficient	Corr coef
-0.70	-0.016	-0.99**	-0.019**			-0.96*	-0.018*	-
-.66	-.008	-.19	-.004	-.75	-.010	-.12	-.005	-
-.83	-.027	-.24	-.007	-.86	-.030			
+.98**	+.058**	-.64	-.041	-.90*	-.025*	-.30	-035	-
				-.98**	-.029**			
-.46	-010	-.82	-022					
				-.93*	-.017*			
-.74	-.018	-.09	-.0002	-.95*	-.054*	-77	-024	-
-.72	-.011	-.32	-.008	-.26	-.003			
-.63	-.023	-.89*	-.019*					

t at 1 percent level
t at 5 percent level

the moisture range would show
a regression of bulk density on
moisture content even though no
shrinking or swelling of the soil
were involved.

Regressions for three
degrees of soil moisture are
illustrated in figure 4. The
"wet soil" regression was computed from 24 samples taken on
January 3, 1952, at one site and
at various depths to 42 inches.
The "intermediate soil" line
was based on 22 samples taken
on October 25 and 26, 1951,
from the same depths. The
"dry soil" line was computed
from 16 samples taken by 3-inch
depths to 12 inches on July 16
and 17, 1951.

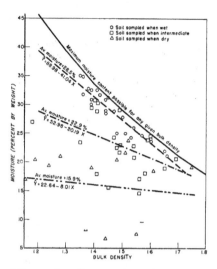

Figure 4. --Regression of moisture
on bulk density under wet, intermediate, and dry conditions.

Individual points deviated
little from the wet soil line, more
from the intermediate soil line, and most from the dry soil line. A t
test was made to determine the significance of the regression coefficients
from zero. For wet soil this test showed significance at the one-percent
level; for intermediate soil, at the 5 percent level; and for dry soil, no
significance. As these data were obtained at different depths, the variation in bulk density is greater than would be expected at any one depth;
however, the data do illustrate that, where bulk density variation exists,
a significant moisture content-bulk density regression may be secured
independent of shrinking or swelling of the soil.

As stated previously, of the 90 regressions based on random field
sampling at Vicksburg, 28 were statistically significant. However, these
regressions are affected by the fact that soil weight per unit volume is a
common element in both variables, and by the correlation between moisture content at a given tension and amount of pore space independent of
soil shrinking or swelling. It is believed that the effect of moisture content on bulk density of these soils is small enough to be disregarded with
respect to gravimetric sampling.

If random field sampling will not give a valid ⸣
change of bulk density with change in moisture conten
accomplished? Several approaches are possible. A
field sampling technique has been suggested: intensiⱱ
of an area when it is wet and again when it is dry to p
bulk density for each condition. The difference betwɛ
ages would be a measure of the swelling or shrinking
curred. A second approach involves the measuremeɪ
elevation of the ground surface with change in moistu
A third possibility lies in methods of determining fiel
situ, such as the gamma-ray scattering method now ⱦ

Summary

It has been shown that even though a change in
may really change the volume occupied by a given soi
bulk density will also change), this change will not af
by electrical-resistance methods of inches of water i
or layer. This is so because the product of variable
variable volume or thickness of a given soil mass reɪ
independent of moisture content.

It is further indicated that even with gravimet
unwise to attempt to adjust bulk density for variation
content (especially percent by weight) because bulk dɛ
mined under dry conditions are likely to be inaccurat
such inverse relationships as have been found are duɛ
the shrinking and swelling of soil.

Literature Cited

(1) Bethlahmy, N.
1952. A method for approximating the water
Trans. Amer. Geophys. Union 33: 6ᴵ

(2) Carlton, P. F., Belcher, D. J., Cuykendall, T.
1953. Modifications and tests of radio-active
uring soil moisture and density. Teᴵ
194, Tech. Development and Evaluat
Aeronautics Admin., Indianapolis, I

(3) Reinhart, K. G.
    1953. Installation and field calibration of fiberglas soil-moisture
    units. Southern Forest Expt. Sta. Occasional Paper
    128, pp. 40-48.

(4) Soil Science Society of America.
    1948. Committee Reports, Terminology, Soil Physics. Soil
    Sci. Soc. Amer. Proc. 13: 573.

(5) Woodruff, C. M.
    1936. Linear changes in the Shelby loam profile as a function of
    soil moisture. Soil Sci. Soc. Amer. Proc. 1: 65-70.

W. M. Broadfoot [1]

The usual laboratory procedure in determining s tension values is to use "undisturbed" soil cores for ter of water and bulk soil samples for higher tensions. Lo usually obtained with a tension table (2) and the higher t pressure plate apparatus (3).

In tension analysis at the Vicksburg Infiltration . soils had higher moisture contents at 1/3 atmosphere te cm (. 06 atmosphere). This was noted particularly in fı and was thought to be related to the difference in relatiͷ the soil samples used under the two tensions. Accordir son was made of the moisture contents of core and bulk various tensions.

Experimental Procedure

The procedure used in preparing samples, and t during the determination, varied for two sets of sample cedure used at first (called A for convenience) was late favor of procedure B, which seemed to yield more accu However, the results from procedure A are comparablє of this study and are therefore included.

Procedure A. --Five pairs of samples, core anc obtained from each of three soil types: Commerce silt͈ burg silt loam, and Bosket very fine sandy loam. The c obtained in stainless steel rings, 2-3/4 inches in diamє in depth, which were driven into the surface 3-inch soiⅼ Coile-type driving head or sampler (1). Each bulk sarr close or adjacent to its corresponding paired core.

Bulk samples were prepared in the laboratory by hand crushing and passing through a U. S. Standard 9-mesh sieve. The soil was placed in the rings, one end of each ring having previously been covered with a filter paper and cheesecloth held in place with a rubber band. The paper insured against loss of fine particles of soil while the cloth held the soil in place. The cores were also covered by cloth and paper. Samples thus contained in rings were placed in a shallow pan of water and allowed to soak until saturated.

The paired samples were run simultaneously at tensions of 5 and 60 cm. of water on the tension table, at 0. 1, 1/3, and 1 atmosphere pressure in pressure cells using common asbestos board as a membrane, and at 3 and 15 atmospheres pressure with Visking sausage casing as a membrane. After they came to equilibrium at each of the tensions, the samples were weighed. After this cycle, as a check on the first determinations, the same samples were rerun at 5 and 60 cm. water tension. Oven-dry weight was obtained after the second 60 cm. water tension determination, and moisture percentage was then calculated for all tensions. Controls or blanks were run on the filter paper, cheesecloth, and rubber bands to determine the moisture content of these materials at the various tensions.

From 6 to 24 hours were allowed for moisture to reach equilibrium in samples on the tension table, and from 24 to 48 hours for samples in the pressure cells. The tension table was covered tightly with oilcloth to prevent evaporation.

Procedure B. --Soil-moisture values secured from procedure A, even though useful for comparative purposes, were somewhat greater at high tensions than data previously obtained on two of the same soils. This indicated that the paper-plus-cheesecloth combination was so thick that it caused the water columns or film to break when the moisture content was reduced. When the water columns were broken, equilibrium was reached prematurely.

Procedure B differed from A in that only one thickness of cheesecloth was used to hold the sample in the ring, and it was removed entirely during the determinations at 3 and 15 atmospheres. In these high-tension runs the soil was directly against the Visking sausage membrane. Moisture content was determined simultaneously for paired samples of soil at the same tensions as stated above.

Sixteen paired samples, taken from various depths of an unidentified loess silt loam near Poplar Bluff, Missouri, were used in the comparison by procedure B. The cores were taken in brass rings, 2

inches in diameter and 1/2 inch in depth, using a core sampler de
scribed on page 10 of this Occasional Paper.

Results and Discussion

Results of the comparison are shown in table 1. Moisture
tents of the bulk samples at 5 cm. of water (.005 atmosphere tens
were significantly higher than those of the cores, ranging from 34
centage points higher for the Commerce silty clay to 10 points hig
the very fine sandy loam soil. As tension was increased to 60 cm
atmosphere), the difference decreased, but was still significantly
The difference in the means remained significant throughout the ir
mediate tension range of .1 to 1 atmosphere, except for the very :
sandy loam. For the sandy soil, bulk samples were significantly
in moisture content only up to .1 atmosphere tension.

At 3 and 15 atmospheres, it was only in the Commerce silt
that bulk samples had significantly higher moisture contents than t
cores. At the same high tensions, cores of the very fine sandy lo
the silt loam from Missouri tended to have slightly higher moistur
tents than the corresponding bulk samples. Differences, however
not significant.

The bulk samples that were repeated at 5 cm. water tensic
dropped considerably in moisture content. The Commerce silty c

Table 1. --Comparison of mean moisture content in percent by weight for core and bulk samples at di
tensions

Tensions (atmospheres)	Procedure A									Procedure B; l silt loam (Mis		
	Commerce silty clay			Briensburg silt loam			Bosket very fine sandy loam					
	Core	Bulk	Difference	Core	Bulk	Difference	Core	Bulk	Difference	Core	Bulk	Diff
	- - Percent - -			- - Percent - -			- - Percent - -			- - Percent		
0.005	36	70	34**	43	66	23**	36	46	10*	40	60	2(
.06	32	48	16**	38	50	12**	31	37	6*	30	44	14
.1	32	42	10**	34	43	9**	28	35	7**	27	35	٤
1/3	30	40	10**	32	41	9**	25	28	3	25	31	٤
1	30	38	8**	32	37	5**	27	25	-2	23	27	4
3	28	32	4**	26	30	4	17	15	-2	17	15	-٤
15	23	29	6*	20	24	4	12	10	-2	12	10	-٤
1/ 005	35	47	12**	41	49	8**	39	39	0	35		
1/.06	32	40	8**	38	42	4**	35	34	-1			

**Difference in means significant at 1 percent level.
* Difference in means significant at 5 percent level
1/Second run.

- 24 -

went from 70 percent to 47 percent, the Briensburg silt loam from 66 to 49 percent, and the Bosket very fine sandy loam from 46 to 39 percent. The corresponding cores remained about the same on the second run as on the first. The difference in moisture content between the rerun core and bulk samples remained highly significant in the Commerce and Briensburg soils but was not significant in the Bosket soil.

The samples repeated at 60 cm. water tension showed the same moisture content for the cores as was obtained on the first run, but a drop in value on the bulk samples of eight percentage points on the heavier soils. Again the difference was highly significant between the core and bulk for the Commerce and Briensburg soils, and was not significant for the Bosket.

Conclusions

At tensions up to 1 atmosphere, bulk samples retain more water than core samples. There is less difference in sandy loam soils than in heavier soils. At higher tensions up to 15 atmospheres it makes little difference whether cores or bulk samples are used in tension analysis, inasmuch as the values obtained show no consistent difference.

Because the core samples represent the "undisturbed" or field condition of the soil, it seems that cores should be used in tension analysis at from 0 to 1 atmosphere.

Literature Cited

(1) Coile, T. S.
     1936. Soil samplers. Soil Sci. 42: 139-142.

(2) Leamer, R. W., and Shaw, B.
     1941. A simple apparatus for measuring noncapillary porosity on an extensive scale. Jour. Amer. Soc. Agron. 33: 1003-1008.

(3) Richards, L. A.
     1949. Methods of measuring soil moisture tension. Soil Sci. 68: 95-112.

Edwin R. Ferguson and William B. Duke[1]

The King-tube (or the modification by Veihmeyer) is w
for obtaining soil samples for moisture content determination
samples are usually collected from various depths and are oft
frequent intervals. To expedite soil-moisture studies being c
east Texas, two devices were developed: a measuring trough
the soil core into depth increments, and a box for storing and
the soil sample cans.

Measuring Trough

The trough (fig. 1) is designed to stand in a slanting pc
that the soil core can be slid from the King-tube with a minim
turbance. The profile position of the core can then be easily
by a yardstick that is permanently fastened to the inside of the

The trough is simple and inexpensive to construct. Th
materials is given below. All lumber is surfaced on four side
sions are nominal.

        2 trough sides, 1" x 4" x 40"
        1 leg, 2" x 4" x 5"
        1 leg, 2" x 4" x 12"
        1 foot, 1" x 4" x 12"
        1 1/4" plywood cover, 5-1/2" x 5-1/2"
        1 3" strap hinge
        1 yardstick

1/   East Texas Branch, Southern Forest Experiment Statioi

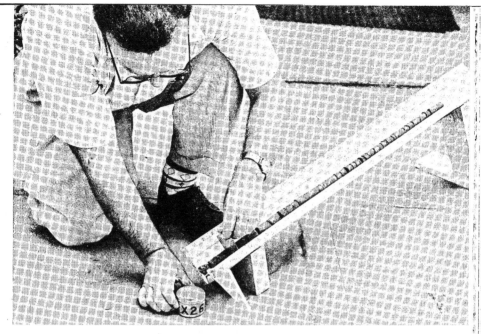

Figure 1.--Measuring trough in use.

Figure 2.--Plywood box for carrying soil sample cans.

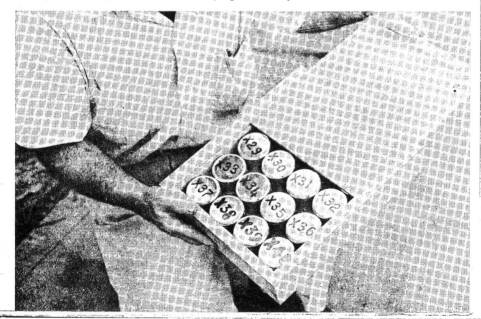

screw; the gate has sufficient play so that it rotate
pieces are notched to conform to the outside angle
short piece is permanently fastened about three inc
(zero) end. The long leg is attached to the trough
hinge. The piece of 1 x 4 x 12 is fastened on the b.
stabilize the trough. For carrying and storage, th
forward under the trough.

To use the trough, the head of the King-tub
against the gate, and the point of the tube is slowly
same time, the soil is pressed gently with the thun
from the constricted end. The core slides free in
upper portion of the core resting against the gate.
sharp withdrawal of the tube along the V toward the
trough will slide the core out. Depths at which the
color and texture can be easily read from the fixed

Soil cores at high moisture contents someti
length within the tube. In this event, the core can
desired lengths by assuming equal compression thr
tifying profile characteristics at known depths. Af
divided into the desired segments, the gate is flipp
ments are slipped into sample cans.

To prevent the core from drying excessivel
the trough, the work should be done rapidly, and p
direct sunlight and wind.

Box

The other device is a box with a sliding cov
sample cans (fig. 2). This box was designed to ca
cans. Besides facilitating the handling of samples
reduced the need for can replacement by providing
impacts.

The bill of materials for the box is as follows. Lumber is surfaced on four sides. Dimensions are actual.

2 sides, 3/4" x 2-1/2" x 13-3/4"
1 back, 3/4" x 2-1/2" x 9-7/8"
1 front, 3/4" x 1-7/8" x 9-7/8"
1 lid pull, 3/8" x 3/4" x 9-7/8"
1 bottom, 1/4" plywood, 11-1/2" x 13-3/4"
1 lid, 1/4" plywood, 10-7/16" x 13-1/4"
1 handle, piece insulated wire, 8-1/2" long

The sides and back have a groove (9/32-inch deep and 5/16-inch wide) cut 3/8-inch from the top edges, in which the lid slides freely. The parts can be fastened together with finishing nails or woodscrews.

TERMINAL PANEL FOR ELECTRICAL SOIL-MOISTURE I

B. D. Doss and W. M. Broadfoot

Since soil-moisture measurement with electrical i
come into fairly wide use, various devices have been deve
vide a rapid and reliable connection of the soil-moisture n
wires from the underground units. Among these devices,
housed multiple switches and a portable multiple plug for
fiberglas instrument have recently been described. [1] The
have conveniently served the purpose for which they were
ever, it has been found that the same objectives can be ob
terminal panel that is much simpler in construction, sture
not require shelter from weather.

Description

The terminal panel is made of 1/4-inch Plexiglas,
wide and long enough to accommodate the number of termi
(fig. 1). Holes 3/16 inch in diameter are drilled in the Pl
distances of 3/4 inch apart vertically and 1 inch apart hor
Brass machine screws 3/16 inch in diameter and 3/4 inch
fastened in the holes with nuts, and lead wires from the m
are soldered to the screw heads. Temperature leads are
the left side, ground wires to the center, and moisture wi
side of the panel. Screw heads and wire connections on th
panel are painted with bakelite resin varnish for waterpro
sulation. The terminal plates are mounted with screws on

A dual clamp--consisting of two battery clamps fa:
gether with Plexiglas--connects the meter with the termin
ends of the machine screws).

---

[1] Palpant, E. H., Thames, J. L., and Helmers, A. E.
shelters for use with soil-moisture units. Southern Fore:
Occas. Paper 128, pp. 21-30. 1953.

The moisture and ground leads from the meter are attached to this clamp. When readings are made, the meter dial knob is set on "moisture" and the clamp is attached to the unit terminals on the left side of the panel for temperature and subsequently to the terminals on the right for moisture readings. The dual clamp can be easily moved from one pair of terminals to another.

This terminal plate is being extensively used at the present time. Its simplicity of construction, sturdiness, and freedom from wear (it has no moving parts) have proven to be distinct advantages.

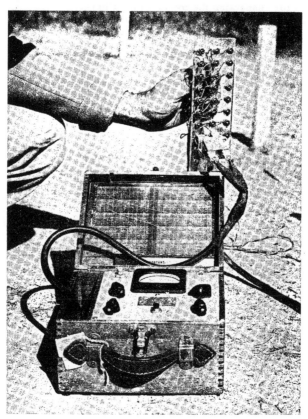

Figure 1. --Terminal panel with fiberglas soil-moisture meter.

PHONE-JACK TERMINALS FOR SO

Frank Woods and Walt H

In west Florida, triple-contact phon
used for connecting the meter to the fibergl.
places the three alligator clamps with which
The jacks are mounted on a Plexiglas termi

Field use has proven the plug and ja
Only one operation is required to make thre
no confusion of leads. However, in making
determinations on each unit, it is necessary
temperature switches to the proper position
to operate, and are less subject to deterior.
switches. They have no moving parts and a
each time a plug is inserted. The 1/4-inch
are recommended; they can usually be obta
shops.

The Plexiglas terminal panel must
support, because a strong pull is sometime
plug. This panel should be coated with a n
decrease the possibilities of short circuits.

A standardized wiring scheme for p
if the same meter is to be used to read sev
standardized scheme could be as follows: t
middle of plug--moisture; base of plug--gr

1/ East Gulfcoast Branch, Southern Fores

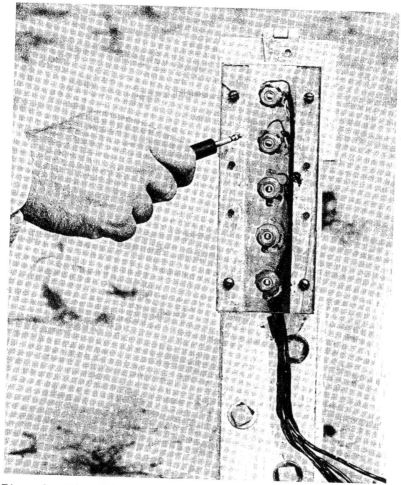

Figure 1. --Phone-jack terminals.

COMPARISON OF LABORATORY AND FIELD CALIBRATION
OF FIBERGLAS MOISTURE UNITS

Charles A. Carlson

Several methods have been used for calibrating fiberglas soil-moisture units. Two laboratory methods, using prepared soil or cores, and one field method are described in the manual of instructions for the instrument (3). Subsequently, in a detailed study of the calibration method, the designers of the unit concluded that "field and laboratory calibration are in good agreement when the laboratory calibration is made in a natural soil core...." (4) In rocky soils a combined method has been used in which laboratory curves were adjusted on the basis of a few field measurements (1).

In a study of the soil-moisture regime in three soils in the vicinity of Vicksburg, Mississippi, preliminary work indicated that laboratory calibration with cores did not agree with field calibration (2). The comparison was not precise because the same moisture unit was not used in the field and laboratory calibrations of any one soil layer. Termination of this study, however, provided an opportunity to make a more rigid test, a comparison of laboratory and field calibrations of several of the units used in the field study.

Methods

In 1951, fiberglas soil-moisture units were installed in Loring silt loam, Collins silt loam, and Commerce clay soils at three-inch depth intervals in quadruplicate stacks. Periodically for six months, duplicate soil-moisture samples were taken at random within a 6- by 6-foot plot surrounding each stack. The soil-moisture contents were plotted against corresponding resistances of the units, and a curve was drawn. Details of installation and field calibration are described elsewhere (5).

Units from single stacks in the Loring and Commerce soils, and from four stacks in the Collins soil, were used for laboratory calibration. Calibration cans were made from eight-ounce moisture cans, 3 inches in diameter and 2 inches high, by cutting disks 2-1/4 inches in diameter from the bottom and lid and placing a hundred-mesh brass

screen backed by 1/4-inch hardware cloth on the inside of the can
against the rim of the openings.

In order to facilitate removal of the units, a hole was dug at the
position of the auger hole that had been used to install the units originally. This hole was filled with water several times so as to wet the
soil about the units. A calibration can was then hammered into the soil
surrounding each unit; the wires of the units passed through a hole in
the screen in the base of the can. The can with enclosed soil core and
unit was then dug out, trimmed, and capped with a lid. Units disturbed
in this operation were not used in the laboratory calibration. Twenty-
two cores were obtained, five at the 4-1/4-inch depth, six at the 7-1/2-
inch depth, five at the 10-1/2-inch depth, four at the 15-inch depth, and
two at the 21-inch depth. Fourteen of these were from the Collins soil
and four each from the Loring and Commerce soils. In order to determine the effects of swelling, if any, half of the cans were bound with
wire.

The cores were saturated from the bottom and left, not quite
submerged, in distilled water overnight. Concurrent weights and resistances were measured once or twice a day as the cores air-dried.
When a unit reached a resistance of about one hundred kilohms, the
core was sealed in a sixteen-ounce can to prevent further drying. After
all samples were dried, the cores were resaturated and the drying
process repeated. Three successive drying calibrations were conducted.

After the third cycle, the cores were slowly wetted by a daily
addition of ten cubic centimeters of distilled water to the top surface of
each core. Weights and resistances were measured just before the
water was added. Following the addition, the cores were equilibrated
in sealed cans until the next measurement. This process was continued
until water dripped from the core base, after which the cores were saturated by standing them in water. Oven-dry weights of the soil cores
were then determined and soil-moisture contents computed for each
weighing of the cans.

The laboratory and field curves of each unit were plotted for
visual and numerical comparisons. For the latter, resistances were
selected at the low end, midpoint, and high end of the resistance range
as reference points for computing moisture content differences between
curves, using the field curves as standards. The moisture content differences at each given resistance were averaged by soil type, by depth,
and for all samples.

One stack of units in the Collins silt loam
the laboratory before it was installed in the field
nary test the core container was smaller and had
cheese cloth. The daily drying procedure include
in a humid chamber maintained at 98 percent rela
brations derived from this earlier test were com
the present study.

Results and Discussion

Table 1 gives average moisture content di
resistances, between the 22 sets of laboratory an
most of the combinations averaged, the laborator
(indicated by positive values) than the field values
0. 5 kilohm, and dryer (negative values) at a res
The lack of agreement between laboratory and fie
shown in figure 1, a typical family of calibration

Generally, the first-cycle laboratory cur
field curve than the second-cycle curve (table 1).
laboratory curve was from 0. 5 to 2. 8 percent wei
cycle curve at 0. 5 kilohm resistance; and from
drier at 50 kilohms resistance. The second- and
usually fell close together.

At high moisture content, resistances ten
successive drying cycles. This may have been d
loss of dissolved salts with repeated saturations.
contents, the resistance frequently dropped some
drying cycle. The reason for this is not known,
night saturation did not wet the soil fully after it
dried for the first time.

The displacement of laboratory curves di
At 5 kilohms the first-cycle laboratory values fo
were, on the average, 1. 4 percent drier than the
silt loam and Commerce clay averaged 3. 4 and 1
For these soils, therefore, a single correction c
adjust laboratory to field curves.

Moisture content differences varied by de
tory curves tended to be moister than the field c
depth, particularly at low resistances. The moi

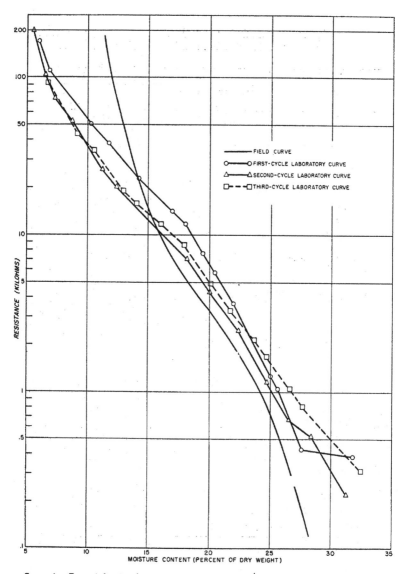

Figure 1-- Typical family of curves, Collins silt loam, 7 1/2- inch depth

Table 1.--Average soil-moisture content differences (as percent of dry weight) between calibration curves at given resistances

Item	Comparisons	Resistance		
		0.5 kilohm	5 kilohms	50 kilohms
	Number	- - Percent - -		
Difference of first-cycle laboratory curves from field curves				
By soil				
Loring	4	-2.2	-1.4	0.1
Collins	14	2.6	3.4	- .2
Commerce	4	.2	1.1	-1.3
By depth (in inches)				
4-1/2	5	- .1	- .4	- .5
7-1/2	6	1.6	2.2	- .3
10-1/2	5	.8	2.2	-2.0
15	4	3.7	4.5	.6
21	2	4.0	1.8	...
Cans tied with wire	10	1.7	2.0	- .7
Cans not tied	12	1.2	2.2	- .2
All samples	22	1.5	2.1	- .4
Difference of second-cycle laboratory curves from field curves	22	2.8	1.2	-2.4
Difference of 1951 second-cycle curves from 1953 second-cycle curves	4	2.5	- .4	- .8
Difference of wetting curves from previous drying curves	22	-3.0	-3.2	[1]-1.4

1/ Average at 20 kilohms.

ᵉsistance of 0.5 kilohm ranged from −0.1 percent at a 4-1/2 inch depth
ᵗ 4.0 percent at 21 inches (table 1). For units installed near the soil
ᵃrface, laboratory calibrations agreed fairly well with field curves;
ᵗr units placed deeper, the laboratory calibration was not satisfactory.
ᵗhe agreement at the surface may have been due to better aeration of
ₜe laboratory cores and of the surface soil in the field. Under these
ₒnditions, carbon dioxide can diffuse readily to the atmosphere. At
ₜwer depths, diffusion is restricted and the carbon dioxide level in-
ᵣeases, thus bringing more salts (the bicarbonates) into solution (6),
ᵗd consequently lowering resistances at given moisture contents. This
ₒuld explain the greater discrepancy of the laboratory curves with in-
ᵣease of soil depth. Field conditions of aeration at lower depths would
ₑ difficult to duplicate in laboratory samples.

Another possible cause of differences in calibrations is the
ₛwelling of cores. This swelling has been observed frequently and may
ₑ expected because in the laboratory the soil is not confined as it is at
ₗwer depths in the field. To determine the effect of swelling, ten of
ₜe calibration cores were bound with wire, so as to secure the lids,
ₜd 12 were left untied. No swelling or lifting of the lids was observed
ₙ any of the cans during the test. Results indicated that tying cans
ᵢth wire had little effect (table 1). Apparently swelling was not a
ᶜtor in this test.

With the earlier laboratory calibrations, before the units had
ᵉen installed in the field, the cores had more opportunity to swell
ᵇcause the ends of the cans were covered only by cheesecloth. This
ₛ evident in the comparison between second-cycle curves (table 1).
ₛelling, and increased water retention of 2.5 percent, is indicated at
ₑ lowest resistance of the earlier calibration. The agreement of the
ᵣves at high resistances is notable (−0.4 percent at 5 kilohms and
ₜ8 percent at 50 kilohms), considering that the laboratory calibra-
ₙs were made by two different methods, 2-1/2 years apart. This
ₐreement confirms the earlier conclusion that equilibration of labora-
ᵣy cores is not necessary (2). Typical curves of the two calibrations
ₑ shown in figure 2.

Further complicating the use of laboratory calibration is the
ᵣror introduced at high resistances by the moisture gradient formed
ᵢthin the core (2). This gradient was produced because the ends of
ₑ core dry more rapidly than the center (where the moisture unit was
ᵗuated). The unit consequently had a lower resistance than was
ₛtified by the gross moisture content of the sample. As a result, the
ₗboratory curve was drier than the field curve in the high resistance
ₐnge.

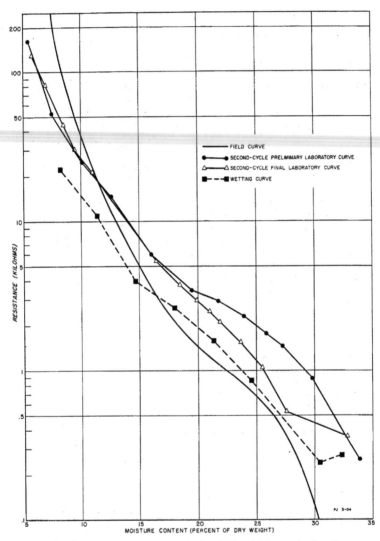

Figure 2 -- Calibration curves before and after field installation, and with wetting
Collins silt loam, 4½-inch depth

The wetting curves were displaced to the left of the drying curves but rejoined at saturation (fig. 2 and table 1). This hysteresis may explain part of the overall difference between the standard laboratory and field curves. Laboratory curves generally are drying curves throughout. Field curves are based on soil samplings interspaced by rains and drainage and depletion periods, and thus may have been a combination of drying and wetting curves. In this study the laboratory wetting curve tended to follow the field curve at low resistances, whereas the drying curve followed more closely at high resistances.

Conclusions

Laboratory calibration of fiberglas soil-moisture units in soil cores did not coincide with the field calibration. Probable causes for the discrepancy are leaching of some of the solutes from the sample, change in aeration and the resulting change in carbon dioxide and bicarbonate concentration in solution, hysteresis between wetting and drying of the soil, moisture gradients in the laboratory cores, and, in some procedures, swelling. It is doubtful that field conditions can be so closely duplicated in the laboratory that field curves can invariably be reproduced. Field calibration is therefore recommended because of the inherent errors in laboratory calibration. Field calibration also reflects more nearly the average moisture content over an area, as opposed to the point sampling associated with laboratory calibration (5).

The laboratory curves do give the general shape and range of the field curves and can be used directly for studies of average field moisture content in which an accuracy of about 3 percent is satisfactory. Where soil-moisture studies are conducted in rocky soils or under pavements, laboratory calibration must be used. But where soil samples can be secured, field calibration is recommended.

Literature Cited

(1) Bethlahmy, N.
  1952. A method for approximating the water content of soils.
    Trans. Amer. Geophys. Union 33: 699-706.

(2) Carlson, Charles A.
  1953. Moisture equilibration in natural cores during laboratory
    calibration of fiberglas soil-moisture units. Southern
    Forest Expt. Sta. Occas. Paper 128, pp. 31-39.

Academic Press, Inc., New York.

A CORE METHOD FOR DETERMINING
THE AMOUNT AND EXTENT OF SMALL ROOTS

Charles A. Carlson

Several root extraction methods utilize soil cores for the determination of root concentration (1, 2, 4). These techniques are relatively quick and simple as compared with those designed to measure the
full extent of individual root systems (3). As commonly applied, however, methods generally involve root-washing techniques that lose unknown amounts of small roots   Since the small unsuberized roots
absorb most of the water and nutrients, they should be measured. A
method using cores was therefore devised to catch the small roots--
those 0. 02 to 2. 0 millimeters in diameter.

The method entails the procurement of a known volume of soil
at a prescribed depth; separation of roots from soil by suspension in a
dispersing solution; collection on a 100-mesh sieve which eliminates
the fine soil particles; decantation to separate roots from coarse soil
particles; measurement of fresh weight, fresh volume, and dry weight;
and ashing the sample to determine the correction for adhering soil
particles. The average diameter of fresh roots is determined with a
microscope, and aggregate length and area of root surface are computed.

Procedure

Core samples were taken from a hole dug with a 5-inch posthole digger to within one inch of the desired depth. A core sampler
with a removable sleeve (about 2-3/4 inches in diameter and 3 inches
high) was driven into the soil at the bottom of the hole. After extraction,
the ends of the core were trimmed flush with the sleeve, and the core
was sealed in a 16-ounce can. Samples at various depths and locations
were taken. Litter and organic matter were scraped off the soil surface
to prevent contamination of the top sample.

Each sample was soaked for 3 days in a 5-gallon bucket containing about 1 gallon of a solution of 0. 02 N sodium hydroxide and 0. 005 N
sodium oxalate. The sample was broken by hand and the mixture stirred
occasionally to break the aggregates and to suspend the soil particles.

The suspension was then diluted with 2 gallons of water, and poured slowly through a 100-mesh sieve. A small stream of water was played on the material to help wash the fine particles through the screen. Dilution and use of the water stream were necessary to prevent clogging of the fine sieve holes. Rhizomes, trash, and roots greater than about 2 mm. in diameter were picked from the sieved material and discarded.

A small stream of water was employed to push the residue on the sieve into a beaker. Floating debris, most common in surface samples, was skimmed off. The roots were then stirred and, after a few seconds during which the sand settled to the bottom, were decanted back to the 100-mesh sieve. Decantation was repeated three or four times to complete the separation of roots from sand. The roots were then transferred to a wide-mouth picnometer for volume determination. Next, the roots were placed on a small 100-mesh screen mounted on a vacuum flask. A rubber stopper was pressed against the roots to help expel excess water, and the fresh weight determined. The sample was then dried in an oven at 65°C. and weighed.

Finally, the sample was put into an evaporating dish and fired in a muffle furnace at a dull red heat, about 650°C. The ash was digested thoroughly in boiling concentrated hydrochloric acid, and the residue washed by decantation, dried, and weighed. The fresh weight and the volume of roots were corrected for residue by subtracting the weight of ash or the calculated volume. A specific gravity of 2.65 was assumed for the residue.

To measure root diameter, a subsample of fresh roots was mounted on a glass slide and all roots measured that crossed a selected transect of the slide. Roots were magnified twelve times; measurements were made by using a graduated eyepiece in a microscope. Four or five subsamples were measured with two transects per slide. A total of 200 to 300 root diameters was measured per sample. The average diameter was determined, and the total surface area and the aggregate length of the roots in the sample were calculated.

The effect of screen size on the collection of roots was checked. Samples were passed through a nest of three sieves--12, 28, and 100 meshes per inch. The root diameter distribution was determined for the material caught on each sieve.

The effects of the chemical dispersing agent on the roots were tested by dispersing three cores in the chemical solution and three matching cores in tap water, and determining the average diameter, moisture content, and density of the fresh roots for each treatment.

- 44 -

Discussion

The washed root samples apparently included most of the fine roots. Few fragments were visible in the water that passed through the 100-mesh sieve. However, since the finest roots and the root hairs are less than 0.149 mm. in diameter (i. e. , the size of the sieve holes), material could be lost even with the 100-mesh sieve.

To the eye, the bulk of the material on the sieve resembled debris. Under the microscope, most of it was seen to be root segments, but some small seeds or decayed organic matter was present and introduced error into the root determination. Because of the opportunity for such contamination, this method is not adapted to soils high in partially decomposed organic matter.

Numerous strands or filaments, approximately 0.010 to 0.020 mm. in diameter, were also visible through the microscope. Since some of these strands were branched, indicative of fungi, none less than 0.020 mm. were measured for diameter. Thus, the root hairs, which have a nominal diameter of 0.015 mm. , were excluded from these measurements. Weight and volume determinations, however, included the filaments.

The diameter distribution of roots passing through the nest of three sieves is given in table 1. The diameter distribution of the roots was about the same for each sieve; no segregation occurred. Although the majority of the roots caught on any one sieve had diameters as small as or smaller than the openings in the 100-mesh sieve, the lengths of the segments were greater than the openings. The length apparently impeded passage through the sieve. The smallest-meshed sieve, used by itself, retained more roots than either of the larger sieves because it caught shorter segments.

The chemical dispersing agent had no apparent effect on the root characteristics. Diameter, moisture content, and density were the same with or without the chemical treatment (table 2).

Root concentrations for these samples averaged 1.7 percent volume of fresh roots per volume of soil. Aggregate length, as calculated from average root diameter, was about 90 meters per 100 cc. of soil. Aggregate surface area of the roots was 416 square cm. per 100 cc. of soil. Thus, though the root volume was small, the aggregate length and surface area per unit volume of soil were considerable.

Root diameter (millimeters)	Material collected			Treatment	Av dia
	--on 12 mesh	--through 12 on 28 mesh	--through 28 on 100 mesh		
	- - - Percent - - -				
0 02 - 0 10	46	38	32	Oxalate-hydroxide	
11 - .20	30	31	46	Sample 1	
21 - 30	16	19	16	2	
31 - .40	6	4	4	3	
Greater than .40	2	8	2		
				Tap water	
				Sample 1	
Average root diameter	0 16 mm.	0 19 mm.	0 16 mm.	2	
				3	

Data obtained by this proposed method were
with those from the study by Dittmer. For the com
data were converted to measurements of 100 cc. of
give comparable results, considering that the metho
cludes roots and filaments smaller than 0. 020 mm.
of the probable contamination by fungi.

Table 3. --Root characteristics by two sampling methods

Item	Sample	Average diameter	Aggregate length	A
	Number	Mm.	Cm.	
Vicksburg method [1]				
Herbaceous on clay soil		0. 135	17, 228	
Herbaceous on silt loam soil		. 160	9, 020	
Hardwoods on silt loam soil	8	. 186	5, 682	
Dittmer's method [2]				
Oats				
Roots	2	. 225	658	
Root hairs	2	. 013	117, 900	
Rye				
Roots	2	. 245	921	

In summary, this method appears useful in studies with a large number of samples in which volume or weight of small roots is desired. Measurements for the average diameter are tedious but necessary if length or surface area is needed.

Literature Cited

(1) Dittmer, H. J.
1938. A comparative study of the subterranean members of three field grasses. Science 88:482.

(2) Gist, G. R., and Smith, R. M.
1948. Root development of several common forage grasses to a depth of eighteen inches. Jour. Amer. Soc. Agron. 40: 1036-1042.

(3) Pavlychenko, T. K.
1937. The soil-block washing method in quantitative root study. Canad. Jour. Res. 15: 33-57.

(4) Ward, H. S., Jr.
1949. Reactions of adapted legumes and grasses on the structural condition of eroded Lindley-Weller soils in southeastern Iowa. Ecol. Monog. 19: 145-171.

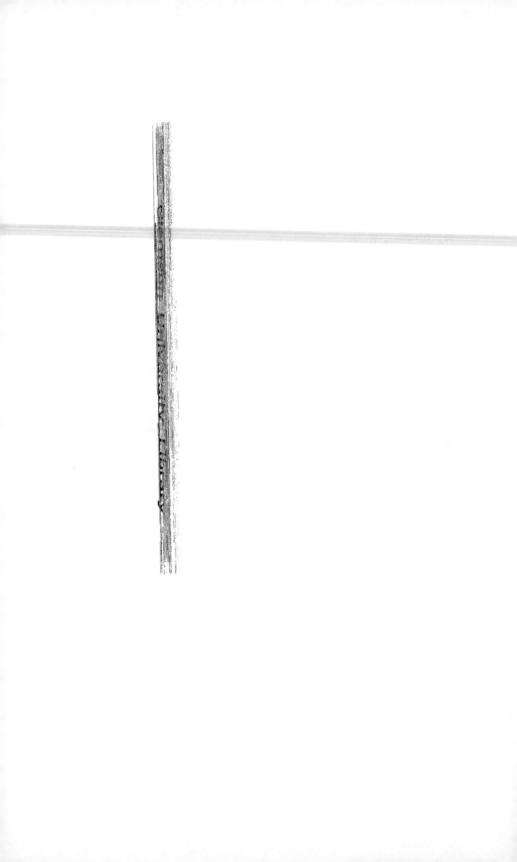

SOME NEW
PINE POLLINATION TECHNIQUES

Philip C. Wakeley and Thomas E. Campbell

SOUTHERN FOREST EXPERIMENT STATION
Philip A Brieglab, Director

Forest Service, U.S. Department of Agriculture

SOME NEW PINE POLLINATION TECHNIQUES

Philip C. Wakeley
Southern Forest Experiment Station
and
Thomas E. Campbell
A. J. Hodges Industries, Incorporated

The authors are cooperating in experimental breeding of southern pines on the A. J. Hodges Industries Experimental Area near Many, in Sabine Parish, Louisiana. They began controlled pollination in 1954 with the techniques described by Cumming and Righter (1) 1/, except that, in common with several other investigators, they substituted synthetic sausage casings for cloth-and-plastic pollination bags. Their work at Many during the spring of 1954 and discussions with fellow workers elsewhere have brought to light new information and further modifications of technique which it seems desirable to share promptly with other agencies and individuals interested in forest tree improvement.

Stages of Development of Female Strobili

Figures 1-3 are offered as guides--more detailed than those in previous publications (1, 2)--to the critical stages at which female "flowers" (strobili) must be protected by bags, or at which pollen must be applied. The stages are designated according to the system of Cumming and Righter (1), who describe them as follows:

(1) Buds small, (2) buds large, (3) buds opening, (4) flowers partly open, (5) flowers maximum, (6) flowers closed, (7) cones enlarging. They add that stages 1-3 are determined by the size or condition of the bud and stages 4-7 by the position of the cone scale relative to the axis of the conelet. They describe the cone scales (not the spines) at stage 5, the optimum for pollination, as being roughly at right angles to the axis. They describe the cone scale at stage 6 as so

---

1/ Underscored numbers in parentheses refer to Literature Cited, page 13.

Right, stage 2-3, bud
large to bud opening;
left, stage 3, bud
opening.

Stage 3-4, bud opening
to flower partly open.

Stage 4, flower partly
open.

Stage 5, "flower max-
imum" (optimum stage
for pollination).

Stage 5-6, flower
closing and becoming
non-receptive to pollen.

Figure 1.--Developmental stages
of female strobili of longleaf pine,
Pinus palustris, according to the
designations of Cumming and
Righter (1).

tightly closed that pollination is no
longer possible. The stages I, II,
III, and IV of the earlier article by
Snow and co-workers (2) correspond closely to stages 2, 3, 5, and
6 of Cumming and Righter.

In the 1954 breeding at
Many, on three species of southern
pine and on a natural hybrid between two of the three, visible
openings between scales were found
superior to scale angle as evidence
that flowers were in stage 5. Stage
6 was more clearly marked by the
closing of these openings and by an
obvious thickening of the tips of the
cone scales, which made them look
like little pillows (fig. 3, left), than
by scale angle. In stage 3, the
spines of the cone scales were more
or less rudimentary, depending on
species; in stage 4, they were fully
formed in all species, but the openings between scales that were
characteristic of stage 5 had not
yet appeared. For easiest recog-
nition, especially in the smaller-
flowered species, these details of
development require a 12-power
hand lens, but in practice, subject
to occasional verification with the
lens, most of them can be recognized with the naked eye.

Plainly discernible changes
in color often accompany the
changes in size and in scale position and thickness shown in figures 1-3. These color changes
differ with species, and to some
extent with trees within species.
As a rule, female strobili of the
southern pines become either darker,
or more distinctly tinged with green,

lination. Flower at left intact; note that the edges of the scales, below the spines, are thin and sharp, and that wide spaces between the scales offer easy access to pollen. Flower at right bisected longitudinally to show more clearly the wide spaces between scales.

and no longer receptive to pollen. Flowers have increased in diameter since stage 5. Note in intact flower to left that ends of scales have thickened so that they press closely together, with no spaces for pollen to enter; in longitudinally bisected flower to right, that gaping openings shown between scales in figure 2 have disappeared.

as they develop. Longleaf strobili, for example, usually pass from pale pink at stage 3 through brighter pink at stage 4 and bright rose at stage 5 to reddish purple at stage 6 and dark purple at stage 7.

Rates of female flower development vary greatly with species, individuals within species, season, protection or non-protection by bags, application or non-application of pollen, and probably other influences. To combine sure protection against contamination by wind-borne pollen and assurance of normal development after pollination, Cumming and Righter advocate bagging at not later than stage 2, and removal of bags after stage 7 is reached. Snow and co-workers considered bagging of slash pine in northern Florida satisfactory well into the equivalent of stage 3. On the A. J. Hodges Industries Experimental Area in 1954, flowers were bagged through stage 3 in crosses designed to produce bulk populations of hybrids from which seedlings suspected of being contaminants could be culled. For more rigorous experiments, however, an attempt was made to bag no flowers beyond stage 2 or very early stage 3.

Right, stage 2-3, bud
large to bud opening;
left, stage 3, bud
opening.

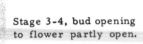

Stage 3-4, bud opening
to flower partly open.

Stage 4, flower partly
open.

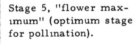

Stage 5, "flower max-
imum" (optimum stage
for pollination).

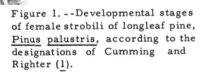

Stage 5-6, flower
closing and becoming
non-receptive to pollen.

Figure 1.--Developmental stages
of female strobili of longleaf pine,
Pinus palustris, according to the
designations of Cumming and
Righter (1).

tightly closed that pollination is no
longer possible. The stages I, II,
III, and IV of the earlier article by
Snow and co-workers (2) correspond closely to stages 2, 3, 5, and
6 of Cumming and Righter.

In the 1954 breeding at
Many, on three species of southern
pine and on a natural hybrid between two of the three, visible
openings between scales were found
superior to scale angle as evidence
that flowers were in stage 5. Stage
6 was more clearly marked by the
closing of these openings and by an
obvious thickening of the tips of the
cone scales, which made them look
like little pillows (fig. 3, left), than
by scale angle. In stage 3, the
spines of the cone scales were more
or less rudimentary, depending on
species; in stage 4, they were fully
formed in all species, but the openings between scales that were
characteristic of stage 5 had not
yet appeared. For easiest recog-
nition, especially in the smaller-
flowered species, these details of
development require a 12-power
hand lens, but in practice, subject
to occasional verification with the
lens, most of them can be recognized with the naked eye.

Plainly discernible changes
in color often accompany the
changes in size and in scale position and thickness shown in figures 1-3. These color changes
differ with species, and to some
extent with trees within species.
As a rule, female strobili of the
southern pines become either darker,
or more distinctly tinged with green,

 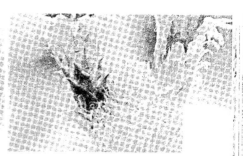

Figure 2. --Female strobili of long-leaf pine in stage 5, optimum for pollination. Flower at left intact; note that the edges of the scales, below the spines, are thin and sharp, and that wide spaces between the scales offer easy access to pollen. Flower at right bisected longitudinally to show more clearly the wide spaces between scales.

Figure 3. --Female strobili of longleaf pine in stage 6, closed and no longer receptive to pollen. Flowers have increased in diameter since stage 5. Note in intact flower to left that ends of scales have thickened so that they press closely together, with no spaces for pollen to enter; in longitudinally bisected flower to right, that gaping openings shown between scales in figure 2 have disappeared.

as they develop. Longleaf strobili, for example, usually pass from pale pink at stage 3 through brighter pink at stage 4 and bright rose at stage 5 to reddish purple at stage 6 and dark purple at stage 7.

Rates of female flower development vary greatly with species, individuals within species, season, protection or non-protection by bags, application or non-application of pollen, and probably other influences. To combine sure protection against contamination by wind-borne pollen and assurance of normal development after pollination, Cumming and Righter advocate bagging at not later than stage 2, and removal of bags after stage 7 is reached. Snow and co-workers considered bagging of slash pine in northern Florida satisfactory well into the equivalent of stage 3. On the A. J. Hodges Industries Experimental Area in 1954, flowers were bagged through stage 3 in crosses designed to produce bulk populations of hybrids from which seedlings suspected of being contaminants could be culled. For more rigorous experiments, however, an attempt was made to bag no flowers beyond stage 2 or very early stage 3.

Synthetic Sausage-Casing Bags

All bagging on the Hodges Experimental Area
with "3-1/4-inch diameter, clear, high-stretch, regul;
ture-permeable synthetic sausage-casing. The partici
came in approximately 23-inch lengths. Casings were
inch lengths for use on longleaf, loblolly, and Sonderegg
of 9 or 8 inches (before folding the ends to close ther
for all but the topmost, very vigorous twigs of shortle
inch diameter" (which refers to the dimensions of the
from such casings; the flattened tube of the fresh, dry
4-7/8 inches across) was satisfactory for all species, b
growing twigs of loblolly and Sonderegger pines 30-inch
15-inch lengths would have been better than the 11-1

Fresh casings were soft and flexible, and were ε
stapled, and attached to the twigs. They soon stiffened
to sun and air. Casings that had been stored at rc
for a year were stiff, brittle, and much less easily fol
than fresh casings.

One end of each casing was stapled shut in the off
lation. The quadruple reverse fold shown in figure 4, A
after another worker 2/ had demonstrated that the ove
shown in figure 4D was not pollen-tight at the corners
finement 3/ consists of inserting cross-wise staples (fi
bag fully expanded and away from the flowers after ι

Prior to installation of the bags, the twigs were
and wrapped with cheap non-absorbent cotton, essentiι
by Cumming and Righter (1). Ordinary pipe cleaners

2/ Bruce Zobel, Texas Forest Service, College Statio
3/ Shown to the authors by Robert G. Hitt, Departm
University of Wisconsin.
4/ Called to the authors' attention by Harold J. Derr,
Experiment Station, Alexandria, Louisiana.

- 4 -

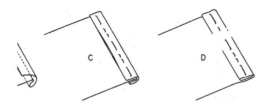

stages in making and stapling pollen-tight,
nd of synthetic sausage-casing pollination
fold in the end of sausage casing is not pol-

ςs to twigs than did the paper-covered wires
pply houses for fastening plants to stakes.

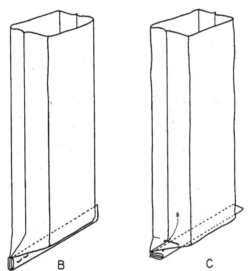

e-casing bag. **A,** With one end folded and
d. **B,** Expanded to maximum rectangular
o it and shaping it with the fingers. **C,** Fixed
by folding end seam into plane at right
astening bottom corners of expanded bag to
aple (s) in each side. Bags are prepared as
B and C are carried out in the tree, just

Disposable Pollen Extractor

To meet immediate needs, some attention was focused on dev
opment of a pollen extractor less elaborate and expensive than
large-scale apparatus described by Cumming and Righter (1). Labo
tory space was not yet available for the special room, sink, steriliz
and large funnels and bags described by those authors or for the blov
and manifold they subsequently developed to aerate their extract
There was need for separate extraction of numerous individual lots
pollen smaller than those for which the large funnels and bags w
especially designed. Lastly, under the climatic conditions of the Sou
difficulty had been encountered with molding of catkins and pollen
quantities as great as those for which Cumming and Righter's equ
ment was best fitted.

The unit extractor developed at Many in 1954 consists of
following (fig. 6):

1. A kraft paper bag of 8-pound capacity.

2. A screen-wire cage, 5 inches deep and of a diameter to per
easy insertion into the paper bag.

3. A plastic funnel 2-1/2 inches from rim to outlet, and with a r
2-7/8 inches in diameter; the opening of this funnel is cove
with fine voile or batiste.

4. Plastic adhesive tape (1 inch wide) to shape the bag, surgi
adhesive tape (1/2-inch wide) to fasten funnel in bag, string
suspend the extractor, and absorbent cotton to plug outlet
funnel.

Both the small plastic funnels (usually obtainable for 5¢ api
at dime stores) and the 8-pound kraft bags vary somewhat in dimensi
from source to source. With due allowances for such variations, c
struction of the extractor is as follows:

a. For the cage to hold the catkins, cut a 5- by 18-inch rectan
of 16-mesh galvanized screen wire and shape it into a cylinder 5 inc
deep and 5-2/5 inches in diameter by overlapping the ends 1 inch
stapling them together. Cut two disks of screen wire each 6-2/5 inc
in diameter. Around the edge of each make cuts about 1 inch apart
extending 1/2 inch toward the center of the disk Bend the tabs

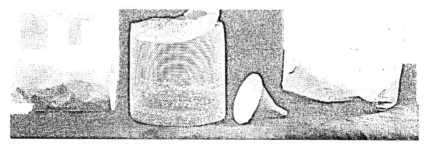

Figure 6. --Component parts of unit pollen extractor: 8-pound capacity kraft paper bag (left, uncut; right, with end and corners trimmed for shaping to fit funnel); screen-wire cage containing catkins (top to be stapled shut before insertion in kraft bag); and 2-7/8-inch plastic funnel covered with voile.

formed up at right angles to the main disk. Insert one such disk 1/2 inch in the bottom of the cylinder and staple each tab to the side of the cylinder. Insert the other disk in the top of the cylinder and staple halfway round, leaving the other half unfastened and bent upward (fig. 6) to permit pouring catkins into the cylinder.

b. To cover funnel with voile, stretch the voile tightly over a sheet of waxed paper on a smooth table top, coat rim of funnel with quick-drying acetate cement, and press rim down on voile till cement dries; trim off the voile close to the funnel with a razor blade.

c. To shape the kraft bag, spread to maximum width and flatten out. Cut open end square and mark the mid-point. From points 5 inches below the square-cut corners of the open end, to points 1-3/8 inches on either side of the marked mid-point of the open end, draw two pencil lines. Cut off the corners of the bag along these lines (bag to right in fig. 6).

d. Place catkins in the wire cage and staple the cage shut. Place cage in the bag. Draw cut edges of bag together, and seal into position

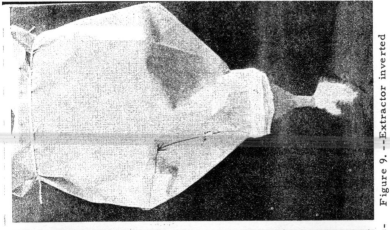

Figure 9.--Extractor inverted and hung up to permit catkins to dry. Note cotton plug in outlet of funnel; this is essential to prevent loss of pollen as it is shed from the catkins.

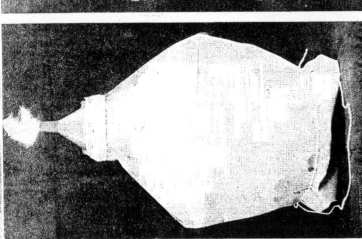

Figure 8.--Extractor completely assembled, with funnel turned vertically, pulled up as far as it will go into the narrowed mouth of the bag, and fastened firmly in place with two turns of 1/2-inch surgical adhesive tape outside the bag.

with strips of 1-inch plastic tape on both inside and outside of the bag (fig. 7); this is most easily done by passing the mouth of the bag over a smooth, slightly rounded strip of wood nailed to a table and projecting 1 foot beyond its edge.

e. Insert funnel in mouth of bag by tipping funnel as shown in figure 7. (If bag has been cut and taped correctly for size of funnel used, it will admit funnel easily in this position but will not permit it to be withdrawn when turned vertically.) Straighten funnel to vertical position, pull upward snugly into tapering mouth of bag, and fasten securely in place by two turns of 1/2-inch surgical adhesive tape, centered on rim of funnel, around the outside of the bag.

f. Plug the outlet of the funnel with absorbent cotton, tie the bottom of the bag firmly around the lower part of the cage with a double turn of soft cord, and to this cord fasten a bridle by which to hang up the extractor (fig 8); invert and hang the extractor (fig. 9) in a warm, dry place.

At Many in 1954, best results were obtained with this extractor by placing in the cage a 1-inch layer of unwashed catkins from which pollen was just beginning to be shed. Such a charge usually yielded practically all its pollen in 24 to 48 hours.

It was demonstrated, however, that for rigorous experiments the extractor could be used with the washing technique described by Cumming and Righter (1). Catkins not quite ready to shed were caged and washed, and cage and catkins were inserted under water in an uncut 8-pound kraft paper bag, essentially as described by those authors for insertion in the cloth bag they used. The mouth of the bag was closed under water, withdrawn from the water still closed, and taped shut. All free water was shaken out of the catkins and cage and drained out through the folds closing the mouth of the bag, without exposing the catkins to the open air with its possible load of pollen. In this process, the kraft bag itself absorbed most of any water clinging to the sides of the cage.

The wet kraft bag was then removed, and replaced with a cut and fitted bag, as described under (d) above, in a pollen-free atmosphere. A pathologist's inoculation chamber or a chemical hood might be used to provide such an atmosphere. In practice, some pollen was transferred in a shower bath in which the air had been washed by running the hot shower for several minutes.

the 1954 season, no trouble was experienced witl
pollen in this extractor, with either unwashed o
only extraction failures resulted from misjudgin{
atkins or the heat to which to expose them. Ful
actors hung in an unused greenhouse released po

:xtraction of pollen, the kraft bag, the voile on tl
plug are thrown away. Cage and funnel are wash{
and, as a double precaution against contaminatio
.lcohol. Fresh voile is cemented onto the funn{
·epared, and the next batch of pollen extracted.

l as described, the kraft-bag extractor may yiel{

r quantities of pollen can be extracted in 10- o
ynthethic sausage-casing of the type used for bagg

ethod is to staple the bottom of a casing shut (poll
l of catkins in the casing, staple a pollen-tight f
id hang the casing by its top in a warm place; pol
laken into the bottom of the casing is then withdra
eedle of which is thrust through the casing $\underline{6}/$. A
)f this method is lack of means of sifting the polle
ragments of catkins, or the small larvae frequen
·acted pollen, may clog the needle.

ernative method is to fill a segment of casing n{
iple pollen-tight seams at both ends, and hang it i
en is discharged, one end of the casing is out of
tightly over a funnel of suitable size $\underline{7}/$, through v
shaken out into a vial. A funnel covered with vo
/iously described, frees the pollen of larvae ai

---

suggested by R. E. Schoenike, Southern Fores1
Crossett, Arkansas, in 1953.
the authors' attention by François Mergen, Sou
ment Station, Lake City, Florida.
by Robert G. Hitt, Department of Genetics, Uni
adison, Wisconsin.

The moisture permeability that gives synthetic sausage casings their advantage over moisture-proof substances for bagging female strobili apparently makes them equally good containers in which to extract pollen. All three of the co-workers cited (footnotes 5, 6, 7) report good yields of pollen from casing extractors, with no molding. The casings are, moreover, inexpensive and extremely simple to prepare.

Pollinizer

For some time, glass or preferably steel and plastic hypodermic syringes of 10 cc capacity equipped with long No. 16 veterinary needles and with special rubber bulbs instead of plungers (1) have been standard pollinating equipment.

A unit pollinizer of this type costs $3.00 to $4.00. Glass syringes are easily broken. The needles must be thrust into corks or the pinched ends of empty .22 long rifle cartridges for safe transportation up the tree. Long needles sometimes become clogged with pollen and must be cleared with a special rod or the wire of a twig tag.

In an attempt to find a cheaper and better pollinizer, 1-ounce rubber ear syringes (or, for larger capacity, bulbs of babies' enema syringes) were tried (fig. 10). These retail for 39¢ to 49¢, and, with needles at 25¢, reduced the price of the individual pollinizer to one-fourth or one-sixth of that paid for glass syringes with special bulbs instead of plungers.

The neck of the ear syringe was cut off at a point such that the orifice left formed a tight joint around the shank of the veterinary needle below the knurled portion (fig. 10). With the enema syringe, the hard-rubber nose was pulled out, and the knurled portion of the needle seated snugly in the neck of the bulb.

A 3/4-inch No. 16 veterinary needle was found preferable to longer needles. The bulb was filled with pollen before the tree was climbed. For transportation up the tree, the needle was inserted in the bulb with the point inward and held in place with 1-inch-wide plastic adhesive tape. For pollination, tape and needle were removed and the needle reinserted point outward. Discharge of pollen was best with the needle pointed horizontally or somewhat downward, depending on whether the bulb was nearly full or nearly empty. Very slight compression of the bulb injected ample pollen into the bag. The 1-ounce ear syringe carried enough pollen for 100 or more bags.

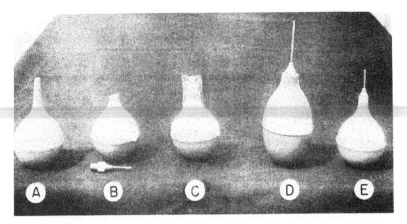

gure 10.--A.--Unmodified 1-ounce rubber ear syringe. B.--Ear
ringe with neck cut off to permit snug insertion of shank of the 3/4-
h No. 16 veterinary needle lying in front of bulb. C.--Ear syringe
h needle taped in place, point inward, for carrying up tree. D.--
bies' enema bulb with (longer) No. 16 needle in place for pollen in-
tion. E.--Ear syringe with 3/4-inch No. 16 needle in place for pol-
injection. The bands of surgical adhesive tape on the bulbs are for
rking, in pencil, the kinds of pollen contained.

The low cost of this pollinizer permitted use of separate units
many different lots of pollen. Any cleaning and sterilizing of polli-
ers while in the tree, and usually any refilling of syringes in the
e, was therefore unnecessary. Ordinarily, no glassware of any
d had to be carried up trees. After pollinizers had been used and the
edles had been turned point inward and taped into place, the bulbs
ld be dropped to the ground without injury or loss of pollen. Bulbs
needles were thoroughly cleaned with hot, soapy water before re-
, and further washed with isopropyl alcohol to kill any adhering
ins of pollen. It was found, however, that this alcohol caused rapid
terioration of the rubber bulbs.

When pollen was carried into the field in 10-cc. vials with small
cks, the close fit between shoulder of bulb and neck of vial made it easy
transfer pollen from vial to bulb (by gravity and slight suction, figure
or back from bulb to vial (by gravity alone) with little or no spilling
no appreciable danger of contamination, even in a high wind.

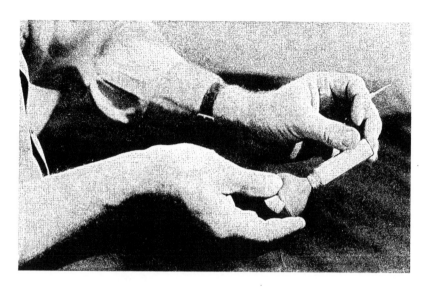

Figure 11.--Ear syringe pollinizer, with needle removed, being filled with pollen from small vial by gravity and slight suction.

Literature Cited

(1) Cumming, W. C., and Righter, F. I.
    1948. Methods used to control pollination of pines in the Sierra·
    Nevada of California. U. S. Dept. Agr. Cir. 792. 18 pp.,
    illus.

(2) Snow, A. G., Jr., Dorman, K. W., and Schopmeyer, C. S.
    1943. Developmental stages of female strobili in slash pine.
    Jour. Forestry 41: 922-923, illus.

SOIL MOISTURE
AFFECTED BY STAND CONDITIONS

Ralph C. Moyle and Robert Zahner

SOUTHERN FOREST EXPERIMENT STATION

Philip A Briegleb, Director

Forest Service, U. S. Department of Agriculture

SOIL MOISTURE AS AFFECTED BY STAND CONDITIONS

Ralph C. Moyle and Robert Zahner

In areas of light rainfall, soil moisture is apt to be the factor that most limits tree growth. To date, however, little thought has been given the problem in areas such as south Arkansas and north Louisiana, where rainfall averages 50 inches or more per year.

Seemingly, an average of nearly an inch of water per week should more than suffice for maximum growth of forest stands. The difficulty is that rainfall is not uniform throughout the year in this territory. Summer droughts from two to six or more weeks in length are fairly regular. Narrow growth rings and mortality of trees have raised the question of just how adequate the moisture supply is.

The seasonal trend of soil moisture in forests has received limited attention. Gaiser (2)[1], working on three upland oak sites in southeastern Ohio, found that water available for plant growth dropped to a critical level by mid-summer. The water regime under a young pine plantation in South Carolina was studied by Hoover, Olson, and Greene (3). They reported that water was removed from a 60-inch depth at about the same rate as from the surface depths, even though roots were concentrated near the surface. Roots removed water down to the wilting point regardless of the depth or root concentration.

Boggess (1), in south Illinois, has shown strong correlation between tree growth and available soil moisture. Both pine and hardwood radial growth virtually ceased by the end of June, when soil moisture neared the wilting point. Intermittent summer showers failed to replenish moisture (in the root zone as a whole) enough for growth to continue.

---

[1] Underscored numbers in parentheses refer to Literature Cited, P. 14.

It would seem that soil moisture becomes
the season when other environmental factors are n
good forest growth. There is a need for detailed n
tions during this period of limited water supply. T
concerned with the effects of various forest conditi
depletion of soil moisture.

Study Areas and Methods

During the very dry summer of 1953, soil
periodically measured under six different forest st
southeastern Arkansas. All of the areas studied w
type, a Leshe[2] silt loam, on the Crossett Experin
Topography also was similar on all areas.

Soil and topographic descriptions

The Leshe soil series is developed on hig]
alluvium composed of sediment from the Permian
red and yellow soil zones. Topography is of the "t
type, with imperfect surface drainage. The micro
up to 4 percent. The dark grey silt loam surface ε
yellow silt loam subsoil at from 8 to 12 inches, wit
silt loam pan development at about 24 inches. Inte
slow. Table 1 summarizes soil profile data for th
lolly pine site index for Leshe soil on imperfectly (
about 90 feet at age 50.

Table 1.--Physical properties[1] of Leshe silt loam

Soil depth (inches)	Composition			Bulk density	Moisture equiva- lent
	Sand	Silt	Clay		
- - Percent - -					- - - Ac
0- 6	21	66	13	1.37	1.6
6-12	20	63	17	1.47	1.7
20-26	20	59	21	1.53	1.8
42-48	20	59	21	1.64	2.0

1/ Averages of the 6 study sites.
2/ Moisture equivalent minus wilting point.
3/ Per 6-inch layer.

2/ Series tentatively assigned, pending correlatior

- 2 -

nd descriptions

Six different stands, on three separate areas, were studied.
two three areas consist of an all-aged loblolly and shortleaf pine stand
h a typical understory of hardwoods, a non-commercial all-aged
dwood type, and a relatively even-aged young hardwood type with
nmercial possibilities. Within the pine type two stand conditions
·e measured, the pine type itself and a clearcut bare area. The all-
d hardwood type furnished three stand conditions: an area that had
eived timber stand improvement in 1951, an identical area that
eived the same treatment in 1953, and an untreated area. The
ng hardwood type provided only the one condition of even-aged
dwood.

)le 2 summarizes stand data for the three areas.

)le 2.--Stand data per acre for the three study areas

Area type and pecies group	Stems	Average d.b.h.	Largest d.b.h.	Basal area
	Number	Inches	Inches	Sq. ft.
-aged pine				
Pine	95	13.4	26	88
Hardwoods	50	6.2	8	13
Total	145			101
l hardwoods				
Pine	5	7.0	10	1
Hardwoods	150	9.3	16	70
Total	155			71
ung hardwoods				
Pine	0	...	...	...
Hardwoods	310	6.3	9	67

All-aged pine.--The Pine site is in an area typical of much of
managed pine forest in southeast Arkansas and northern Louisiana
. 1). The dominant trees range up to 26 inches d.b.h. The basal
a is about 70 square feet per acre, and the board-foot volume is
) 8 M per acre. The understory hardwoods are chiefly sweetgum

- 3 -

Figure 1. -- The all-aged loblolly-shortleaf pine area.

and southern red oak, of all sizes up to about 20 to 30 feet in height and 8 inches d.b.h. Other oaks, dogwood, hawthorn, and vaccinium species are also present.

During the winter of 1953, a 30- by 30-foot plot within the pine area was completely cleared of all vegetation. All overstory and larger understory trees within 50 feet of the plot center were also removed. Understory hardwoods on the plot were cut and the stumps chemically treated to prevent sprouting; the foliage of lesser vegetation was chemically sprayed. The resulting plot was completely bare of vegetation. The soil was exposed by removing the surface litter. This plot is called the Bare site.

All-aged cull hardwoods.--This stand is typical of much of the badly cutover upland pine-hardwood forest of the area (fig. 2). Most pine was removed at least 40 years ago and the small cull hardwoods that remained after logging grew into the present all-aged hardwood type, with a moderately developed hardwood understory. The principal species are southern red oak, post oak, and sweetgum, the largest of which have diameters up to 16 inches. A few scattered pine seedlings and saplings are also on the plots. About four large loblolly pine and shortleaf pine per acre occupy the area, but do not occur on the plots themselves.

Within the all-aged cull hardwoods area, three 50- by 50-foot plots were chosen for similarity of stand conditions. In 1951, one plot had been given a timber stand improvement treatment in which all hardwoods four inches d.b.h. and over were frilled and treated with a water emulsion of 2,4-D; the remaining stand consisted of a few small hardwoods and scattered pine. In 1953, a second plot was treated identically. The third plot was left as an untreated control. The three plots are called, respectively, TSI 1951, TSI 1953, and Untreated Hardwood. In contrast to the other two plots, the TSI 1951 plot had developed some scattered pine reproduction by the summer of 1953.

Young even-aged hardwood.--This area is being managed for the commercial production of upland hardwood sawlogs (fig. 3). The dominant stand varies from 30 to 35 years of age. The area is well stocked, and diameters range from 3 to 9 inches. Principal species are southern red oak and post oak. Understory hardwoods are present but not abundant.

Figure 2. --The cull hardwoods area.

Figure 3. --The young upland hardwoods area.

In this even-aged hardwood stand, a representative 50-foot plot was selected. This plot is called the Young Har site.

Measurements

Soil moisture was measured on all six sites from M September 24, 1953. Moisture content was determined at fo 0-6 inches, 6-12 inches, 20-26 inches, and 42-48 inches. O 1953 and the Young Hardwood plots[3/], four randomly selecte were sampled gravimetrically at each depth at two-week inte On the other four sites fiberglas units were employed accord method described by Reinhart (4), except that calibrations of were made on 18- by 18-foot areas. Fiberglas unit readings four depths were recorded daily.

Bulk density and wilting point determinations were 1 three to four times for each of the four depths at each site. density was measured with 71.0 c.c. undisturbed soil cores. point was determined at 15 atmospheres pressure on disturb samples in a pressure-membrane cell.

Rainfall

Gross daily rainfall was recorded for the pine and t hardwoods areas. Rain gauges were placed in open areas nc from the study sites. Rainfall measured at the cull hardwoo was considered applicable to the Young Hardwood site, three of a mile away.

Rainfall data are summarized in table 3. All of the (9.50 inches for the hardwood areas and 10.06 for the pine) 1 to May 20, when soil-moisture observations began. This ex more than five inches during May, on top of normal winter a1 precipitation, was slow in draining off. On the other hand, (1.4 inches fell on either area from May 20 to July 20, and it seven separate showers. A severe drought was therefore fe the latter part of June and the first 20 days of July. Both ar corded about normal precipitation for August, although the f. weeks of this month were dry. September again showed sev(conditions.

─────────────────────

3/ C. X. Grano collected the data on these two sites.

Table 3.--Rainfall data for May through September, showing normals
for the Crossett area and the 1953 record for two of the
study areas

| Month | Rainfall | | | | |
| | Normal [1] | Cull hardwood area | | All-aged pine area | |
		Recorded 1953	Deviation from normal	Recorded 1953	Deviation from normal
	- - - - - - - - - - - - Inches - - - - - - - - - - - -				
May	4.14	9.50	+5.36	10.06	+5.92
June	4.07	.93	-3.14	.65	-3.42
July	4.62	3.86	- .76	5.26	+ .64
August	3.21	2.62	- .59	2.91	- .30
September	2.39	.31	-2.08	.26	-2.13
Total	18.43	17.22	-1.21	19.14	+ .71

1/ From official records of U. S. Weather Bureau.

Analytical methods

Soil moisture observations were calculated in terms of inches
of available water per 6-inch layer of soil for each of the four depths.
Available water is taken as that field moisture content over and above
the moisture content at wilting point.

Available water in the upper 48 inches of soil on each site was
calculated as the sum of water available in the four 6-inch layers, plus
an interpolated amount of water in the intervening layers.

Results and Discussion

The three forested sites showed similar trends of soil
moisture depletion. Where vegetation was modified, however, the trend
and degree of depletion were respectively modified.

TSI sites

Comparisons of soil moisture on the three sites of the cull
hardwood area are presented in figures 4 and 5. Figure 4 shows the
available water in the whole upper 48 inches of profile on the three sites.
Soil water was probably adequate for growth throughout the summer on

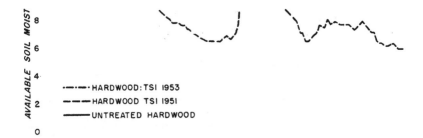

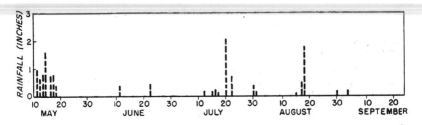

Figure 4.--Summer trends of soil moisture in the upper 48 inches of the TSI and Untreated Hardwood sites. 1953.

both TSI sites, whereas by the first of July soil water may have been critical on the Untreated site. This difference in available water was due, of course, to transpiration by the trees on the Untreated site.

The differences between the TSI 1951 and TSI 1953 sites are confined to the surface layers rather than the subsoil (fig. 5). Along with the pine seedlings a rather dense ground cover of herbaceous vegetation and blackberry vines (evidently an invasion after the 1951 treatment) removed water from the surface 18 inches of soil on the TSI 1951 site. Lesser vegetation had not had time to develop on the TSI 1953 site, and little water was removed. These results suggest that reproduction

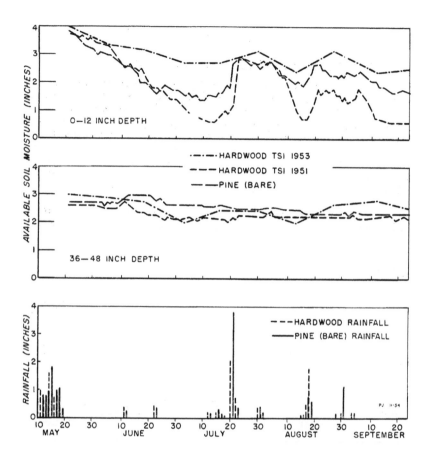

Figure 5. --Summer trends of soil moisture in the 0-12 inch depth
and the 36-48 inch depth of the TSI and Bare sites. 1953.

established in the first year after timber stand improvement should
grow better than similar regeneration during later years.

The summer rains of July and August did not appreciably increase the total water on the Untreated and TSI 1953 sites. The latter
was already wet, and on the former the rains did not reach the 20-inch
sampling depth.

Untreated Hardwood, Young Hardwood, and Pine sites

Results on the Untreated Hardwood, Young Hardwood, and Pine sites show that well-stocked forests, regardless of composition, rapidly deplete the soil of water during the summer. Transpiration in a few weeks reduces soil moisture to a point where growth undoubtedly is restricted. On all three sites, soil moisture dropped from saturation in late May to near the wilting point by late June. At the beginning of June the Leshe silt loam soil still had about 11 inches of available water stored in the surface 48 inches. Transpiration removed about 4 inches of this in only two weeks, and up to 7 inches in four weeks.

Figure 6 compares the all-aged Pine and the all-aged Untreated Hardwood sites. The moisture depletion curves for the entire profile

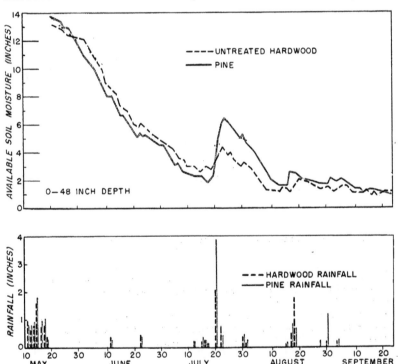

Figure 6. --Summer trends of soil moisture in the upper 48 inches of the Young Hardwood, Untreated Hardwood, and Pine sites. 1953.

- 12 -

are essentially identical, except for the influence of the rain of July 20. The additional rain on the Pine site wetted the soil to the 20- to 26-inch depth, whereas this depth was not affected on the Untreated Hardwood site. Prior to this rain, both sites were dry, the trees having used up most of the available water by early July.

The young even-aged hardwoods depleted the soil of water almost to the extent that the all-aged cull hardwoods did. Trees and shrubs on both sites removed most of the available water by early July (fig. 6), and there is little difference between the two areas. The additional inch of water that was apparently removed from the Untreated site may not be significant for the methods employed.

TSI 1953 and Bare sites

Although moisture was depleted rapidly on the untreated forest sites, it remained relatively high throughout the summer on the TSI 1953 and Bare sites. The TSI 1953 site consistently had the higher water content (fig. 5), undoubtedly because surface evaporation was reduced by the undisturbed litter and shade from the dead hardwoods. Again, the difference was confined to the surface 12 inches of soil.

Summary and Conclusions

During the summer of 1953, soil water depletion was measured under six different forest conditions in south Arkansas. Three forested sites were left undisturbed, while three others were variously treated to remove vegetation.

Where pine or hardwood stands with a stocking of 70 to 100 square feet of basal area were undisturbed, water was removed from the ground rapidly with the onset of hot dry weather. On plots where large cull hardwoods were deadened, and where all living vegetation was removed, soil water remained relatively high throughout the summer.

The summer was a dry one, and rainfall only moderately affected the total moisture on the undisturbed sites. Evapo-transpiration quickly depleted the soil of water added by these showers.

Soil water depletion was greater where all vegetation had been removed than where only culls had been deadened, but this difference was apparent only in the surface layers. Below the effective zone of evaporation on these sites, soil water remained at a rather constant high level.

The study indicates that very serious consideration should be given to methods of stand treatment which will both conserve moisture and permit the water supply to be used by the more desirable species. During the summer, droughts occur nearly every year throughout the western portion of the shortleaf-loblolly pine-hardwood type, and lack of soil moisture undoubtedly limits tree growth.

Literature Cited

1. Boggess, W. R.
   1953. Diameter growth of shortleaf pine and white oak during a dry season. Univ. Ill. Agr. Expt. Sta. Forestry Note 37, 7 pp.

2. Gaiser, R. N.
   1952. Readily available water in forest soils. Proc. Soil Sci. Soc. Amer. 16: 334-338.

3. Hoover, M. D., Olson, D. F., Jr., and Greene, G. E.
   1953. Soil moisture under a young loblolly pine plantation. Proc. Soil Sci. Soc. Amer. 17: 147-150.

4. Reinhart, K. G.
   1953. Installation and field calibration of fiberglas soil-moisture units. U. S. Forest Serv., South. Forest Expt. Sta. Occas. Paper 128, pp. 40-48.

1955

O
C
C
A
S
I
O
N
A
L

P
A
P
E
R

1
3
8

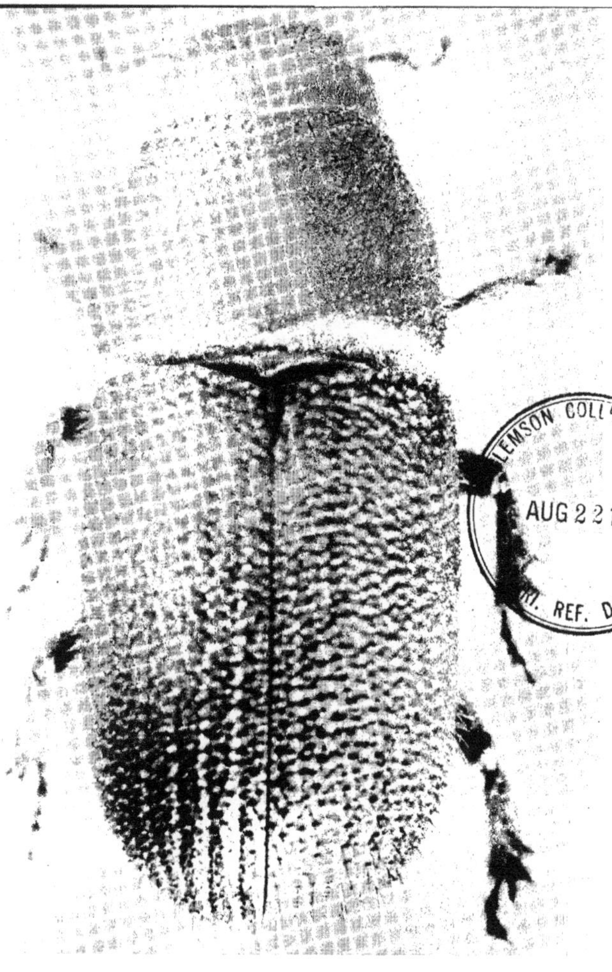

THE BLACK TURPENTINE BEETLE, ITS HABITS AND CONTROL

R. E. Lee and R. H. Smith

SOUTHERN FOREST EXPERIMENT STATION

Philip A Briegleb, Director

The photo on the cover shows an adult black turpentine beetle enlarged about 25 times. It was made by the Southern Regional Research Laboratory (New Orleans, La.), Agricultural Research Service, U. S. Department of Agriculture.

THE BLACK TURPENTINE BEETLE,
ITS HABITS AND CONTROL

R. E. Lee[1] and R. H. Smith[2]

SUMMARY

The black turpentine beetle, long considered relatively harmless, has since 1949 been killing large numbers of pines throughout the Deep South, causing substantial losses to both timber growers and turpentine farmers.

Presence of this large, black or reddish-brown bark beetle is indicated by tubular masses of reddish pitch on fresh stumps and on the lower trunks of standing pines. Adult beetles deposit groups of eggs along tunnels in the inner bark. The larvae feed on the inner bark and often consume enough of it to girdle the tree. All species of southern pine are attacked.

The worst outbreaks have occurred in stands disturbed by either natural or man-made causes. The insect does not always confine itself to weakened trees, but kills apparently healthy ones as well. Other bark beetles, stain, and decay follow the turpentine beetle and develop so rapidly that the killed trees must be salvaged very promptly.

Tests with various insecticides show that spraying with solutions of benzene hexachloride (BHC) in fuel oil will prevent attack on stumps and trees. BHC also kills broods beneath the bark, though it may not save heavily infested trees. The effectiveness of stump treatments in reducing beetle populations has been demonstrated on large-scale control operations. It is not practical to use fire in controlling the beetle in stumps.

Attacks can be prevented or minimized by reducing or avoiding stand disturbances as much as possible, removing or spraying with BHC the trees injured in logging, promptly salvaging dead and dying

---

1/ Southern Forest Experiment Station.
2/ Southeastern Forest Experiment Station.

trees, spraying stumps with BHC following salvage of infested trees o
cutting of green trees in an outbreak area, and spraying seed trees in
outbreak areas.

When insecticides are used to control the beetle, trustworthy
should be assigned to the sprayers, the insecticide should be prepar
carefully, and the bark should be covered thoroughly with the spra
Spraymen should avoid all unnecessary contact with the insecticide.

THE BEETLE AND ITS HABITS

The black turpentine beetle (Dendroctonus terebrans (Oliver))
one of two Dendroctonus species described in 1836 when the generic n
was proposed. Dendroctonus means "killers of trees." The name is
propriate, for the 23 or more species that today represent the genus
clude some of the most destructive insect enemies of North American
forests (2). 3/

Description and range. --Adults of the black turpentine beetle
larger than those of most of the other species of Dendroctonus, varyi
length from 1/5 to 1/3 of an inch. They are hard-shelled, almost cy
lindrical in shape, and reddish-brown to black in color. The crea
white larvae are legless and grub-like, attaining a maximum length
about 3/8 inch. The insect ranges from New Hampshire south through
Appalachians and throughout the range of the southern pines.

Host and site preferences. --Within the beetle's range, all spec
of pine, and red spruce as well, are attacked (1). The insect readil
breeds in fresh stumps and in the lower trunk and roots of weakened ti
Populations sometimes build up rapidly, and then pines which appear t
healthy may be attacked. The beetle prefers large trees, seldom a
tacking those less than 3 inches in diameter.

Beetle activity usually centers in areas where some disturbanc
has occurred, and tree-killing commonly begins within one year of
disturbance.

H. R. Johnston and R. C. Morris, in an unpublished report of
Southern Forest Experiment Station, have pointed out that the most se
infestations occur on poorly drained sites. However, trees on uplan
areas are by no means immune.

Signs of attack. --The most obvious signs of attack are tubula
masses of pitch on the lower trunk of trees and on stumps (fig. 1). A

3/ Underscored numbers in parentheses refer to Literature Cited, p.

part of a third each year. Broods   turpentine beetles.
overlap; with a mild winter the
insect may remain active throughout the year. It is most active, however,
from early spring to late fall.

When the beetle attacks, it bores through the outer bark to the
living inner bark, where it seeks to excavate a longitudinal gallery (usually parallel to the grain of the wood) in which to breed. As the tunnel is
extended, the beetle must remove the resin that flows into it from severed bark
tissues. It does this by pushing tiny bits of resin and bark to the entrance
hole, where a pitch tube is formed. When, on occasion, the beetle is overcome by the resin flow, it is said to be "pitched out." The quantity of gum
produced by a tree, however, does not appear to affect its resistance to
attack, for heavy gum producers such as longleaf and slash pine are killed
as often as are loblolly and shortleaf pines.

The exact role of male and female in the construction and maintenance of the gallery is not known. It is thought that the male initiates attack and begins excavating the tunnel. Subsequently, one and occasionally
two females join and assist him.

The gallery may exceed 20 inches in length. It is widened at
points, and at each of the wide spaces the female deposits from 70 to 200
or more eggs. The eggs are laid on a soft cushion of pulverized bark
which probably serves as an incubator and which may also protect the eggs
from predators (2).

In the Gulf States, during the warmer months, the eggs hatch after about ten days. The larvae feed gregariously on the inner bark. When several broods occur at approximately the same height on the trunk, they may girdle the tree.

When the larvae cease feeding, they form cells at the ends of their galleries and transform into pupae (fig. 2). In 10 to 14 days the pupae become adults. The young adults remain under the bark for a time, gradually changing from a very light amber to the color of the mature beetle. They then emerge to begin the life cycle anew--either by flying to new stumps or trees or by infesting green portions of the tree in which they developed.

Figure 2. --Pupae and pupal cells of the black turpentine beetle.

Attacks on roots. --The beetle also attacks the roots of pines. In 1951, the soil was washed from the roots of four trees in Mississippi, and in 1953 a similar study was made on eight pines in east Texas. 4/ Root attacks were found as deep as 5 feet underground. They appeared to be limited only by root diameter: roots less than 1 inch thick were not attacked. The larger laterals and taproots of heavily infested trees had been extensively damaged. However, there appeared to have been little successful development of brood except on roots in the upper eighteen inches of the soil.

Because of the thinness of root bark, adult galleries were furrows through the bark rather than tunnels beneath it. The furrows were deep enough to cause the roots to "bleed, " but the soil prevented the resin from flowing as it does above ground. As a result, galleries along the roots were covered by a hardened mixture of soil and resin.

4/ Acknowledgment is made to the Kirby Lumber Corporation, Houston, Texas, on whose lands the 1953 work was done, for providing a centrifugal pump and other assistance.

These studies suggested that frequency of attack on roots varies directly with frequency of attack above ground. Work in 1954 showed that this relationship does not always hold--several trees were found with heavy root attack and only light trunk attack.

The relationship of root attack to trunk attack needs intensive investigation. Does the initial infestation always occur above ground, or are the roots sometimes attacked first? Can the beetle spread from one tree to another along the roots? What effect do rainfall and the soil water table have upon the development of brood on roots? The answers might have an important bearing on control methods.

Tree-killing ability. --Prior to about 1949, the black turpentine beetle was regarded as a pest chiefly because of its habit of killing large patches of inner bark at the base of healthy pines (1). The aggressive tendencies that it has recently displayed were first observed by foresters and turpentine farmers in the Gulf States.

A report (unpublished) by Lloyd F. Smith, of damage on the McNeill Experimental Forest in south Mississippi, did much to focus attention on the insect. In June 1950, entomologists began keeping case histories on 82 of the infested longleaf pines at McNeill. By June 1951, 61 of the infested trees had died. Though limited in scope, this pioneering study nevertheless dramatized the deadliness of the beetle's new phase and yielded clues as to the conditions associated with attack.

Meanwhile, reports of unusual tree killing were accumulating to reveal outbreaks all across the lower South, from the Atlantic Coast into Texas.

In Louisiana, trees in a 125,000-acre tract of loblolly and spruce pine (Pinus glabra Walt.) became infested in 1949, and over 1,000,000 board feet of timber, plus additional pulpwood, was killed during the year. Further killing occurred in 1950. By 1951 trees were dying on over 40 percent of the area, and a volume of 2,000,000 board feet and 14,000 cords was killed before the end of the year. Through a well-organized program for controlling the beetle and salvaging dead trees, the owners averted much of the direct loss, but some 5,000 cords of pulpwood were too inaccessible and scattered to salvage.

In northern Florida, slash pine on pond sites has suffered a great deal. On areas up to 5 acres in size, losses have amounted to 30 to 50 percent of the volume, while losses of 5 to 10 percent have not been uncommon on larger areas. Severe infestations of loblolly and longleaf pine also have been observed in Alabama, Georgia, Mississippi, and east Texas.

Figure 3 --Tree heavily attacked by black turpentine beetles and associated insects. White deposit at base of tree is wood dust excavated by ambrosia beetles, which cause severe staining of wood.

On some intensively man forests losses have threatened to the cutting budget, and the grow stock on many areas has been ex sively reduced. There have bee serious effects. The stumpage for the salvaged timber is usuall than for regular stumpage becau scattered distribution of the infe trees makes them costly to log. over, unless salvage is prompt, decay, and other insects increas loss rapidly (fig. 3). When infe follow improvement cuttings, lo greater because only high-qualit ber is attacked. Turpentine far also have suffered. Heavily infe trees yield little resin and usual within a few months. Young tre served for future turpentining, a as for crop trees, have been kil

There is much to be lear garding the role of this heretofo relatively harmless insect. Wh great need exists for careful stu the host and environment, the be must also be investigated much thoroughly than it has been. Ger changes, such as occur in other sect species, may explain the be new role--which may be only ter but which could also be permane

The fact that this "well-k insect suddenly became primary trates why the so-called lesser pests can never be taken for gra and points up the need for intens research on such insects.

INVESTIGATIONS OF CONT

Research was begun in 1 determine methods of controlling

black turpentine beetle. These investigations were along two lines: (1) treatment of valuable individual trees that were only lightly infested, and (2) reduction of the general population in outbreak areas. J. F. Coyne was the first to attempt chemical treatment of living infested trees, and R. J. Kowal developed the stump-treating method that is now being used to reduce beetle populations (3, 4).

Treatment of individual trees. --Initial attempts to control the beetle were aimed primarily at saving living infested trees. In 1950, and again in 1951, formulations of benzene hexachloride (BHC) were applied as sprays to the infested trunks. These tests showed that water emulsions containing 0.40 percent of the gamma isomer gave poor results. Solutions containing 0.25 percent of the gamma isomer in No. 2 fuel oil [5] killed the beetle brood but took effect so slowly that in severe infestations they failed to prevent tree mortality. Recent tests in Florida, however, indicate that increasing the concentration to 0.5 or even 1.0 percent kills the brood more rapidly and thus saves trees which otherwise might die (5).

The need for a method of protecting valuable uninfested trees, such as naval stores trees, seed trees, and trees in seed orchards, soon became apparent. It was found that a solution of 0.50 percent gamma isomer of BHC in fuel oil prevented attack on green stumps and trees for at least seven months. This treatment was successfully applied on the Osceola National Forest in Florida. Most of the beetle damage on that area had been confined to slash pine on pond sites where the trees were being worked for naval stores. The aim of the control program was to remove the susceptible trees and to keep the beetle population at a low level by treating all freshly cut stumps with BHC in oil. To insure a future stand of timber on the logged areas, the basal portions of seed trees, of which there are 3 to 4 per acre, were sprayed with the BHC solution. When more than 10,000 of these seed trees were examined recently, none were found to be attacked.

General reduction of beetle population. --In stumps and in trees mechanically injured, struck by lightning, or similarly damaged, the beetle breeds and develops large populations which later attack apparently vigorous trees. Two series of tests were established in 1951: (1) preventive, aimed at treating stumps immediately after cutting to prevent attack and, if possible, to kill the attacking beetles; (2) remedial, directed toward killing the broods already established in stumps.

5/ Directions for formulating the 0.25 percent gamma isomer solution of BHC are given on page 12.

In the preventive series, residual chlorinated hydrocarbon insecticides were applied to freshly cut stumps. The following chemicals were tested on two-acre areas: 2 percent chlordane in fuel oil; 2.5 percent DDT in fuel oil; 10 percent gamma isomer of BHC in dust form; 1 percent gamma isomer of wettable BHC in water; and 0.25 percent gamma isomer of BHC in fuel oil. Liquids were applied by spraying to the drip-point dosage--approximately 1 gallon per 100 square feet of bark. Dusts were applied at the rate of 1/2 pound per 100 square feet of bark. All stumps were approximately 11 inches high and 12 inches in diameter.

For the remedial tests, insecticides were applied to stumps about two months old and heavily infested with the beetle. Formulations and procedures were virtually the same as for the preventive tests except that BHC dust was omitted.

BHC gave superior results in both series of tests. In the preventive test, dust and wettable powder gave 98 and 96 percent control respectively; the fuel oil solution of BHC gave 92 percent control. Chlordane and DDT exerted only 66 and 47 percent control. In remedial tests, only BHC in fuel oil gave a creditable performance, exerting its chief influence upon the population as the beetles attempted to emerge through the bark. These tests indicated that under usual conditions, in which both infested and uninfested stumps exist, BHC in fuel oil is the best material.

In 1952 tests on 800 stumps in Florida and 400 in Mississippi confirmed this suggestion. In both States the oil solutions of BHC were superior to aqueous preparations or dusts for both preventive and remedial purposes.

Because of the thick bark on slash pines in Florida, all oil formulations for the Florida tests were in a 0.5 percent gamma isomer concentration. The addition of tall oil, turpentine, dipentine, and trichlorobenzene did little to increase the effectiveness of the basic oil solution. However, in Mississippi both 0.25 percent and 0.5 percent gamma isomer oil formulations were tested on longleaf pine stumps without appreciable difference in the results.

During the large Louisiana outbreak in 1949, efforts were made to control the beetle through prompt removal of infested trees. By 1950, some parts of the area had had as many as three salvage cuttings and the outbreak was still uncontrolled. In 1951, treatment of the infested stumps with BHC oil solutions was undertaken coincident with the salvage operation, and by the end of the year the beetle population had declined markedly. Likewise, tree killing by this insect remained at a very low level on the area in 1952.

Use of fire. --In 1951 a small study was conducted to see if fire could be used to control the beetle. An area containing a number of heavily infested stumps was burned over by a "hot" fire, dried logging slash being piled on some stumps to increase the heat. The fire failed to destroy the infestation, and it was concluded that a burn hot enough to control the beetles would seriously injure residual trees on the area.

RECOMMENDATIONS FOR CONTROL

The following recommendations for controlling the beetle are necessarily general and must be applied with care. The amount of tree mortality within a stand varies with the beetle population and also with the vigor and possibly the species of the trees involved. Consequently, some modifications of these measures may be warranted, depending upon individual circumstances and management policy.

Indirect measures to prevent outbreaks. --Prevention by careful management is always better than cure by use of insecticides. Therefore:

1. Unnecessary disturbances in pine stands should be avoided. Woods operations should be conducted so as to prevent excessive injury to residual trees.

2. Stands disturbed by logging, turpentining, fire, hail, wind, lightning strikes, and other insects should be kept under observation for possible build-up in beetle population, especially during drought periods. Provision should be made for at least two inspections during the first year following the disturbance.

3. If disturbances occur, the more seriously damaged trees should be removed as quickly as possible. Skinned trees along skid trails, and others injured through logging, should be sprayed with BHC Many poorly drained areas can be logged only when dry. Because of the preference of the beetle for low, wet sites, special care should be taken to avoid logging damage to the residual stand.

4. Thinning or other conditioning of a stand prior to naval stores operation should be done, if possible, two years before working of the crop trees begins. If stands are thinned only one year before they are worked, the stumps should be treated with BHC. Thinning preferably should be done in the winter. As soon as facing is completed, the worked-out trees should be removed.

5. To prevent attack of high-value trees in an outbreak area, the lower boles should be treated with BHC. The height to which the

ayed varies with the tree species and area. **Where**
ccur up to 10 or 12 feet above ground, preventive
end this high on the trunks. In most areas a high
tacks occur on the basal two feet of a tree, so that
: of six feet should suffice. Recent tests in Florida
aying in turpentine orchards can be confined to the
ash pine and three feet on longleaf (on turpentine
? to keep the amount of spraying to a minimum
int may reduce gum flow). The spray deposit re-
six to seven months, after which respraying may be
is still a large beetle population in the area.

ures to control outbreaks. --For areas where out-
ed in spite of preventive measures or before the
 the following suggestions are offered:
iat are dead or dying as a result of beetle attack
as promptly as possible. A conservative guide to
ees is to take only those with ambrosia beetle dust
r these trees nearly always die. In some cases **the**
ford to return to the area frequently and therefore
move other heavily attacked trees which have not
· ambrosia beetles. If this is the case, care should
in that the turpentine beetle attacks are successful,
in the tunnels, and have not been "pitched out." In
he trees should be cut close to the ground in order
mum number of beetles from the woods. The logs
nptly and the slabs should be burned before the new
es. Stumps should be sprayed with BHC. In fact,
ce that salvage will be delayed until after beetle
tter to spray the trees before cutting begins. This
antage of eliminating the need for burning of the

l treatment of living infested trees is not to be
criminately. However, to save especially valuable
 trees) that have become infested, prompt spraying
iC may prove effective, particularly if the trees have
icked (fig. 4).
 other than salvage, should not be conducted in the
of outbreaks unless provision can be made for
tumps (fig. 5).

Valuable pine being
arrest existing in-
d to prevent future

tump spraying with
le.   Nozzle is held
p and its delivery
t low to permit
aying with minimum
nical.

Materials and preparation. --No. 2 fu
carrier for BHC. It is safe for use in most
other light oils should never be substituted fc
trees are to be treated, as they may kill the

The BHC may be purchased either in
or as a technical grade chemical. The perce
specified for each product indicates the amou
that it contains.

Preparation of the solutions is simple
common sources of error--faulty measureme
the technical BHC completely. Commercial
for this purpose, require only the addition of
slight stirring. Sprays made from technical
deal of agitation but are less expensive. Pei
in the sun for several hours before use aids

Concentrations recommended. --Most
BHC for control of the black turpentine beetl
solution of the gamma isomer in No. 2 fuel o
will give effective control of the insect in rel
or stumps and where the broods have develop
the bark to loosen and become partly dried.
green stumps or trees to prevent attack or to
broods, a 0.50 percent solution is considera
thick-barked slash pine in Florida, it may be
percent solution (5).

Following are the amounts of various
that should be dissolved in 50 gallons of No.
lution containing 0.25 percent of the gamma
percent solution is desired, these amounts s
or four.

Spray application. --Common garden sprayers with an approximate capacity of 3-1/2 gallons are satisfactory for general use. The sprayer should have an oil-resistant hose. A small nozzle opening works best. A 50-gallon drum mounted on a jeep can be used to refill the sprayer.

Thorough coverage of the bark surface, especially in crevices, is necessary for good control. The spray should be applied until it begins to run off. To assure maximum benefits from each dollar spent, dependable men should be assigned to spray crews. The quality of the labor largely determines the success of the operation.

Many woodland grazing areas have been treated without apparent harm to cattle. Cattle are not attracted to stumps or trees treated with BHC.

Spray crews should take reasonable precautions to protect themselves. Oil solutions can be absorbed through the skin, and the BHC may irritate tender skin. After using the chemical, it is advisable to bathe in warm, soapy water. Things a sprayman will not have to worry about, however, are mosquitoes, chiggers, and ticks!

Cost. --The cost of treating stumps or trees depends upon the degree of refinement and the quantity of insecticide purchased, the source of supply, the cost of fuel oil, the cost of labor, and the distribution and number of trees or stumps to be treated.

On the average, four trees 8 inches in d.b.h. (or two trees 16 inches d.b.h.) can be sprayed to a height of 10 feet with one gallon of the fuel oil solution. About ten stumps, averaging 16 inches in diameter and 12 inches high, can be treated with one gallon of the solution. Trees or stumps with rough, thick bark may require twice this amount. Because the bark is thicker at the base, more liquid is needed to spray a stump of a given surface area than to treat a similar area higher on the bole. When fuel oil and insecticide are purchased in large quantities, a gallon of finished 0.25 solution can be prepared for about 17 cents.

Sources of supply. --It is impractical to furnish a complete list of suppliers of BHC and sprayers. The following list is provided for the convenience of the reader. Many of these companies have local distributors. No discrimination is intended against firms not listed, and no guarantee of reliability is implied.

<u>BHC</u>

Ashcraft-Wilkinson Company, Atlanta 3, Georgia

Chapman Chemical Company, 707 Dermon Building, Memphis 3, Tennessee

Dow Chemical Company, Midland, Michigan

E. I. duPont deNemours and Company, Agricultural Chemical Division, Wilmington 98, Delaware

Ethyl Corporation, 100 Park Avenue, New York 17, New York

Florida Agricultural Supply Company, Box 658, Jacksonville, Florida

Mathieson Chemical Corporation, Box 991, Little Rock, Arkansas

Pennsylvania Salt Manufacturing Company, Bryan, Texas

Thompson-Hayward Chemical Company, 2915 Southwest Boulevard, Kansas City 8, Missouri

<u>Sprayers</u>

John Bean Manufacturing Company, Box 890, Lansing 4, Michigan

H. D. Hudson Manufacturing Company, 589 East Illinois Street, Chicago 11, Illinois

The F. E. Myers and Brother Company, 241 Fourth Street, Ashland, Ohio

D. B. Smith and Company, 440 Main Street, Utica 2, New York

LITERATURE CITED

1. Beal, J. A., and Massey, C. L.
    1945. Bark beetles and ambrosia beetles (Coleoptera: Scolytoide Duke Univ. School Forestry Bul. 10, 178 pp.

2. Hopkins, A. D.
    1909. Bark beetles of the genus <u>Dendroctonus</u>. U. S. Dept. Agr Bureau of Entomology Bul. 83, Pt. I, 169 pp.

3. Johnston, H. R.
    1952. Insect control. practical methods for the control of insects attacking green logs and lumber. South. Lumberman 184(2307): 37-39.

4. Kowal, R. J., and Coyne, J. F.
    1951. The black turpentine beetle can kill trees. AT-FA Jour. 13(9): 7, 14-15.

5. Smith, R. H.
    1954. Benzene hexachloride controls black turpentine beetle. South. Lumberman 189(2369): 155-157.

FORAGE WEIGHT INVENTORIES
ON SOUTHERN FOREST RANGES

R. S. Campbell and John T. Cassady

SOUTHERN FOREST EXPERIMENT STATION

Philip A Briegleb, Director

Forest Service, U.S. Department of Agriculture

OCCASIONAL PAPER 139   1955

FORAGE WEIGHT INVENTORIES
ON SOUTHERN FOREST RANGES

R. S. Campbell and John T. Cassady
Southern Forest Experiment Station

A first step in the management of a forest range is to inventory the forage--to determine the kind and amount of plants edible to livestock on various parts of the range. This information is needed to plan the proper number and distribution of animals and the season of grazing. The forage inventory method described here has been used on several experimental forests of the Southern Forest Experiment Station since 1944, and on national forests in the Gulf States and Arkansas since 1951.

The method may be illustrated by the situation of a farmer who has 10 tons of hay. He reckons that if he allows 20 pounds of hay per cow per day he can winter about 11 cows for 3 months. The farmer's problem is fairly easy, for the hay is cut and stored in the stack or in the barn. On the range, cattle do the harvesting, and the stockman must determine (that is, estimate by sampling) the amount of green grass on an area and then calculate the approximate number of cattle the area will support during the grazing season.

As described here, forage inventories for cattle range management will determine only the amount of green grass. On longleaf pine forest ranges in central Louisiana and south Mississippi, grass makes up about 95 percent of the cattle diet in spring, summer, and fall. Browse and weeds (forbs) contribute so little that it is unnecessary to sample them. On some livestock ranges, and especially on deer ranges, edible portions of weeds and browse are of real importance and should be included in the inventory.

For spring-summer-fall ranges, it is usually best to make the inventory in July, after the main grass growth is complete but before leaves begin to toughen and flower stalks become abundant. Usually about 70 percent of the year's growth will have been made by this time. However, on ranges grazed mainly during one season, such as winter or spring, the inventory should be timed to measure forage production during that season.

Relative forage production during the year of the inventory must be considered in applying the results over a period of years. In central Louisiana, a drought year produced 40 percent less range forage than

FORAGE WEIGHT INVENTORIES
ON SOUTHERN FOREST RANGES

R. S. Campbell and John T. Cassady
Southern Forest Experiment Station

A first step in the management of a forest range is to inventory
the forage--to determine the kind and amount of plants edible to livestock on various parts of the range. This information is needed to plan
the proper number and distribution of animals and the season of grazing.
The forage inventory method described here has been used on several
experimental forests of the Southern Forest Experiment Station since
1944, and on national forests in the Gulf States and Arkansas since 1951.

The method may be illustrated by the situation of a farmer who
has 10 tons of hay. He reckons that if he allows 20 pounds of hay per
cow per day he can winter about 11 cows for 3 months. The farmer's
problem is fairly easy, for the hay is cut and stored in the stack or in
the barn. On the range, cattle do the harvesting, and the stockman must
determine (that is, estimate by sampling) the amount of green grass on
an area and then calculate the approximate number of cattle the area will
support during the grazing season.

As described here, forage inventories for cattle range management will determine only the amount of green grass. On longleaf pine
forest ranges in central Louisiana and south Mississippi, grass makes
up about 95 percent of the cattle diet in spring, summer, and fall.
Browse and weeds (forbs) contribute so little that it is unnecessary to
sample them. On some livestock ranges, and especially on deer ranges,
edible portions of weeds and browse are of real importance and should be
included in the inventory.

For spring-summer-fall ranges, it is usually best to make the
inventory in July, after the main grass growth is complete but before
leaves begin to toughen and flower stalks become abundant. Usually
about 70 percent of the year's growth will have been made by this time.
However, on ranges grazed mainly during one season, such as winter or
spring, the inventory should be timed to measure forage production during
that season.

Relative forage production during the year of the inventory must
be considered in applying the results over a period of years. In central
Louisiana, a drought year produced 40 percent less range forage than

inventory is above or below average in forage production. A
range should be inspected each year to show whether utilizat
measurably above or below proper use. These check-ups w
determine if adjustments are needed in application of the for
ventory data.

Basic Steps in Forage Inventory

The basic steps for making a range forage inventory
immediately below. Detailed directions for carrying out sor
steps, as well as shortcuts to some of the field and office we
discussed later.

1. A sample plot 3.1 feet square is a convenient size
with the abundant grass on southern ranges (2), because of e
version to pounds per acre. The number of plots to be taken
on the degree of accuracy desired. Plots are taken in group
systematically selected sampling points.

2. The green grass or current season's growth is cl
weighed in grams per plot. With training, the examiner ma·
grass weight on the stalk with reasonable accuracy.

3. A few samples of the grass are saved and dried in
convert the plot weights to an air-dry basis (similar to hay).

4. The air-dry weight of green grass in grams per F
plied by 10 to convert to pounds per acre.

5. "Usable grass" is then calculated. On most sout
types, about 60 percent of the annual growth of native grasse
forest ranges should be left ungrazed in order to protect the
the soil. Therefore, under proper range use, only 40 perce
grass is available for grazing in most types.

6. The "usable grass" per acre is multiplied by the
acres in the range unit to obtain the pounds of usable grass (
"hay" available under approximately proper use.

1/ Underscored numbers in parentheses refer to Literature
page 15.

7. The total usable grass in the range unit is divided by 600 (20 pounds per cow per day times 30 days) to obtain the number of cow-months of grazing.

8. The total cow-months are divided by the number of months in the grazing season to determine the number of cows to be grazed.

Details of Field and Office Procedure

Preliminary Office Work

Inventory map. --Prepare for field use a forest range type map of the range unit to be inventoried (fig. 1). A convenient scale is 2 inches to the mile. Show main drainages, survey corners, fences, and roads for control. Outline with penciled dots each forest range type that is to be sampled. For practical purposes, small types of only a few acres should be lumped with larger adjacent types unless very high forage values are involved, as in the carpetgrass type. See Appendix I for suggested list of types to be recognized and sampled separately. Compute and record on the map the acreage in each type.

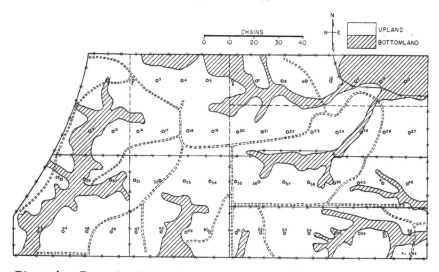

Figure 1.--Example of forest range type map of 1,200-acre area, showing sampling grid for forage inventory.

Sampling points. --Lay out tentative sampling
them on the map, systematically covering the entire
a car is to be used in making the inventory, care mu
for sampling across the drainages rather than to tak
or close to roads and main livestock trails.

In forage inventories, it is desirable to come
20 percent accuracy (two standard errors), which is
as adequate in timber cruising for management plan

Where sampling is on a grid, as in figure 1,
garded in summarizing the data, unless the type bre
management or research purposes. Indications are
sampling, and where type breakdowns are not desire
sampling points (195 plots) are needed to sample any
less of size) that has more than one broad type. Fo
than 8 sections, however, a minimum of 8 sampling
should be taken.

Where sampling is mostly from roads (i. e.,
sampling should be planned by types. For example,
standard errors on the burned and unburned wiregra
Apalachicola National Forest in north Florida requi
pling points (150 plots) in each type. The carpetgras
more uniform, required only about 20 sampling point

Experience in 5 Gulfcoast states suggests the
minimum numbers of sampling points (3 plots at each
types:

Type	San
Bluestem	
Wiregrass	
Carpetgrass	
Loblolly-shortleaf pine	
Upland hardwood	
Bottomland hardwood	
Switchcane	
Brush	
Swamp	

- 4 -

Equipment. --Here is a check list of equipment needed for field work:

1. Transportation
2. Plot frame, 3.1 feet square, made of welding rod or heavy wire
3. Spring balances, 500 gram
4. Sheep shears
5. Paper sacks, 16-pound kraft quality
6. Map, for control in inventory
7. Record forms and holder
8. Pencils, with erasers
9. Forest Service compass and Jake stick
10. Water
11. Lunch
12. Back packsack

Field Procedure

A two-man crew is recommended, but the inventory can be made efficiently by one man working alone.

1. Take three plots at each sampling point. The three plot values will be recorded on a single form. If travel is cross-country, then plots may be 1/4 chain apart on line. If travel is by car along roads, then the first plot should be 2 chains from road, and the other 2 plots 1 chain a-part, on a line perpendicular to the road. It is best to keep the 3 plots of each sampling point within the same broad type. If a new type is en-countered, set up a new sampling point and take 3 plots.

2. Locate the first plot by pacing the predetermined distance, then lay plot frame on ground directly ahead of last pace. If plot lands in a stream or hole with no grass, record grass weight as 0. If plot falls with a large tree or stump in it, place frame close beside the object and proceed as usual. Plots in brush areas should be taken as they fall, with appropriate notes as to accessibility of the forage to grazing animals.

3. Fill in appropriate headings on form: cover type, location, date, examiner, etc.; see sample form on page 6. The headings consti-tute a list of items to be considered, and need not be filled out completely in routine forage inventory.

4. Note previous grazing, if any, and record estimated utilization to nearest 10 percent.

Field crews should train themselves to determine utilization by the standard ocular-estimate-by-plot method developed by Pechanec and Pickford (8). In training, simulate grazing on an ungrazed plot by clipping and retaining the plant material removed. Estimate the percent-

RANGE FORAGE WEIGHT INVENTORY

Forest _Apalachicola_ District _Wakulla_
Allotment _Simmons_ Date _July 29, 1952_
Examiner _Stradt - Bryan_ Sampling point No. _26_
Location _see map_
Forest type _slash pine_ Age _____ Density _____
Forage type _wiregrass_
Main species _Three-awn, muhly, panicum_
Grazing: Animals _cattle_ Intensity _moderate_
Burning _Area burned Jan., 1951_
Soil and condition _sandy loam_
Wildlife _____

| | Plots | | | |
	A	B	C	
Clipped	✓			Total
Estimated		✓	✓	
Grasses:				
Green wt. (grams)	75	90	45	X
Utilization (%)	15	10	20	X
Total production (grams)	88	100	56	244

Weeds _____
Browse _____
Litter _____
Grass: wet _____ damp _____ dry ✓

Remarks: _Area unburned in current year. Very dry season._

(Form to be half letter size)

- 6 -

age weight removed. Then clip the remainder of the herbage, weigh
both parts, calculate the actual percent utilization, and note the error
of the original estimate. Training should continue until utilization can
be determined within 10 percent.

Below is a descriptive utilization scale for approximating
utilization percentage on bluestem grasses. The scale must, of course,
be tempered according to time of year and general productivity of the
grass growth in the particular year.

Description	Utilization Percent
Utilization perceptible but very patchy, leaf tips bitten off occasional grass tufts.	10
Utilization obvious but still patchy; i. e., leaf tips bitten off 1/4 to 1/2 of grass tufts.	20
Leaf tips bitten off 1/2 to 3/4 of grass tufts, but many tufts or patches still ungrazed.	30
Leaf tips bitten off more than 3/4 of grass tufts, occasional tufts or patches ungrazed, considerable nuzzling for new green growth at base of tufts on unburned range.	40
Leaf tips bitten off nearly all grass tufts, average stubble about 4 to 5 inches high. Some evidence of trampling.	50
Nearly all grasses closely grazed, especially on burned range. Average stubble only 1 to 3 inches high. Trampling.	60-95

When a forage weight inventory is made in midsummer on a
range that has been grazed for the preceding 3 or 4 months, some of the
utilization will be obscured by subsequent regrowth of the grass. Therefore, if any utilization can be detected in midsummer, it represents at
least 10 percent utilization of the main forage grasses.

5. Remove litter and last year's grass from plot by gleaning
and separating it with fingers from the current or green grass. The inventory measures only the current year's grass production, some of
which may be cured or dry.

6. Clip remaining grass as close as possible to the ground, in paper sack, and weigh in grams. Deduct tare for sack and reco1 green weight on form. If any old leaves, needles, or previous year grass growth remain, pick them out before weighing. When weeds browse are included in the inventory, clip palatable weeds and weig rately from grass. Strip leaves and tender twigs from palatable br plants up to 4-1/2 feet above ground and weigh separately. Proper factors for these kinds of plants are at present considered to be abc percent.

7. Proceed to next sampling point and repeat; use a new fo1 each sampling point.

8. On the forms, note any range conditions that may be use determining good management, such as livestock concentrations or grazed areas, extraordinary feeding habits, signs of deer, areas ir cessible to livestock or too distant from water, or recently burned

9. Save samples of green grass (100 to 200 grams) during f fourth, and last hour of the work day for dry-weight determination. each sample is taken, record exact green weight, date, time of day plot number or location on the paper sack. Three such green grass ples for each main type are ordinarily sufficient. Take all samples office for drying. If some of the plots are taken when grass is wet dew or rain, use one of the following special procedures:

a. Moisture classification of plots. Classify each plot into
   one of three broad categories, and record the classifica
   (1) wet, when droplets of water still cling to the leaves,
   (2) damp, when there are no drops, but a thin film of
       moisture, damp to the touch, remains,
   (3) dry, when no moisture can be felt on the leaves.

   It is necessary to save a sample of grass in each
   moisture class.

b. Ocular estimate of grass weight. After some experienc
   in clipping and weighing grass on plots, the examiner ca
   train himself to estimate grass weight within about 10
   percent of the actual weight. If clipping is interrupted b
   a brief shower, the worker may continue after the rain t
   estimating the weight of green grass without the surface
   moisture. When the leaves no longer feel moist, he
   should resume the standard method of clipping and weigh
   ing the grass.

- 8 -

For accurate research work, the wet grass on each plot can be clipped, weighed, and all of it kept and air-dried at the office. Weighing will then give the exact air-dry weight in grams per plot.

Determining Air-dry Percentages of Grass

Air-dry the grass samples. Keep samples in original sacks and avoid loss of grass or mixing between samples. After about 10 days, weigh a few samples and record weights on sacks. About three days later, repeat on same samples. If there is no appreciable change, weigh all samples and record air-dry weight on each original sack. A fan blowing over the sacks (with tops open) will hasten drying time to less than a week. Be sure to subtract tare and record net weight for both "green" and air-dry weights. Tare weight of small sacks may be determined by weighing 10 empty sacks together and dividing by 10.

Group the data for each type or moisture classification and calculate average air-dry percentages to nearest 1 percent.

$$\frac{\text{Air-dry weight x 100}}{\text{Green weight}} = \text{air-dry percent}$$

For example, if total air-dry weight of 3 samples is 300 grams, and total green weight is 650 grams, then

300 x 100/650 = 46 percent air-dry.

Final Office Calculation

1. Correct grass production for utilization on each plot in space provided on inventory form, using table on page 18. For example, if actual weight of grass on plot is 40 grams, and utilization is 20 percent, then calculated total grass production is 50 grams.

The values in the table were calculated as follows:

$$\frac{\text{Grams of grass, green weight}}{\text{Percent not grazed (decimal)}} = \text{Total grass production}$$

Using the same example as above:

40/0.8 = 50 grams

Percent not grazed is 100 minus utilization percentage
20% utilization = 80% ungrazed (decimal value = 0.8).

2. For each sampling point, record total grass product
grams (green weight) by adding together the three plots.

3. Sort field record sheets by types, conditions, or oth
fications to which identical air-dry weight percentages and prop
factors may be applied.

4. Determine total green weight of grass in grams for
in type or other classification: add together the total productio
each field form.

5. Calculate for type or other classification the averag
weight of grass in grams per plot: divide total from No. 4 abov
number of plots. Remember that there are 3 plots per field sh

6. Calculate for each type or other classification the a
air-dry weight of grass in grams per plot: multiply average gr
weight per plot by appropriate air-dry percentage as previousl
termined. When this has been done, the various classifications
combined by types.

7. Determine average air-dry grass for each type in po
per acre: multiply average air-dry weight per plot in grams by
obtain pounds per acre.

8. Determine average usable grass in pounds per acre:
average air-dry grass (pounds per acre) by appropriate proper-
factor (40 percent for most types).

9. Calculate total usable grass in type: multiply averag
grass (pounds per acre) by number of acres in type.

10. Add all types to obtain total pounds of usable air-dry
(hay) in entire range unit being inventoried.

11. Calculate estimated grazing capacity in cow months
range unit: divide total pounds of usable air-dry grass by 600 (p
required per cow per month).

12. Determine rate of stocking or grazing capacity for th
unit: divide number of cow months of feed by number of months
approved grazing season (number of cows for the season grazed

An example of field plot records and computations is shown in the field sheet, page 6. Examples of final office procedure for several types are shown in figure 2. For research purposes, more detailed plot records and calculations may be made as needed.

| Item | Wiregrass | | Carpetgrass | | Brush | Swamp | Total or average |
	Unburned	Burned	Woodland	Roads and headquarters			
Range forage inventory. Simmons Pasture, east allotment, Wakulla Ranger District, Apalachicola National Forest. July 28-30, 1952.							
Grass on plots							
Green weight (grams)	33,613	7,544	582	2,642	60	1,764	
Plots (number)	447	95	13	10	77	49	691
Average per plot (grams)	75.2	79.4	44.8	264.2	.8	36.0	
Air-dry (percent)	60	50	40	40	50	30	
Air-dry weight per plot (grams)	45.1	39.7	17.9	105.7	.4	10.8	
Air-dry grass per acre (pounds)	451	397	179	1,057	4	108	
Proper-use factor (percent)	10	40	100	100	10	60	
Usable grass per acre (pounds)	45	159	179	1,057	.4	65	
Acres in type	9,976	2,100	600	100	1,000	7,485	21,261
Total usable grass (pounds)	448,920	333,900	107,400	105,700	400	486,525	1,482,845
Grazing capacity							
Cow-months (grass ÷ 600)	748	556	179	176	1	811	2,471
Cows yearlong (cow months ÷ 12)	62	46	15	15	...	68	206
Acres per cow yearlong	161	46	40	7	...	110	103
Cows for 8-month season (Cow months ÷ 8)	94	70	22	22	...	101	309
Acres per cow, 8-month season	106	30	27	5	...	74	69

Figure 2. --Example of summary work sheet.

Shortcut Methods

For speeding up the forage inventory or for quick approximation of grazing capacity, various shortcuts may be employed, although they may involve some loss in accuracy where the assumed air-dry weight of 50 percent differs from the actual air-dry weight.

The first shortcut method can be used in types where the proper-use factor is about 40 percent. The plots are inventoried in the field as in the standard method but grazing capacity is estimated from weight of the grasses. The shortcut calculation for the burned type illustrated in figure 2 would be as follows:

Grass on plots
    Total green weight (grams)
    Number of plots
    Average grams per plot
Pounds of green grass per acre
Acres in type
Total green grass, pounds   1, 66
Grazing capacity
    Cow months (green grass ÷ 3, 000) 2/
    Acres per cow month (2, 100 ÷ 556)
    Cows, 8-month season (cow months ÷ 8)
    Acres per cow, 8 month-season

The second shortcut method gives a quick approximat capacity for a given type or area. Several plots should be in and the average green weight of grass per acre calculated. ' proximate number of acres required per cow month can be re table 1.

The rough approximations of grazing capacity in table on the following formula:

$$\frac{\text{Green weight of grass per cow month}}{\text{Green weight of grass per acre}} = \text{acres per co}$$

Example:

$$\frac{3,000 \text{ lbs.}}{250 \text{ lbs.}} = 12.0 \text{ acres per cow month}$$

Table 1 may be applied directly for bluestem and burn types. The inventory on the Apalachicola National Forest ind other types differences in air-dry weights and proper-use fac fig. 2) may necessitate certain adjustments:

_____

2/ This provides 3, 000 pounds of green grass per month, or per day. A cow would utilize about 40 percent of it, or 40 p day.

- 12 -

Wiregrass, unburned, multiply acreages in table by 3.5.
Carpetgrass, divide acreages in table by 2.
Swamp, add 10 percent to acreages in table.

Table 1. --<u>Approximate grazing capacity in relation to</u>
<u>amount of grass per acre (green weight)</u>

Grass per acre (pounds)	Acres per cow month	Grass per acre (pounds)	Acres per cow month
100	30.0	800	3.75
120	25.0	833	3.6
150	20.0	882	3.4
171	17.5	938	3.2
200	15.0	1,000	3.0
214	14.0	1,034	2.9
231	13.0	1,071	2.8
240	12.5	1,111	2.7
		1,154	2.6
250	12.0		
261	11.5	1,200	2.5
273	11.0	1,250	2.4
286	10.5	1,304	2.3
		1,364	2.2
300	10.0	1,429	2.1
316	9.5		
333	9.0	1,500	2.0
353	8.5	1,579	1.9
375	8.0	1,667	1.8
		1,765	1.7
400	7.5	1,875	1.6
429	7.0		
462	6.5	2,000	1.5
500	6.0	2,143	1.4
545	5.5	2,308	1.3
		2,500	1.2
600	5.0	2,727	1.1
632	4.75		
667	4.50	3,000	1.0
706	4.25		
750	4.00		

Grazing Management

Management features to keep in mind in applyir
ventory are as follows:

Fenced control of cattle. --Fences are indispen
range management. Cattle can be grazed during suita
proper numbers only if the range has the necessary bo
fences.

Season of grazing. --The upland types generall)
ably good grazing for 3 to 4 months from late March o
through mid-July, with only fair grazing in summer ar
poor grazing in winter (1, 3, 4, 6, 9). The bottomlanc
fair grazing in winter.

Grazing damage to reproduction. --Usually catt
reproduction results from concentrations of animals.
cal to exclude grazing when the trees are vulnerable, t
damage should be minimized by: (1) grazing moderatel
relatively uniform distribution over the range; (2) excl
pecially in winter; and (3) preventing or remedying an)
cause unusual concentrations of cattle.

Distribution of grazing. --Where prescribed bur
for silvicultural or other purposes, it becomes the ma
bution of grazing. Since cattle concentrate on burned a
age burned should be sufficiently large to provide the t
for the herd (7). If possible, the burns should be locat
courage animals to use all parts of the pasture and avc
grazing.

On unburned ranges, the principal means of dis
through wise placement of salt or mineral supplement,
et cetera.

LITERATURE CITED

) Bond, W. E., and Campbell, Robert S.
   1951. Planted pines and cattle grazing--a profitable use of south-
         west Louisiana's cut-over pine land. La. Forestry
         Comm. Bul. 4, 28 pp., illus.

) Campbell, Robert S., and Cassady, John T.
   1949. Determining forage weight on southern forest ranges.
         Jour. Range Mangt. 2: 30-32, illus.

) _____ and Cassady, John T.
   1951. Grazing values for cattle on pine forest ranges in Louisiana.
         La. Agr. Expt. Sta. Bul. 452, 31 pp., illus.

) _____ Epps, E. A, Jr., Moreland, C. C., and others.
   1954. Nutritive values of native plants on forest range in central
         Louisiana. La. Agr. Expt. Sta. Bul. 488, 18 pp., illus.

) Cassady, John T.
   1953. Herbage production on bluestem range in central Louisiana.
         Jour. Range Mangt. 6: 38-43, illus.

5) _____ and Campbell, R. S.
   1951. Pine forest ranges in Louisiana. U. S. Forest Serv. South.
         Forest Expt. Sta., 7 pp., illus. [Processed.]

7) Halls, L. K., Southwell, B. L., and Knox, F. E.
   1952. Burning and grazing in Coastal Plain forests. Ga. Coastal
         Plain Expt. Sta. Bul. 51, 33 pp., illus.

3) Pechanec, Joseph F., and Pickford, G. D.
   1937. A comparison of some methods used in determining percent-
         age utilization of range grasses. Jour. Agr. Res. 54:
         753-765.

9) Shepherd, Weldon O., Southwell, Byron L., and Stevenson, J. W.
   1953. Grazing longleaf-slash pine forests. U. S. Dept. Agr.
         Cir. 928, 31 pp., illus.

APPENDIX I

Broad Forest Range Types

In the upland forest types, three main grass subtypes may be recognized:

Bluestem: proper-use factor 40 percent.

Wiregrass: proper-use factor about 10 percent if unburned, 40 percent if burned during preceding winter or spring.

Carpetgrass: because of its ability to withstand repeated close grazing and to make regrowth, this type may be credited with 100 percent proper-use factor for current forage production.

The upland forest range types usually recognized are as follows:

Grassland: heavily cutover forest land with trees absent or very scattered. Usually clear-cut and non-restocking longleaf pine stumpland.

Open forest: heavily cut over, and partially restocked. Trees prominent, but not sufficiently dense to reduce grass cover conspicuously.

Longleaf pine: second-growth or planted longleaf pine with more than 300 pine and hardwood trees per acre; enough litter to reduce grass cover distinctly.

Slash pine: similar to longleaf pine, but usually with better timber stocking and frequently with more underbrush and less grass.

Upland hardwood: good upland hardwoods or scrub oaks, similar to pine types but with very few or no pine trees.

Loblolly-shortleaf pine-hardwood: many variations, but generally with less grass than longleaf and slash pine types. May be separated into open and heavily wooded if types are sufficiently distinct.

Brush: dense underbrush, such as ti-ti or gallberry, usually
with very low grass production.

In the bottomland forest types, the principal forage species are
lush sedges, rushes, weeds, and wetland grasses, especially panicums.
A proper-use factor of 60 percent may be applied in areas that are
accessible during dry periods when cattle can get into the bottomlands
to graze.

The bottomland types are as follows:

Bottomland hardwood: creek bottoms and alluvial soils where
bottomland hardwoods predominate. Make separate
subtype for switchcane areas.

Swamp: poorly drained bottomland and swamp sites.

Total Grass Production on Grazed Plots, Calculated From Weight Measurements and Estimated Utilization

Actual weight of grass on plot (grams)	Utilization percent																
	5	10	15	20	25	30	35	40	45	50	55	60	65	70	75	80	85
	Calculated total grass production--grams																
5	5	6	6	6	7	7	8	8	9	10	11	12	14	17	20	25	33
10	11	11	12	12	13	14	15	17	18	20	22	25	29	33	40	50	67
15	16	17	18	19	20	21	23	25	27	30	33	38	43	50	60	75	100
20	21	22	24	25	27	29	31	33	36	40	44	50	57	67	80	100	133
25	26	28	29	31	33	36	38	42	45	50	56	62	71	83	100	125	167
30	32	33	35	38	40	43	46	50	55	60	67	75	86	100	120	150	200
35	37	39	41	44	47	50	54	58	64	70	78	88	100	117	140	175	233
40	42	44	47	50	53	57	62	67	73	80	89	100	114	133	160	200	267
45	47	50	53	56	60	64	69	75	82	90	100	112	129	150	180	225	300
50	53	56	59	62	67	71	77	83	91	100	111	125	143	167	200	250	333
55	58	61	65	69	73	79	85	92	100	110	122	138	157	183	220	275	367
60	63	67	71	75	80	86	92	100	109	120	133	150	171	200	240	300	400
65	68	72	76	81	87	93	100	108	118	130	144	162	186	217	260	325	433
70	74	78	82	88	93	100	108	117	127	140	156	175	200	233	280	350	467
75	79	83	88	94	100	107	115	125	136	150	167	188	214	250	300	375	500
80	84	89	94	100	107	114	123	133	145	160	178	200	229	267	320	400	533
85	89	94	100	106	113	121	131	142	155	170	189	212	243	283	340	425	567
90	95	100	106	112	120	129	138	150	164	180	200	225	257	300	360	450	600
95	100	106	112	119	127	136	146	158	173	190	211	238	271	317	380	475	633
100	105	111	118	125	133	143	154	167	182	200	222	250	286	333	400	500	667
105	111	117	123	131	140	150	161	175	191	210	233	262	300	350	420	525	700
110	116	122	129	138	147	157	169	183	200	220	244	275	314	367	440	550	733
115	121	128	135	144	153	164	177	192	209	230	256	288	329	383	460	575	767
120	126	133	141	150	160	171	185	200	218	240	267	300	343	400	480	600	800
125	132	139	147	156	167	179	192	208	227	250	278	312	357	417	500	625	833
130	137	144	153	162	173	186	200	217	236	260	289	325	371	433	520	650	867
135	142	150	159	169	180	193	208	225	245	270	300	338	386	450	540	675	900
140	147	156	163	175	187	200	215	233	255	280	311	350	400	467	560	700	933
145	153	161	171	181	193	207	223	242	264	290	322	362	414	483	580	725	967
150	158	167	176	188	200	214	231	250	273	300	333	375	429	500	600	750	
155	163	172	182	194	207	221	238	258	282	310	344	388	443	517	620	775	
160	168	178	188	200	213	228	246	267	291	320	356	400	457	533	640	800	
165	174	183	194	206	220	236	254	275	300	330	367	412	471	550	660	825	
170	179	189	200	212	227	243	261	283	309	340	378	425	486	567	680	850	
175	184	194	206	219	233	250	269	292	318	350	389	438	500	583	700	875	
180	189	200	212	225	240	257	277	300	327	360	400	450	514	600	720	900	
185	195	206	218	231	247	264	285	308	336	370	411	462	529	617	740	925	
190	200	211	223	238	253	271	292	317	345	380	422	475	543	633	760	950	
195	205	217	229	244	260	278	300	325	355	390	433	488	557	650	780	975	
200	211	222	235	250	267	286	308	333	364	400	444	500	571	667	800		

CCASIONAL PAPER 140

1955

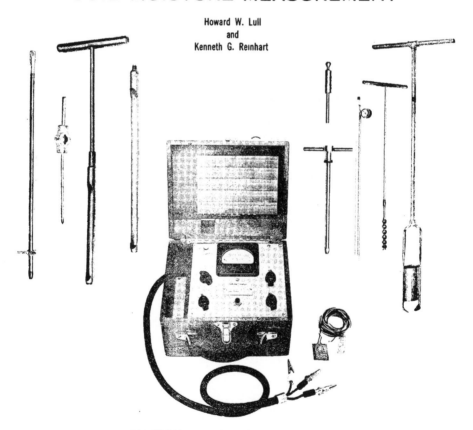

SOIL·MOISTURE MEASUREMENT

Howard W. Lull
and
Kenneth G. Reinhart

SOUTHERN FOREST EXPERIMENT STATION
Philip A. Briegleb, Director
Forest Service, U. S. Department of Agriculture

CONTENTS

SOIL-MOISTURE MEASUREMENT

Howard W. Lull and Kenneth G. Reinhart

In agriculture, forestry, and engineering, considerable attention is devoted to the amount of moisture in the soil and its influence on such factors as crop yields, forest growth, and soil strength. Soil-moisture records are not easily secured. By conventional methods, soil sampling is laborious, and drying of samples is time-consuming. Sampling results are often exceedingly variable, for soil moisture changes from day to day, point to point, and depth to depth. Certainly, it is as difficult a factor to measure as any encountered in hydrologic research.

Because of the importance of soil-moisture records and difficulties in securing them, considerable time and energy has been spent in developing new ways of measuring moisture. This paper reviews most of these methods, describes and compares the most commonly used instruments, and discusses soil-moisture expression, variation, and sampling. Wherever possible the authors have relied on first-hand experience gained at the Vicksburg Infiltration Project. [1]

Development of Methods

Current methods of estimating soil-moisture content range from the age-old one of feeling the soil to the recent employment of radio-active materials and associated intricate and expensive instrumentation. In the pages immediately following, the various methods are briefly reviewed in rough chronological order. Later sections will describe in detail the three now in common use--gravimetric, electrical resistance, and tensiometer--and will summarize experience so far with nuclear instruments.

---

[1] A cooperative soil-moisture study conducted by the Southern Forest Experiment Station, Forest Service, U.S. Department of Agriculture, for the Waterways Experiment Station, Corps of Engineers, U.S. Army. Project headquarters are at Vicksburg, Mississippi.

Consistency Tests

Since ancient times, estimates of soil consistency have been us
to judge the best times for plowing and cultivating and the amount (
moisture available for plant growth. An early recorded observation ha
been cited by Keen from Fitzherbert's Boke of Husbandry, published i
1523: to find out if the soil was ready to be sown to peas and beans, th
husbandman was directed to walk on the plowed land "and if it synge o
crye, or make any noise under thy fete, then it is to wete to sowe: an
if it make no noyse, and wyll beare thy horses, thanne sowe in the nan
of god" (43). 2/

Farmers still pick up and squeeze a handful of soil before worl
their fields. Such simple moisture estimates can be very accurate.
1937, Stoeckeler reported that forest nursery personnel had been trai
to gage the total moisture content of the soil within 2 percent (78). M
recently, Diebold has described a series of manual strength tests i
estimating readily available moisture in medium- to fine-textured so
(32).

Gravimetric Method

One can only guess when the first soil-moisture measurements
were made--perhaps in the eighteenth century. Russell gives results
soil-moisture studies started at Rothamsted in 1843 (68). One of tl
illustrations in a recent edition of his text compares moisture und(
cropped soil and under a fallow area after a prolonged drought in 187
Whenever made, the first measurement probably consisted of weight
a sample before and after drying--essentially, the gravimetric methc

This was the method used by the Department of Agriculture w
it started about 1894 to study "the circulation of water in some of
important types of soil in the United States" (88). Samples were tak
to a depth of 12 inches with a 15-inch brazed brass tube 7/8-inch
diameter. After both ends of the tube had been sealed with rubber ca
the 75- to 100-gram sample was carried to the laboratory and dried
110°C. for 20 hours. Moisture content was expressed, contrary t
present-day practice, in percent by weight of the original moist
sample (89).

2/ Underscored numbers in parentheses refer to Literature Cited, p. '

Dr. F. J. Veihmeyer states in correspondence that the idea of using a tube for sampling soil is old, antedating the development of the well-known King tube in 1890. He also notes that by 1897 the Department of Agriculture was reporting soil moisture on a dry-weight basis, as is done today, and that as early as 1892 Hilgard, in a report of the California Agricultural Experiment Station, gave moisture determinations with respect to both weight and volume.

Since these early beginnings, efficient types of drying ovens, scales, and sampling instruments have been developed. The gravimetric method is still the most widely used technique for obtaining soil-moisture records. Moreover, as the only direct way of measuring soil moisture, it is indispensable for calibrating instruments used in the indirect methods.

The gravimetric method has several serious disadvantages. Much labor is required to secure the samples and considerable time is needed for drying them. Repeated sampling destroys the experimental area. Evaluation of moisture differences between successive samples is complicated by the fact that no two samples can be taken from exactly the same point. The elimination of these disadvantages has been the goal of many investigators.

Soil Points

From about 1920 to the present, researchers have studied the possibility of estimating soil moisture from changes in the weight of porous blocks or points placed in the soil. Livingston and Koketsu, who developed the original blocks, found that permanent wilting began when a block failed to absorb more than 85 milligrams of water after two-hour exposures in soils of various textures (47). Wilson noted that plants remained healthy as long as there was about 500 milligrams' increase in weight of a point set at a soil depth of 6 centimeters for one hour; the critical value for beginning of drought was about 100 milligrams (91). Stoeckeler found that specific gains in weight of soil points could be used to indicate the moisture condition of a light sandy loam (78). Davis and Slater have described a soil point made up of a porous chamber within which is a close-fitting plug that can be removed for weighing (29). Recently, Dimbleby has developed a porous clay pencil which is stuck into the soil. Comparative moisture contents are estimated after one hour from the color change as moisture moves up the pencil (33).

Water-Absorbing Liquids

The techniques using water-absorbing liquids and the several other "rapid" methods were developed mainly to avoid the 24-hour

oven-drying period that is more or less standard in the gravimetr
pling method. Some of these methods show promise, but they fre
require more of the investigator's time than do weighing procedur
associated with oven-drying.

In 1936, Begemann described a method of determining soil
moisture content by adding xylol to a sample, distilling off tl
and water, and measuring volumetrically the distilled water (4).
process takes about 45 minutes. In several comparisons, va
from oven-drying ranged from 0 to 0.4 percent. Bidwell and St
reported a similar method in which toluene was used as the imm
liquid (8). The time required for the individual moisture determ
is the obvious limitation of this method.

In 1927-1931 Bouyoucos described the alcohol method, the
several methods developed by him. The procedure consisted of
chanically dispersing a 25-gram soil sample in 75 milliliters of p
methyl alcohol, filtering the solution, and determining its specifi
with a special alcohol hydrometer (9). Moisture content was dete
by calibrating the hydrometer in known mixtures of alcohol and wa
The method was   accurate, but Smith and Flint (74) noted tha
quired 3 to 5 times as much operator's time as is needed in oven-

Heat of Solution

This method utilizes the principle that a mixture of water .
centrated sulphuric acid produces heat in proportion to the quantit
Emmert employed it by using 2 milliliters of acid and 1-gram por
soil, and measuring the rise in temperature of the solution (35). S
calibration curves have to be derived for each soil type. The met
reported accurate to about 0.5 percent by weight. Extreme care
exercised in using the correct volume of acid and correct weight c

Heat Diffusion

The rate of transmission of heat through soil varies with
moisture content. For dry soils the heat conductivity is low be
the particles make only point contact. As water is added, the a
conduction is greatly increased, and the temperature different
around a heat source is consequently reduced.

Recently, more than 50 moisture cells employing this prir
were tested (1). Changes in soil temperature were determined
thermistor attached to a current-measuring circuit. None of the

- 4 -

measured the full range of moisture content. If a successful cell is developed, it will require separate calibration for different soils and densities.

Shaw and Baver noted that heat conductivity-moisture content relationships for clay and sand samples gave entirely different curves that reflected differences in pore space distribution (71). They also found that concentrations of salt ranging from 100 to 10,000 parts per million had no effect on the readings.

Momim measured heat diffusion with a thermometer inserted to a soil depth of two feet (50). He passed a fixed current through a heating element wound on the bulb and recorded the time required to raise the temperature $5^{\circ}C$. This device was field-calibrated.

Calcium Carbide Method

In 1940, White-Stevens and Jacob described an interesting method based upon the reaction of calcium carbide with free water (87). Ten grams of soil are placed in a specially designed container with an equal weight of calcium carbide. When the container is shaken, the soil moisture comes in contact with the carbide and gas is formed. The moisture content is calculated from the equivalence that each 26-gram loss in weight represents 36 grams of water.

Constant-Volume Methods

Papadakis, in 1941, suggested a method based on addition of water to soil samples to make up a constant volume (57). The difference between the weight of the soil-water volume and the weight of an equal volume of water, when multiplied by a factor, gives the oven-dry weight of the soil. The factor has to be determined for each kind of soil by dividing the oven-dry weight (obtained conventionally) of one sample by the difference mentioned above. Fifty-gram soil samples were used and water was added to make 100 milliliters.

For soils that do not vary widely in volume weight, Uhland (81) has suggested the possibility of determining moisture content very quickly by weighing a given volume of undisturbed soil and comparing it to an average oven-dry weight of the same volume.

Resistance to Penetration

Moisture content can be estimated by relating it to the force required to push an instrument, often called a penetrometer, through

the soil. An example is the device developed by All
measure the amount of force required to drive a pair
soil core (3). For soils in which the instrument had
total moisture content could be estimated within 0.5 pe
2 percent of the available moisture capacity. A simil
mation was obtained with an instrument later develop
This instrument measures the resistance offered to
diamond-shaped steel point driven into the soil.

With any penetrometer, considerable difficulty
in stony soils.

Electrical-Resistance Method

The use of soil points and the heat-diffusion me
tempts to measure moisture in situ in order to pres
point and thus provide continuity in both space and tim
records. This is not possible in gravimetric sampli
moisture difference in successive samplings may be d
actual moisture change with time or to variation in mo
tween sampling points.

The most successful of the in situ devices have
trical-resistance instruments which operate on the prin
ance to the passage of an electrical current between 2
in the soil will depend on the moisture content of the s
not new; the Department of Agriculture used it from
Results were unpromising, largely because of the vari
between the electrodes and the surrounding soil partic

The first successful device appeared in 1940,
and Mick embedded the electrodes in a plaster of par
years later a fiberglas unit was described by Colm
Bouyoucos brought out a fabric unit made of nylon (11).

The use of fabric or plaster of paris permits u
contact with moisture, and thus overcomes the major
instruments. Buried in the soil, the porous materia
and dries along with the soil around it, and the char
content affect the electrical conductivity or resistanc
Wires lead from the unit to the surface of the ground,
is read with a meter. To convert resistances to soil-
requires calibration of the units in the soil being studi

Tensiometers

Tensiometers came into prominence about the same time that electrical-resistance units were developed, Richards presenting a definitive paper on them in 1942 (61). Tensiometers measure soil moisture tension at high moisture contents. Essentially they consist of a porous ceramic cup connected to a vacuum gage or mercury manometer. When the system is filled with water, the water in the cup comes into equilibrium with that in the surrounding soil. As the soil dries or wets, water flows from or into the cup: these changes activate the pressure-measuring device. The instruments function to a tension of 0.8 to 0.9 atmosphere. At greater pressures, air enters the system.

Air-Picnometer

With the air-picnometer developed by Russell, soil moisture, as well as bulk density and percentage of air-filled pores, can be measured directly in the field (67). The method consists of taking a soil core and placing it in an air-tight system at barometric pressure. The volume of the system is then changed a known amount and the resultant pressure measured. This value is used with data on the total volume of the system, the volume change, and volume of sample to calculate the percentage of the sample volume that is filled with air. Moisture content can then be computed from the weight of the sample and an estimated specific gravity of 2.6. At the Vicksburg Project these calculations are made through a series of nomographs. According to Russell's data, most deviations of actual from estimated soil-moisture values were less than 2 percent.

This method is not recommended for soils containing over 5 percent organic matter or for stony soils, since the estimated specific gravity cannot be applied. Broadfoot recently developed a convenient sampler for taking cores that fit the air-picnometer (19).

Nuclear Method

In 1950 a method based on use of radioactive material was introduced (5). The technique involves measuring the slowing down of neutrons emitted into the soil, an effect proportional to the concentration of hydrogen atoms. Still in the developmental stage, this new method holds considerable promise.

Summary

Of the 13 methods that have been described, only the gravi1
electrical-resistance, and tensiometer methods are commonly u
Consistency tests and penetrometers are useful for approximation
will not give the accuracy required for most records. Soil points
vide only relative measurements. The heat-diffusion method awai
ble instrumentation. The air-picnometer possesses the advantage
speed but at the cost of accuracy. The other methods--water-ab
liquids, heat of combustion, calcium carbide, and constant volur
have little promise because they take too much time or are inaccu1
The nuclear method requires further development. The commonly
methods also have their disadvantages. In short, a method that wi
accurate, rapid, _in situ_ values has not yet been perfected.

Gravimetric Method

As has been noted, the gravimetric method is the only com
used direct means of measuring soil moisture. Simply, it involves
a soil sample, weighing it, oven-drying it, reweighing, and expre
the original moisture content in percent of oven-dry weight of soi
first step, securing the sample, is the most troublesome, the deg
difficulty depending on the condition of the soil.

Soil-Sampling Conditions

Sampling conditions are ideal when soil is just moist enoug
easy ingress and egress of the sampling instrument, and where s
roots, and organic matter are not a problem. Such condition
seldom encountered.

With free water in the soil, moisture can never be sa
accurately: water will drip off as the sample is removed fron
ground or compaction may squeeze it out. Rapid sampling is esser
to prevent undue losses. Stickiness, often a problem in wet soils
be alleviated by keeping instruments clean.

Where a wet layer overlies a dry one, samples taken from
drier levels may be contaminated. After the upper soil is we
a summer rain, for instance, removal of a sample may permit v
to run down the hole toward drier soil. The same problem is
countered when samples are taken below a perched water table.
instruments add to contamination and should be dried between sam1

When the soil is dry and hard, the principal problem in fine-textured soils is to get the sampling instrument in and out again. In coarse-textured soils particularly, samples may slide out of the instrument as it is withdrawn from the soil. Fortunately, after the soil dries in summer, its moisture content may fluctuate so little that frequent sampling may be unnecessary.

In certain sections of the country, the problem of stoniness outweighs all others. Stones of any size make it difficult to secure samples and to express moisture content in a useful way. Generally speaking, larger soil samples are required in stony areas and moisture content should be expressed in terms of both the total sample (stones included) and of only the material below the size of coarse sand. In forest soils, heavy accumulations of organic matter and large roots often make sampling difficult.

Instruments

The type of sampling instrument to be used is determined largely by the soil conditions most likely to be encountered. All of the instruments are designed so that an equal volume of soil will be taken per unit increase of depth.

Soil tubes. --The most commonly used sampler is the soil tube, often called a King tube or Veihmeyer tube. Essentially, it is a pipe of about 1-inch diameter which can be obtained in lengths from 3 to 20 feet (fig. 1). As modified and improved by Veihmeyer (84), this tube cuts into mineral soil with a minimum of compaction, though wet clay samples more than a few inches in depth and samples high in organic content tend to compress within the tube.

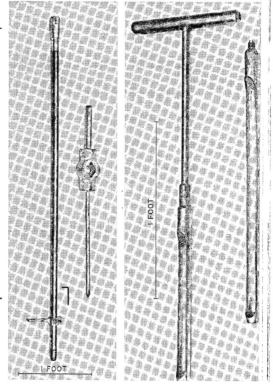

Figure 1. --Two types of soil tubes. Left, soil tube and hammer; the step facilitates sampling of uniform shallow depths. Right, open-side tube.

The tube is driven into the ground with a hammʀ
15 pounds. A specially designed jack is often necessar
tubes driven more than 2 feet into dry soil.

To help prevent the core from falling out whe
withdrawn from the soil, the lower end of the tube i:
stricted. Loss of dry sandy samples can often be prev
compacting the soil in the tube with a rod (84). In very
of core by suction can be avoided by closing the open eı
while withdrawing it, with a rubber stopper or the palm
Tubes may be provided with a variety of points for sa
different textures.

When sampling is especially laborious or when
gradients may lead to contamination of samples, a c
depth of sampling may be taken and apportioned as n
division of the entire core into successive increments
using an inclined measuring trough (37). Under norı
conditions, however, samples of the successive layers
from the hole one at a time; this gives better control
and minimizes exposure and drying.

Soil tubes have several advantages. A sample
several feet can be taken in one operation; as the soil ı
less subject to contamination than samples taken by
side tubes; and, in comparison with other instrumenl
a minimum of site disturbance. In some dry soils, tu
means of securing a sample. Their main disadvantag
amounts of time and labor required and the fact thal
used in stony soils.

An open-side tube useful in moist soils is shc
The open face makes extraction of samples easy. Co
be minimized by pulling the open side away from the wɛ
the tube is withdrawn.

Augers. --The soil auger, perhaps the most fa
instrument, is particularly useful in sticky or in stony
bit will often slide off larger rocks on its downward
samples to depths greater than bit length can be taken ɾ
samplings, contamination can result. When the soil ı:
soil-tube, however, there is little other choice. In s
soils even augers are not satisfactory, and the only reɾ
and shovel.

Two augers are shown in figure 2. According to the Soil Survey Manual (76), the screw auger consists of a 1-1/4 or 1-1/2-inch wood bit, with cutting lips and tip removed, welded to a steel rod with a handle. The worm should be about 7 inches long with distance between flanges about the same as the auger's diameter.

The Iwan or posthole type is particularly useful in gravelly soils or when sizeable samples are desired. The sampler pictured has an inside diameter of 4 inches and a barrel length of 8 inches. Smaller augers are available. As the auger is rotated, the cutting blades loosen the soil and force it into the cylinder. Under favorable conditions, and with shaft extensions of common pipe, this auger can be used to sample to depths of 20 feet or more.

<u>Depth and Frequency of Sampling</u>

The depths sampled depend on the objectives of the study. If the intent is to determine moisture supplies for plant growth, sampling should include all of the root zone. Relatively little sampling is done below 5 or 6 feet.

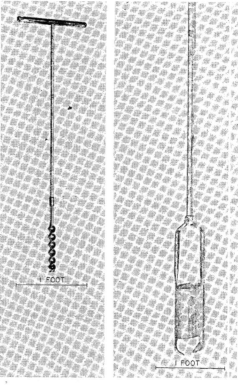

Figure 2. --Two types of soil augers. Left, common type. Right, Iwan or posthole type.

Gravimetric sampling is so arduous that its frequency should be directed by close regard to weather and soil-moisture conditions. For instance, to define the soil drying curve it is necessary to sample daily, or perhaps even more frequently, for the first few days after irrigation or after soaking rains in the growing season. At such times, the rate of change in soil-moisture content is most rapid. As the soil dries, the rate of change decreases and sampling may be less frequent.

Soil-Sample Containers

With a 1-inch diameter soil tube, sample
25 to 50 grams per 3-inch soil depth. Samples
grams have been found unsatisfactory for the acc
of soil-moisture content (52).

Samples may be placed in tinned cans hav
4, 8, 16, or 32 fluid ounces. Roughly, the 4-oun
6-inch soil sample taken with a soil tube or scr
carry cans are easily made of plywood (37).

Cans should be numbered, both on the lid
tare-weighed, usually to the nearest 0.1 gram.

Generally, the seal on these cans is cons
prevent moisture losses for several days. In a
soil-tube cores of silt loam were placed in cans
day of sampling and at various times thereafter
magnitude of loss due to delay in weighing. Th
60 grams of dry soil and 30 percent moisture by
days after sampling, no loss occurred. In 3 day
than 0.1 gram per sample; in 9 days, 0.2 gram; i
or 1 percent by weight. When samples are stor
they either should be weighed before storage or s
electrical or masking tape.

After oven-drying, cans should be thoroug
a large number of cans are in use, a motor-dri
helpful. Tin cans in frequent use often do not la
An acid- and heat-resistant varnish delays rusti
cans more difficult to open. Aluminum cans ma
expensive and dent easily--after which lids will n

Because of convenient packaging, air-tigh
rust, and ease of procurement, 1/2-pint Mason ja
the Vicksburg Project to hold 18-inch soil-tube sa

Oven-Drying and Weighing

As a common practice, samples are ove
110° C. to a constant weight. A 24-hour period is
large numbers of samples are to be dried, fo
desirable; they can dry 50-gram samples in 4 to (

Davisson and Sivaslian recommended that soils be dried in an electric oven at 105° C. in a vacuum over phosphorus pentoxide (30). With this procedure 4 hours' drying was sufficient. Slightly more moisture was removed than with standard oven-drying.

Generally, samples of less than 100 grams are weighed to the nearest tenth of a gram. When many are to be weighed daily, a rapid-reading laboratory scale, such as the Toledo Model 4636BA, is worth the investment. With samples larger than 100 grams, allowable error may permit use of an inexpensive dietetic scale accurate to one gram. To speed up weighing on a torsion balance, Olson and Hoover (51) suggest that samples first be weighed to the nearest 0.5 gram on a direct-weighing scale.

Stout and Holben have described a method of weighing and drying that will give moisture content in 2 to 3 hours (79). In this method, 20-gram soil samples are placed in 4-ounce soil cans of exactly equal tare weight and dried in a forced-draft oven at 130° C. After drying, the samples are weighed on a scale counterpoised to the can weight so that the difference between the dry weight and 20 represents the moisture loss. According to the authors, results checked closely with those obtained by drying overnight at 105° C.

Sen (70) dried 10- to 15-gram soil samples by placing them in a fused silica basin held 2 to 3 inches above the flame of a rose Bunsen burner. The soil, which was stirred constantly, dried completely in 3 to 6 minutes. Moisture contents did not vary from oven-dry values by more than one percent.

Bouyoucos (10) has suggested treating soil samples with alcohol and burning off the alcohol to remove the water. Carter (23) found that it required about five minutes to make a determination in this way and that the moisture percentages obtained were slightly higher than those determined by oven-drying. Considering the time required, he believed the method worth while only for field determinations.

Electrical-Resistance Units

The development of electrical methods has been excellently summarized by Olson and Hoover (53). Herein major emphasis will be placed on the most common instruments and their use. Much of this material is a condensation of earlier papers prepared at the Vicksburg Project (55, 59).

Types of Instruments

Currently, three types of units are in use: the fiberglas unit developed by Colman and Hendrix (27), the nylon unit developed by Bouyoucos (11), and the plaster of paris or gypsum block of Bouyoucos and Mick (17). These three units are shown in figure 3, together with a fiberglas-gypsum unit that was developed by Youker and Dreibelbis (93) but has not come into general use.

Dimensions and details will be found in table 1. In the fiberglas unit, electrical resistance is measured between two monel metal screens separated by two layers of fiberglas cloth, the whole enclosed in three layers of the same material and bound in a monel metal case that is spot-welded at the edges. The fiberglas unit differs from all

Figure 3. --Soil-moisture units and meters.

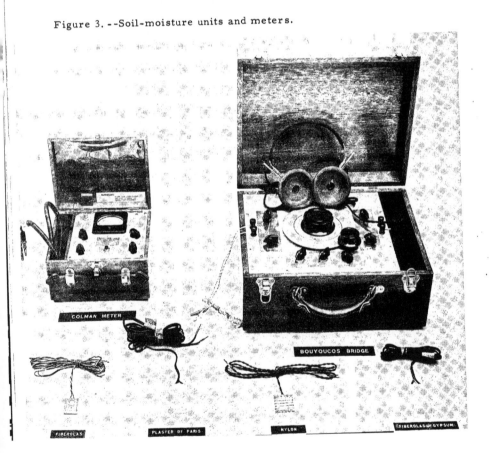

Table 1 --Dimensions and details of electrical soil-moisture units

Item	Type of unit			
	Fiberglas	Nylon 1/	Plaster of paris 2/	Fiberglas-gypsum
Outside dimensions, inches	1.5 x 1.0 x 0 12	1.5 x 1.25 x 0.12	1.7 x 1.25 x 0.7	2 5 x 1.5 x 0.5
Absorbent material between electrodes	2 layers of fiberglas	1 layer of nylon	Plaster of paris	Fiberglas, plaster of paris
Distance between electrodes, inch	. 03	. 03	. 19	. 75
Electrodes:				
Area, square inch	. 39	2. 0	. 46	...
Length, inch	...	...	...	. 25 - . 50
Mesh, wires per inch	60	96	20	...
Area of fabric in unit, square inches	5	5. 6		
Area of absorbent exposed to soil, square inches	. 20	1. 20	8. 38	11. 50

1/ This unit is now available encased in plaster of paris The "nylon-plaster" unit has outside dimensions of 1. 90 x 1. 70 x 46 inches.

2/ Data for unit as now manufactured with wire-screen electrodes.

others in that it contains a thermistor, thus permitting soil-temperature measurements by which resistances may be corrected to a common temperature. The resistance between the electrodes is read with a battery-operated alternating-current ohmmeter.

The nylon unit, similar in shape and size to the fiberglas, consists of two electrodes of fine metal screen to which wire leads are silver-soldered. The electrodes are separated by wrappings of nylon and enclosed in a perforated nickel case with edges mechanically united. A later design has the case and electrodes of stainless steel (13). This unit differs from the fiberglas by having larger grids for electrodes and a much greater area of fabric exposed to the soil.

The plaster of paris block was first constructed simply by embedding two electrodes, each two inches long, in plaster of paris. As now manufactured, the block has electrodes of stainless steel screen (15) and is impregnated with a nylon plastic resin for greater durability (14). The fiberglas-gypsum block is similar to the plaster of paris, but the two electrodes are separated from the plaster of paris by a layer of fiberglas cloth. Recently, Bouyoucos has encased his nylon unit in plaster of paris (16).

Resistances of units other than fiberglas are measured with a modified Wheatstone bridge developed originally for use with the plaster of paris block (18). They may also be determined with the ohmmeter used with the fiberglas unit.

Installation

Careful installation of units is extremely important (59). Units should be installed when the soil is moist enough to pack well; it should be below field capacity but not dry and hard. Unless the excavated soil is repacked to at least its original density, an artificial channel may be created through which water will travel rapidly to the units. Such a channel is indicated when, after a rain, soil samples at a given depth are still dry whereas the resistance of the corresponding unit has dropped significantly. The problem is especially serious when the natural soil has a hardpan or other layer of low permeability that is difficult to duplicate when refilling excavations.

At the Vicksburg Infiltration Project, fiberglas units were first installed in pits. Since this involved considerable soil disturbance and was time-consuming, further installations were in 5-inch auger holes Units are pressed into the sidewall of the hole; to avoid disturbance between units, unit positions were 45 degrees apart (fig. 4). At first, a stick with a notch cut to fit the unit was used to press the unit into the wall of the hole; later a device for mechanically inserting the unit was devised (54).

The units are placed in the soil perpendicular to the ground surface. To prevent water movement along the wires to the units, the wires are led slightly downward before being brought to the surface. Soil is replaced in its original position and repacked to at least its original density. Under favorable soil conditions three stacks of ten units each,

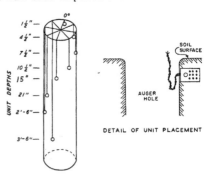

Figure 4. --Installation of fiberglas units in auger holes.

with the deepest unit at 42 inches, have been installed in two man-days.
With recently developed equipment for insertion of plaster of paris
blocks, 7 units were installed in 15 minutes (49).

To facilitate reading, lead wires from the units can be wired
into several different types of terminal panels or switch housings
(34, 56, 92). A simple terminal panel (fig. 5) for fiberglas units
can be made of 1/2-inch Plexiglas, 2-3/4 inches wide and long enough
to accommodate the number of terminals desired. Lead wires from the
units are soldered to brass bolts inserted in the Plexiglas.

Figure 5. -- Terminal panel.

Electrical resistance of a soil-moisture unit is affected by sture content and, to a lesser but appreciable extent, by temperature. s if accurate soil-moisture data are required, corrections for temature are necessary. Figure 6 gives an idea of the extent of the rection (to a common temperature of 60° F.) for one fiberglas installed near Vicksburg (55). The temperatures ranged from 30° 0° F. At high moisture contents the correction in this temperature ge was small, amounting to 0.05 inch at 1.20 inches of water. At sture content of 0.70 inch, the variation was 0.22 inch.

Bouyoucos notes that temperature influence on nylon units is lerate and can be ignored where temperature fluctuations are not it. For careful research where corrections are needed, he

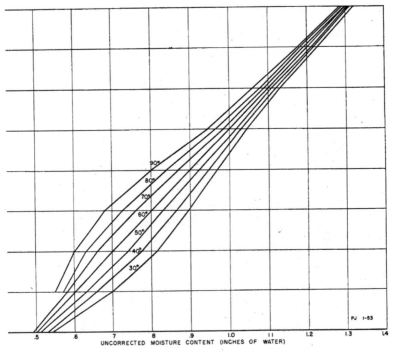

ire 6. --Effect of temperature (degrees F.) on determination of sture content for the 0- to 3-inch soil depth in Commerce clay.

- 18 -

suggests that temperature be measured with a bimetallic dial-indicating thermometer at shallow depths and with liquid electrical-resistance thermometers or thermistors for deeper depths (11). Temperature corrections for plaster of paris blocks can be made from a figure given in the original paper describing the block (17) or with a recently developed slide rule (66).

Determination of Resistance

Dial readings of the meter used with the fiberglas units are usually converted into ohms. On the Vicksburg Project, this was first done by use of tables relating dial reading and temperature of the unit to resistance in ohms adjusted to a common temperature of 60 degrees F. Later, to facilitate plotting of the calibration curves, tables were prepared giving the logarithm of resistance in ohms. Bethlahmy (6) prepared an alinement chart to determine resistances corrected to a common temperature. At the San Dimas Experimental Forest in Glendora, California, dial readings are corrected directly for temperature by use of an alinement chart, and resistances are never directly determined. The Wheatstone bridge used with Bouyoucos units gives readings directly in ohms uncorrected for temperature.

Calibration

Units are calibrated either in the laboratory or in the field. The objective in each situation is to determine the relationship between resistance of a particular unit and actual soil-moisture content. For this relationship, direct measurements of soil moisture are plotted against concurrent resistances over the experienced range of moisture content. Laboratory calibration is accomplished by drying and periodically weighing soil cores in which units have been inserted; resistances of the units are measured during the drying cycles. Field procedures consist of taking gravimetric samples from an area close to the units, and relating their moisture content to measured resistance.

From studies at Vicksburg, laboratory calibration cannot be recommended for general use (20, 21). Laboratory curves not only differed markedly from field calibration curves but also differed from each other in successive drying cycles. Discrepancies probably originate in the treatment of laboratory cores and differences between laboratory and field conditions. Laboratory calibration, however, is necessary when units are to be placed in very stony soils, under pavements, or in other situations where field samples cannot be taken.

In field calibration each stack of units
sampling area. Samples must be taken cl
represent the same moisture condition, yet n
the natural soil and moisture relations at th
large enough to take care of repeated samplin

Sampling areas used by the Vicksburg
arrangement in figure 7 has been found most
cause it avoids disturbing the central plot con
Samples, usually in duplicate, are taken with
foot blocks, one sample from each of two of t]
and blocks are randomly selected.

Successive soil cores are removed fro
mid-point of a core corresponds to the depth c
and the length of the core is generally made t
layer represented by the unit. After samplir
similar soil to prevent movement of water

When rain follows a dry period the d
considerably between points only a few feet
content at the unit may differ from that at

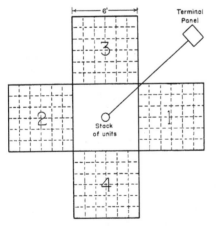

taken t
the cur
during
ing sea
remain
tent. S
rapidly
be requ
section
bration
ional sa
through
as a ch

frequen

Figure 7.--Calibration plot arrange-   several
ment.   tent bet

Inspection of the values for the soil layers just above and just below the values in question may give a clue as to whether the difference is due to error in measurement or truly represents soil conditions.

Calibration curves. --A calibration curve--an example of which is given in figure 8--is prepared for each soil-moisture unit. Each moisture-content value of the duplicate samples and the mean value is plotted from the abscissa against corresponding resistances measured on the ordinate. Resistances can be most conveniently plotted on a 4-cycle logarithmic scale, or, if the logarithm of resistance is used, on an arithmetic scale. On the average, 15 to 20 duplicate samples, taken at moisture contents ranging from saturation to wilting point, adequately define the calibration curve.

Points should be plotted as data are obtained. When sufficient points show reasonable agreement, the calibration curve is drawn. This calls for some judgment: for instance, individual samples of a pair that vary widely should be given less weight than pairs that agree closely. In general, all samples taken when the moisture content varies considerably from depth to depth are subject to doubt.

Conversion of Resistance to Moisture Content

After the development of the calibration curve, resistances, corrected to a common temperature, can be converted to moisture contents. Moisture contents may be read directly from the curve, or a table may be prepared from the curve. With plaster of paris blocks, a slide rule (66) may be used to correct the observed electrical resistance for soil temperature variations and to change the resistance readings to percentage of available moisture. On the Vicksburg Project, moisture content was marked on the abscissa of the calibration curve in both percent by weight and inches depth of water. Thus, after calibration

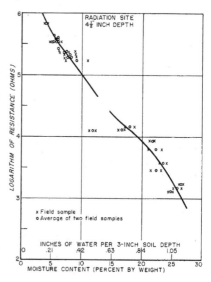

Figure 8. --Typical calibration curve.

- 21 -

data have been plotted in percent by weight, daily resistance values can be readily converted to inches depth for the soil-moisture record.

Recording Instrument

Resistance readings are usually taken daily or less frequently. For more frequent readings, Korty and Kohnke hooked a recording potentiometer, connected with an ohmmeter circuit, to nylon units (44). Readings were recorded at 8-minute intervals. Maximum accuracy was 5 to 10 percent of the resistance being measured--10 percent of the resistance was equal to a 1/2 percent error in soil-moisture content. This instrument responded to sudden changes in soil moisture from rainstorms and was useful in following general soil-moisture conditions. To measure small changes in moisture, more costly devices are necessary.

Other Types of Meters

Since the development of the modified Wheatstone bridge, more simple types of instruments have been devised for use with Bouyoucos units. One is a light-weight instrument that measures the electrical resistance of plaster of paris blocks to passage of a direct current (46). Readings can be taken in about 15 seconds. As a guide to irrigation practice, Bouyoucos developed an alternating-current impedance meter calibrated to read directly in percentage of available moisture in the soil (12).

Tensiometers

Tensiometers measure moisture content through a tension range from 0 to 0.85 atmosphere. In soils of the finest texture this covers about half the moisture range from field capacity to wilting percentage; for coarser, sandy soils it will cover more than 90 percent of the range (62).

Tensiometers come in various lengths from 6 inches to 4 feet or more. They are equipped with a mercury manometer or a Bourdon-type vacuum gage (fig. 9). The manometer type is more accurate. The Bourdon type is simpler to operate; it gives precision of 2 percent of the field moisture range (62).

Field installations have been made to depths of 15 feet (61). The instruments can be placed in soil-tube holes; soil around the top four inches of the hole is tamped to prevent runoff from flowing downward to the cup. According to Richards (61), two men in less than two hours installed 36 instruments at depths from 6 inches to 5 feet. About 8 hours is required for the instrument to reach equilibrium. When first set out, the system should be filled with boiled distilled water. Thereafter, when only small amounts are added, the water need not be boiled. With freezing temperatures, instruments must be brought in from the field; however, a wool sock over the gage is sufficient protection for the Bourdon type at temperatures slightly below the freezing point.

Tensiometer readings are subject to daily variations. Haise and Kelly (38) found that in Yuma fine sand the variation extended to a depth of 48 inches. Daily variations in tension of 350 to 400 centimeters of water at the 6-inch depth were common. These variations are believed due to temperature gradients (between the porous cup and the surrounding soil) which affected vapor transfer and condensation. The effect can be minimized by reading tensiometers at the same time each day, preferably in the morning. Richards and Gardner used a recording vacuum gage to obtain a continuous record of tension (63).

To obtain moisture-content values with tensiometers, calibration is necessary either by taking field samples and relating their moisture content to concurrent tensions, placing a tensiometer in a container of soil that can be weighed from time to time to get moisture changes, or determining a moisture content-tension relation in the laboratory by using tension tables and pressure membrane cells. Scofield, working with the first two methods, secured best results with an open container of growing plants (69). The soil in the container was saturated and weights and tensions were recorded as it dried.

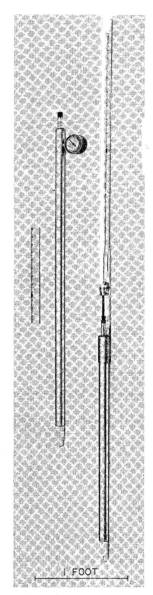

FOOT

Figure 9. -- Types of tensiometers. Left, vacuum gage. Right, manometer.

Nuclear Method

Moisture measurement by the nuclear method depen
a source of fast neutrons, (b) the slowing down and deflectic
neutrons by the water in soil, and (c) a measure of the res(
neutrons. Carlton et al. (22) have stated the theory as f

The measurement of soil moisture is based on the pl
cal laws governing the scattering of neutrons in matt
and, in particular, the scattering of neutrons in soil.
When a fast neutron source is placed in soil, the emi
neutrons collide with the atoms comprising the soil.
result of these collisions, the neutrons are scattered
all directions and some of them return to the vicinity
the source. However, in each collision the neutron l
part of its kinetic energy and is slowed down. The av
age energy loss is much greater in neutron collisions
with atoms of low atomic weight than in collisions inv
heavier atoms. As a result, the number of slow neu
found near the source is a function of the number of a
of low atomic weight present in the soil. If the numb
these atoms in the soil is increased, a greater numbe
slow neutrons will be found near the source.

Hydrogen is the only element of low atomic weight fo
in ordinary soils in appreciable amounts. Therefore
a device for detecting slow neutrons is placed in the ε
near a fast neutron source, the number of slow neutr
counted per unit of time is a measure of the concentr.
of hydrogen atoms in the soil. Since the hydrogen is
ly contained in molecules of free water (moisture tha
be evaporated by heating the soil to a temperature of
C.), the slow neutron count is a direct measure of
moisture content of the soil.

The nuclear method gives the amount of water per un
soil. Since the relationship between neutron count and moistu
is largely independent of the character of the soil, it is possil
calibration curve will suffice for all locations. If moisture
percent by weight is desired, the soil bulk density must be d
Nuclear equipment that utilizes gamma rays produced by col
been devised to measure bulk density.

The equipment used in tests at the Vicksburg Project
signed at Cornell University (5, 22). It included a source of

- 24 -

trons (25 millicuries of radium D-beryllium) and a Geiger-Muller counter tube, both sealed within a moisture probe; access tubes which were driven into the ground and into which the moisture probe was lowered; a scaling mechanism for electronically indicating the pulses received from the counter tube; a timer; and a container of water in which standard readings were taken (fig. 10).

The moisture probe, about 9 inches long, was made of thin-walled brass tubing 1 inch in outside diameter. Within the probe, the neutron source and counter tube were separated by a 1-1/2-inch lead plug. Access tubes were made from stainless steel tubing 1-1/16 inches in outside diameter and with a wall thickness of 1/32 inch. One end of each tube was fitted with a metal point. Tubes were 3 and 5 feet long; longer tubes can be used.

A 13-foot length of coaxial cable connected the scaling unit to the moisture probe. A trailer-mounted gasoline-driven generator provided current for field use.

Before readings were taken in the soil, three 3-minute readings were made in the standard. Then the probe was lowered into the access tube in the soil and a series of three 3-minute readings was taken at various depths to 42 inches.

Figure 10. --Nuclear equipment for measuring soil moisture and soil bulk density.

The instrument was cal-
ibrated by graphically relating
count ratio (count in soil divided
by count in standard) to soil-
moisture percent by volume as
obtained from duplicate soil-
tube samples and cylinder bulk-
density samples (85). Calibration points of the 1-1/2- and
4-1/2-inch soil depths did not
agree with those from lower
depths because of loss of radiation at the ground surface. Separate calibration curves, three
in all, were developed for the
1-1/2-, the 4-1/2-, and 7-1/2-
inch and lower depths (fig. 11).

Fifty-five measured
moisture contents for depths
from 7-1/2 to 42 inches deviated
from the calibration curve by an
average of 2.0 percent by volume or about 1.4 percent of the

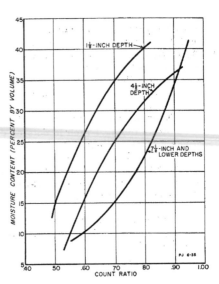

Figure 11. --Sample calibration
curves for nuclear equipment.

oven-dry weight of the soil. For eleven measurements, the average deviation was 1.4 percent by volume at the 1-1/2-inch depth and 2.0 percent by volume at the 4-1/2-inch depth. With additional experience in
handling the equipment, accuracy may be increased.

The above deviations compare well with the findings of others.
Carlton et al. report a deviation of 1.3 percent by volume in laboratory
tests (22). Using similar equipment, also in the laboratory, investigators
at the University of California found an average deviation of 1.4 percent
by volume (41), while in a field study at the University of Saskatchewan
all deviations were within 3 percent by weight and most were 2 percent
or less (45). If an average density of 1.3 grams per cubic centimeter is
assumed, 2 percent by weight is equivalent to 2.6 percent by volume.

Portable meter. --Recently, a portable slow-neutron flux meter
for measuring soil moisture has been devised by Underwood, van Bavel,
and Swanson. This meter has a source-counter assembly which can be
inserted in a 2-inch well or, for surface measurements, laid flat on the
ground (82). It uses a radium-beryllium source of 10 millicuries nominal

strength and a proportional counter. A 10-foot coaxial cable connects the counter to a battery-operated rate meter which is custom built. A bucket of water, with a central cylinder for insertion of the source-counter assembly, is used as a standard. The weight of the assembled equipment, not including the standard, is less than 20 pounds.

An instrument of this type is now being commercially produced. It is called a "direct soil-moisture determinator."

Evaluation of Methods

Each of the four methods just described has certain advantages and disadvantages. Three of the methods, the electrical-resistance, tensiometer, and nuclear, were devised primarily to give in situ records, a distinct advantage for many types of investigations over the gravimetric method, in which the sampling point is destroyed. As will be pointed out later, this is an over-simplification because field calibration recommended for the in situ instruments tends to reduce their point significance, lending the data an areal significance not generally appreciated. Too, with sufficient replication, gravimetric sampling can yield records susceptible of point interpretation. Thus, an evaluation of instruments in respect to their in situ measuring ability is not too meaningful. The exception is the nuclear method: apparently independent of field calibration, it may with further development prove to give true in situ measurements.

In the following sections the methods will be judged as to their accuracy and range, freedom from error due to variations in salt concentration or changes in temperature, ease and speed of measurement, and durability and relative cost of equipment.

Range of Measurement and Accuracy

Electrical-resistance units are most responsive to moisture-content changes from below wilting point to field capacity. The plaster of paris units are not sensitive above field capacity (17). At moisture contents between field capacity and saturation, the resistance change per unit change of moisture for the fiberglas and nylon units is less than at lower moisture levels. Also, field calibration of these units at moisture contents above field capacity is inaccurate because, in sampling, gravitational water may be lost or extraneous free water included. The calibration curve for the nuclear equipment used at Vicksburg tends to level off at very high moisture contents, an indication of decreased accuracy in this range.

Tensiometers are the best method now available for t
soil moisture from field capacity to saturation. On the other
fact that they cannot be used to measure low moisture conte
prevents securing continuous records. Slater and Bryant nc
1,078 instrument days of measurement in Maryland, the tens
were inoperative 73 percent of the time (73).

The accuracy of the various methods is somewhat d
evaluate, since rigorous comparisons have never been made.
and Mick stated that the relative experimental error of lal
calibrated plaster of paris units was from ± 0.1 to ± 1.0 pei
As the blocks are not accurate at high moisture content,
laboratory calibration curves, from Vicksburg Project expe
not agree with those derived by field calibration, comparab
errors could be expected to be somewhat greater.

In a recent comparison at a Vicksburg site, nuclear
and fiberglas units gave very similar accuracy (85). In tl
method there was an average deviation of 1.4 percent by weig
calibration curve. Eight fiberglas units, calibrated in the fiel
an average deviation of 1 percent by weight.

In more than 1,100 observations at three Vicksburg
average deviation of the mean of duplicate gravimetric sam
fiberglas-unit calibration curves was 1.6 percent by weight.
and Hendrix (27) measured deviations from a mean calibratio
five laboratory-calibrated fiberglas units. Deviations were l
0.5 percent from wilting point to field capacity and about 1 pe
bove field capacity. Stackhouse and Youker found close agre
tween the moisture records of fiberglas-gypsum units and the
monolith lysimeter in which they were buried (77).

Plaster of paris blocks, fiberglas units, and nylon ur
with about equal rapidity to changes in soil moisture (55).
plaster of paris and fiberglas units and tensiometers, Ewart
(36) reported that a rapid drop in the resistance of one type
followed by a similar drop in the other. However, the tensic
from 40 minutes to 6 hours slower than the other instrum
cording a drop. On the basis of daily readings, Slater and Br
found response of plaster of paris blocks and tensiometers pi
tensiometers more sensitive at higher moisture contents. Th
also noted that tensiometers frequently leak, and thus give
results.

An undetermined source of error with plaster of paris blocks is the rather high proportion of electrical conduction which follows paths partially outside the block. Slater found that conduction may be confined wholly within the block by locating one electrode centrally in a cylindrical screen that serves as the second electrode (72).

Another source of error in these blocks is hysteresis--i. e., the tendency of the block, at a given soil-moisture tension, to be higher during soil drying than during soil wetting. Tanner and Hanks (80) determined hysteresis over a tension range of 0 to 8 atmospheres and found that unless the blocks are calibrated individually and unless they begin drying.from a definite moisture condition, an estimate of soil-moisture tension from block resistance may be in error by an amount equal to 0.5 to 1.0 times the estimated tension. Hysteresis effects in the blocks tended to compensate for moisture hysteresis effects in the soil when blocks were used for moisture-content determination.

Soil-moisture records secured with tensiometers also exhibit considerable hysteresis (64). Richards points out that drying curves are of chief interest in practical agriculture, since the wetting process is usually of short duration (62).

Effect of Salt Concentration

The salt concentration of the soil solution does not affect the gravimetric, tensiometer, or nuclear methods. For the electrical units, it can be the deciding factor as to whether certain types are used.

Ewart and Baver (36) found that increasing the salinity from 0 to 0.2 percent had little or no effect on moisture readings from Bouyoucos blocks. When the concentration was raised to 0.5 percent, drops in resistance were noticeable. Fiberglas units were significantly affected by an increase from 0 to 0.1 percent. The authors concluded that, in mildly saline regions, salt concentrations are not high enough to affect the gypsum block values.

In successive drying cycles during laboratory calibration, Colman and Hendrix (27) found that sufficient salts accumulated from tap water to reduce the resistance of fiberglas units. They recommended that distilled water be used in laboratory calibrations and that units be field-calibrated if they are to be placed in soils varying in soluble salt content during the year.

According to Bouyoucos (11), nylon has virtually no buffering action, but since the units are calibrated for their respective soil this

is not a hardship. Changes in salt concentration within
are considered unimportant under average conditions. :
of paris block, Bouyoucos and Mick report no effect or
content-resistance relation with soil samples containing a
1,000 pounds of 4-16-8 fertilizer per acre (17).

Weaver and Jamison found that the resistance of
nylon, and plaster of paris units decreased appreciably .
and 0.01 to 0.1 normal solution of sodium chloride at te.
0 and 15 atmospheres. The three solutions were equal
sistance to application of 40, 850, and 1,900 pounds of
4-10-7 fertilizer per acre 6 inches of topsoil (86).

Effect of Temperature

To secure accurate records with electrical uni
must be adjusted to a common temperature. The thermi
glas unit makes it possible to measure soil temperature
other types of units separate temperature-sensing ins
necessary.

As has been mentioned, temperature gradients betw
cup of the tensiometer and the surrounding soil may caus
tension readings. Measurements by the gravimetric r
affected by temperature changes above freezing. The r
reportedly operates independently of temperature (5).

Ease and Speed of Measurement

In effort required to secure frequent soil-moist
ments, the four methods may be ranked (from most to le.
as follows: gravimetric, nuclear, electrical, tensiomete
no question that the gravimetric method requires the mos
great extent the total effort required to use electrical
ometers depends on how much gravimetric sampling is n
calibration. Once calibrated, both types of instruments.
If the nuclear method will require only one calibration fo:
may become the easiest method.

Considering speed of measurement, the same
followed. Time required for sampling with a soil tube c
condition of the soil. For soils easy to sample, 16 to 25
increasing depths to 4 feet can be taken in about one ho
conditions, sampling would take twice that long; in stony
a full day. With the nuclear equipment used at the Vicks

- 30 -

eight measurements required about two man-hours. Less time is required with electrical-resistance units; in a recent test, 1 minute and 26 seconds were required to read 10 fiberglas units with the ohmmeter and convert readings to moisture contents (55). An average of 4 minutes and 3 seconds was required to read resistances of 10 fiberglas units with the Bouyoucos bridge. According to Bouyoucos and Mick (18), a single reading can generally be taken in less than a minute. Tensiometers, with their indicating mechanism, are most rapidly read.

Durability

All of the commonly used instruments are very durable. Occasionally, the points of soil tubes become worn or chipped and must be replaced. Fiberglas units have been used as long as four years without failure. Until recently, a disadvantage of the plaster of paris blocks was their tendency to dissolve when the soil remained wet for long periods of time. This has been corrected by impregnating these blocks with a nylon plastic resin, a treatment which does not change their physical or chemical characteristics (14). The nuclear equipment tested at Vicksburg frequently broke down.

Relative Cost

All of the methods involve equipment that is commercially produced for soil-moisture measurement. The following current prices will vary from time to time and are only for relative comparison.

Gravimetric sampling

Five-foot soil tube with hammer	$25.50
Soil auger and open-side tube	13.50

Electrical-resistance method

Plaster of paris blocks, each	$1.85
Nylon unit (encased in plaster of paris)	3.85
Bouyoucos bridge	200.00
Fiberglas unit	4.60
Alternating-current ohmmeter	160.00

Tensiometer

One- to 4-foot models	$18 to $22

Radioactive method

Direct soil-moisture determinator
(complete with radium-beryllium)
Equipment described on pages 24-25
(not including generator or radium
D-beryllium [3]/)

Summary

Selection of instruments will depend on their r
already considered and on proposed frequency of mea
of record, experience of personnel, and commonsens

As a general rule, the gravimetric method sho
measurement by one of the indirect methods is absolu
satisfy study requirements. The gravimetric method
perience than any of the others. As it provides direc
the time, effort, and possibility of error associated
resistances, tension measurements, or neutron count
content are avoided. Even if one of the other methods i
ous gravimetric samplings are usually required for cal

Considerable knowledge of the soil-moisture
secured by judicious gravimetric sampling, i. e., s
dates with regard to weather conditions and making
points and depths sampled will give a maximum

Electrical-resistance instruments are most us
daily records of considerable duration; in this situati
area disturbance and time and effort involved with the
are usually prohibitive. Daily measurement is nec
moisture content before and after rainfall is needed.
range is to be covered, the fiberglas or nylon-gypsum
mended. If temperature corrections are considered n
glas unit should be used. If the soil has been fertilize
concentration of the soil solution may prevent satisfac
with electrical units.

Tensiometers are most useful in irrigation stu
the soil is dry too frequently to give them any value.

3/ The radium D-beryllium was rented at $24.00 a m

Nuclear instruments have not been tested sufficiently to demonstrate their suitability for routine field use; they do hold promise.

Soil-Moisture Expression

Moisture content can be expressed on either a weight or volume basis. On a weight basis, it is almost always given as a percentage of the oven-dry weight of the soil. On a volume basis, it is usually stated as the percentage of the total soil volume occupied by water.

Values in percent by weight are relatively easy to compute:

$$\text{Moisture content (percent by weight)} = \frac{\text{Weight of water in sample x 100}}{\text{Oven-dry weight of soil in sample}}$$

To obtain values in percent by volume, volume measurements are substituted for weight:

$$\text{Moisture content (percent by volume)} = \frac{\text{Volume of water in sample x 100}}{\text{Total volume of sample}}$$

The volume of water in the sample can readily be determined from its weight (1 cubic centimeter equals 1 gram). The volume of the sample, that is, the volume it occupied in the field, can be determined by using cylinders of known size to secure soil samples. Henrie has pointed out that the inches depth of water in a soil sample taken with a soil tube equals the product of the weight of water in grams times $1/2 D^2$, D being the inside diameter of the cutter in centimeters (39).

Generally, percent by volume can be less accurately and less easily obtained by field measurement than can percent by weight. Therefore the usual practice is to derive percent-by-volume values from percent by weight by using an average bulk density:

$$\text{Moisture content (percent by volume)} = \text{Moisture content (percent by weight)} \times \text{Bulk density}$$

Bulk density, the ratio of the weight of oven-dry soil to the volume it occupied in the field, is generally expressed in grams per cubic centimeter.

Expression of moisture content in percent by weight is satisfactory when moisture contents of soils of similar bulk density are compared. However, values in percent by weight do not indicate how

much water is in the soil. Values on a volume basis are neces
absolute quantities of moisture content are required or for stri
parisons between soils of different bulk densities. They are v‹
in hydrologic studies where soil moisture, precipitation, and s
are expressed volumetrically. Percent-by-volume values do,
possess an added source of error over percent-by-weight, bec
measurements of soil volume are also involved.

Other Means of Expression

Moisture content of the soil can be expressed in seve
ways, all derived from percentages by weight or volume. O
most common and most generally useful expressions is in inch
of water--either per inch or foot of soil, or in a given so
or in the root zone of a specific crop, or in some other spec
depth. The value in inches depth is easily obtained from the
content in percent by volume:

$$\frac{\text{Moisture content}}{\text{(inches depth)}} = \frac{\text{Moisture content}}{\text{(percent by volume)}} \times \frac{\text{Soil depth}}{\text{(in inches)}} \times \text{]}$$

In crop studies, the term available water is sometim‹
This refers to the water available to plants and is the moistur
of the soil (gravitation water excluded) less the amount remain
permanent wilting point. Available water can be expressed in
percent by weight, percent by volume, or inches depth. The p
wilting point must be established for each soil by appropriat‹
laboratory measurements.

Sometimes the amount of available water present in the
given time is reported as a percentage of the soil's availa
capacity, or the difference between field capacity and wiltir

Conrad and Veihmeyer (28) proposed the term relativ
to express the ratio of moisture content to moisture equivalent
expression enables comparisons of moisture conditions betwee
between soil horizons that differ in texture.

The moisture level in a soil is sometimes given in ter
tension, indicating the force by which the water is held. Te
negative pressure, is generally expressed in atmospheres
centimeters of water; for all practical purposes one atmospher
to 1,000 centimeters. The term pF, the logarithm of the
centimeters, has also been used. Moisture tension can be r

4/ Moisture equivalent is the moisture content of soil that has
jected to a centrifugal force of 1,000 times gravity.

directly with a tensiometer. To convert tension values to moisture content, a moisture-tension curve must be prepared for individual soils from concurrent laboratory or field measurements of moisture content and tension.

Stoniness

Stone in the soil complicates soil-moisture expression as well as its measurement. Depending upon the amount and size of the stones and objectives of the study, soil-moisture content can be reported as either a percent of total weight or volume inclusive of stones, or as a percent of net weight or volume exclusive of stones. In any case, it should be made clear which method is being used, why it is being used, and, if stone is excluded, the minimum size considered. Degrees of stoniness should be reported in site descriptions. Nikiforoff (51) and the Soil Survey Manual (76) present discussions and classifications of stoniness.

Where soils are stony, moisture content in percent by weight is often related to the net weight of soil. Olson and Hoover (53) suggest that samples with an estimated rock content of 2 percent by weight or greater be screened through a 2-millimeter sieve after the samples have been oven-dried. The weight of the stones is deducted from the oven-dry weight to obtain the net oven-dry weight. Without sieving, the moisture percent by weight will represent the soil exclusive of material too large to be sampled by the equipment used.

Moisture content in percent by volume exclusive of stone can be determined from the corresponding percent-by-weight value and a bulk density value that is also based upon weight and volume determinations exclusive of stone. Percent-by-volume moisture content based on total soil volume including stone is probably more useful. If stones are small enough to be sampled with the soil-moisture and bulk-density equipment being used, no difficulty will be encountered in determining this value. Otherwise, an estimate must be made, for each soil horizon or layer, of the percent of total volume that is occupied by stones larger than the size included in the samples. Hoover, Olson, and Metz (40) determined bulk density exclusive of particles above 2 millimeters in size and then, after determining average stone content for the horizons sampled, adjusted bulk density to prevailing field values.

Bethlahmy (7) presented a method based upon weighing the stones found in each 6-inch soil layer, and dividing their weight by their specific gravity to obtain their volume. With this estimate, moisture

percent by volume based on total soil volume can be compute
moisture percent based on net soil volume (less stone):

Moisture content =
(percent of total volume)

Moisture content   x (100 - percent stone)
(percent of net volume)

Rowe and Colman (65) estimated stone content by mappi
edges of the stones where they intersected the vertical face of a p
at the sampling plot, and then calculating the percentage of area o
by stones on the pit face. This percentage was assumed equivaler
percentage of soil volume occupied by stones.

Where possible, values of moisture content in inches depth
be based on total soil volume. This facilitates conversion to mea
values of acre-inches or acre-feet of water and comparisons wit
cipitation and streamflow amounts. Inches-depth values based
volume excluding stone give excessively high moisture contents;
area, because that part of the volume occupied by stones cannot
pected to have as much water as the soil that was sampled.

The foregoing is based on the usual assumption that m
change in stone is negligible. Coile (25) has shown that moisture
of small stone, especially that between 2 millimeters and 1/2 i:
diameter, may change appreciably and that the amount of cha:
related to the size and kind of stone. Little error will result fro
source except when determining moisture content in percent of net
or net volume after sieving out material larger than 2 millimeters
then only when the proportion of such material is large.

Use of Soil Bulk-Density Data

As noted, soil bulk density is determined in order to conve
moisture content in percent by weight to percent by volume or
depth. Since bulk density varies from spot to spot and also with d
average value, and its deviation, is usually computed for any one
within a given area.

Bulk density also changes as soils shrink or swell with ch;
moisture content. Because of this, Bethlahmy (7) used bulk-dens:
values adjusted for moisture content when converting moisture va
from percent by weight to inches depth.

At Vicksburg, data from over 1,200 bulk-density samples, taken at different depths, moisture contents, and sites, indicated that the effect of soil shrinking and swelling upon bulk density was small compared to the natural bulk-density variation (60). Hoover, Olson, and Metz (40) recommended taking bulk-density samples near field capacity; for these samples no correction for moisture content was considered necessary. The relation of soil bulk density to moisture content as it affects soil-moisture records has been discussed in detail by Reinhart (60). In considering the effect of shrinking and swelling on the vertical distance between electrical-resistance units, he pointed out that since the mass of soil per unit area of the layer represented by each soil-moisture unit does not change (even though the thickness of the soil layer may change slightly), no bulk-density adjustment should be made. This conclusion is in agreement with that of van Bavel and Gilbert (83).

Bulk-density measurements should be made when the soil is at field capacity (48). Broadfoot (19) has recently described and compared three methods of measuring bulk density that have been used at the Vicksburg Project. In the first of these, a rectangular block of soil 1 foot square by 3 inches thick is carefully excavated and total wet weight and moisture content determined. In the second, a "San Dimas sampler" secures soil cores 2.72 inches in diameter by 3 inches in depth without the necessity of digging a pit. The third system involves the use of an air-picnometer to obtain cores 2 inches in diameter and 1-3/8 inches high (67). Considering such factors as disturbance of sampling area, compression of soil sample, and time and skill required, Broadfoot concluded that the San Dimas and air-picnometer procedures appeared about equally superior to the block procedure. None of these methods can be used in stony soils. Equipment designed to measure bulk density by the nuclear method is illustrated in figure 10(22).

Soil-Moisture Variation and Sampling

Associated with soil-moisture measurements in any area is soil-moisture variation. This variation affects sampling techniques and the accuracy of measurements. An estimate of the amount of variation over a study area can be obtained during the course of an investigation; however, some preliminary idea of the magnitude of the variation likely to be encountered aids in planning sampling procedures.

Moisture content will of course vary by soil depth, but this can be taken care of by separate determinations for each soil horizon or layer. The variation to be considered here is that which occurs at any one time within one layer of an area apparently homogeneous as to soil and vegetation.

Soil variation, in such properties as texture, structure, c
content, and bulk density, will exist over any area mapped as o
type. A certain heterogeneity within a mapped soil type may be e>
according to the Soil Survey Manual (76):

> The soil type is a subdivision of the soil series based on
> the texture of the surface soil.... $\boxed{\text{It}}$ is the lowest and
> most nearly homogeneous unit in the natural system of
> classification. A soil type may include defined variations
> in such characteristics as slope, stoniness, degree of
> erosion, or depth to bedrock or layers of unconformable
> material.

The effect of vegetation on soil-moisture variation is leas
the area is fully occupied with cover of uniform composition and g
when the soil is only partially covered--as with row-crops or sc.
bunchgrasses or shrubs. Variation in vegetal cover causes diffe
in interception, stemflow, and transpiration. Soil-moisture cont·
its variation are thereby influenced.

Uneven disposition of rainfall in the soil profile probably a·
for a major part of soil-moisture variation; variation may be cau
differences in infiltration rates, runoff from parts of the area or
from adjacent areas, and gain or loss of water at a given point t
surface flow. Variation in distribution of rainfall over an area a
sults in variation in soil-moisture content. Where a water tabl
present, even minor differences in elevation over an area may
in large differences in soil-moisture content at any given soil d

In comparing four methods of soil-moisture measuremen·
and Bryant (73) studied variation in two soils at Beltsville, Maryl
each soil, areas 40 feet square were selected for uniformity and
into plots 10 feet square. The areas were at various tim·
mulched, or covered with bluegrass or Italian ryegrass.

Soil samples, representative of a 4-inch intercept cen
the 6-inch soil depth, were taken in duplicate with a soil tube froi
plot about 20 times a season. The standard deviation of moistur·
pling within plots was 0.75 percent by weight on Beltsville silt lo
0.53 percent on Muirkirk sand. These standard deviations incli
the real differences in soil moisture on different parts of a singl
and whatever errors may have been made in the determinatioi

The Vicksburg Project studied the amount of variation ii
moisture content of the 0- to 6- and the 6- to 12-inch layers on a

apparently homogeneous soil and cover in Mississippi, Louisiana, Arkansas, New Mexico, Colorado, and Wisconsin. Each sampling area contained 40,000 square feet (about 0.92 acre), usually as a 200-foot square. Each area was divided into four blocks, each block into four plots, and each plot into twenty-five sampling squares (10 by 10 feet each). At weekly intervals, two blocks, two plots within each of these blocks, and two squares within each of these plots were randomly selected and gravimetrically sampled. From each selected square, one soil-tube sample was obtained from the 0- to 6-inch depth and one from the 6- to 12-inch depth--a total of 8 samples weekly from each depth. Individual samples averaged about 50 grams, dry weight, of soil. At several of the areas, measurements were made for 10 to 16 weeks during the drier part of the year and again in the wet season. The location of study areas, inclusive dates of sampling, and soil and vegetation characteristics are given in table 2.

For each week of sampling at each site and depth, the variance (s^2) associated with any single sample, taken at random anywhere in

Table 2. --Soil-moisture variation study

Site name and location	Soil texture	Vegetation	Season	Weeks	Period covered
				Number	
Mound Madison Parish, Louisiana	Silty clay	Herbaceous--well stocked	Dry Wet	12 12	August 14 to October 29, 1953 January 14 to April 1, 1954
Durden Warren County, Mississippi	Silt loam	Herbaceous--well stocked	Dry Wet	12 12	August 13 to October 28, 1953 January 13 to April 1, 1954
Radiation Warren County, Mississippi	Silt loam	Herbaceous--well stocked	Dry Wet	12 12	August 13 to October 28, 1953 January 13 to April 1, 1954
Headquarters Ashley County, Arkansas	Silt loam	Forest--70-yr -old loblolly pine	Dry Wet	12 12	September 4 to November 20, 1953 December 24, 1953, to March 12, 1954
Pine Flat Bernalillo County, New Mexico	Silt loam	Open forest--pinon pine, oak brush, juniper, perennial grasses, and weeds	Dry Wet	12 16	August 28 to November 5, 1953 November 9, 1953, to April 5, 1954
Mesa Mesa County, Colorado	Silt loam	Mountain meadow--sagebrush and herbaceous, 25 percent bare	Dry	10	August 18 to October 19, 1953
Escalante Delta County, Colorado	Silty clay loam	Desert--5 to 10 percent herbaceous, remainder bare	Wet	10	December 21, 1953, to March 1, 1954
Sortek Oneida County, Wisconsin	Silt loam	Herbaceous--timothy grass, well stocked	Dry	12	August 24 to November 16, 1953

the whole "acre", was computed from the individual mean squ

$$s^2 = \frac{2\text{MS within plots} + \text{MS between plots within blocks} + \text{MS b}}{4}$$

Standard deviations were derived from these varian

The variance was also partitioned into components: w (s_w^2), between plots within blocks (s_p^2), and between bloc Variances for each area were then averaged for the number o the study; results are shown in table 3.

One of the most striking features is the variation from and sometimes from season to season within a site. Th standard deviation (square root of the average variance) for ranges from 0.09 inch (1.1 percent by weight) to 0.46 inch (5. in the 0- to 6-inch depth and from 0.12 inch (1.4 percent) to 0 (4.1 percent) in the 6- to 12-inch depth. These values are mu than the 0.75 percent and 0.53 percent of Slater and Bryar

The magnitude of the variation that was encountered ca cases be related to observed site factors. Escalante, with

Table 3.--Soil-moisture variation at 8 sites

Site name	Season	0- to 6-inch soil depth					6- to 12-inch soil depth			
		Mean moisture content	Average standard deviation (s)	Proportion of total variance			Mean moisture content	Average standard deviation (s)	Proportion total vari	
				s_w^2	s_p^2	s_b^2			s_w^2	s_p^2
		Inches	Inch	- - - Percent - - -			Inches	Inch	- - - Percent	
Mound	Dry	1.42	0.09	78	22	0	1.64	0.15	76	16
	Wet	2.58	.20	87	0	13	2.60	.19	100	0
Durden	Dry	1.47	.34	33	67	0	1.55	.29	43	57
	Wet	2.39	.25	26	41	33	2.39	.14	37	37
Radiation	Dry	.74	.13	94	0	6	1.02	.21	56	24
	Wet	2.17	.15	76	0	24	2.31	.23	40	19
Head-quarters	Dry	.80	.16	70	30	0	.72	.16	89	11
	Wet	2.10	.20	63	37	0	2.05	.15	50	32
Pine Flat	Dry	.74	.14	58	5	37	.90	.12	61	39
	Wet	1.76	.20	51	27	22	1.56	.22	82	0
Mesa	Dry	.84	.13	78	5	17	.88	.15	72	0
Escal-ante	Wet	.98	.46	67	11	22	.88	.33	68	18
Sortek	Dry	1.56	.28	100	0	0	1.53	.33	100	0
Mean		...	.23	68	19	13	...	.22	67	20

variation at both depths, was characterized by scattered clumps of bunchgrasses; the site in effect was a patchwork of herbaceous and bare areas. This would naturally lead to considerable variation.

Mesa also had unevenly distributed vegetation. However, moisture contents were uniformly low through most of the dry period, and variation was much less than at Escalante.

Vegetation at Pine Flat was mixed pinyon pine, juniper, and herbaceous species. Here, variation was large in the wet season but not so large in the dry period.

Soil at Durden, colluvial in origin, was quite variable. Moisture variation in the dry season was large. In the wet season the variation was much less; the water table at this site was quite close to the surface and, as a result, all samples obtained during this period were more or less uniformly wet.

There was also a considerable difference among the sites in the proportion of the total variation which may be assigned to variation within plots, between plots within blocks, and between blocks. The between-block and between-plot variances were probably a measure of the success achieved in laying out uniform plots. At Mound, Headquarters, and Sortek, the percentage of total variation attributable to variation between blocks was small, ranging up to 18 percent. At Radiation, between-block variation was high during the wet period; this may be the result of considerable slope at this site and possible gradation from one soil type to another. At Durden, variation between blocks during the wet period (as compared to total variation) was much greater than during the dry period. It is likely that slight differences in elevation of the blocks affected distance from water table of the samples secured and resulted in this larger block variation.

At four sites, data were collected in the dry season and again in the wet season. At two of these, Mound and Pine Flat, variation for both depths was much larger in the wet season. For one, Radiation, values for each depth were about the same in both seasons. At Durden, probably because of the water-table influence, variation in the wet period was less than under dry conditions.

The variances and standard deviations so far considered give a measure of the reliability of a single sample in determining the true weekly mean. The actual means determined from 8 samples are of course more reliable. The hypothetical standard deviation of each weekly mean can be determined by dividing the standard deviation by 2.8, the square

root of the number of samples. The result will vary somewha
true value because the 8 samples were not completely randomi
if the standard deviation is 0.23 inch, the standard deviation of
mean is only 0.08 inch.

Sampling in this study was from the upper foot of soi
order to get a measure of the variation of deeper layers, ana
made of the duplicate samples taken from the 6- by 6-foot
calibration of fiberglas units at the Mound, Durden, and Radia
Standard deviations in percent by weight of single gravimetric
are given in table 4. For comparison, corresponding standard

Table 4. --Soil-moisture variation by depth: standard deviation (s) of moisture con
of a single gravimetric sample

Soil depth (inches)	Mound		Durden		Radiation		All
	Pairs	s	Pairs	s	Pairs	s	Pairs
	Number	Percent by wt	Number	Percent by wt.	Number	Percent by wt.	Number
Within 6- by 6-foot calibration plots							
0-3	72	2.0	31	1.9	12	1.4	115
3-6	72	1.7	32	1.1	12	1.1	116
6-9	71	2.0	31	1.3	12	1.4	114
9-12	70	2.1	31	1.4	12	1.3	113
12-15	42	2.2	31	1.6			73
15-18	42	2.8	31	1.4			73
12-18	27	2.4	...	...	12	1.6	39
18-21	40	2.4	28	1.6			68
21-24	41	2.4	29	2.2			70
18-24	26	2.3			12	0.8	38
28-32	58	1.9	26	1.2	12	1.1	96
40-44	51	1.6	24	1.3	12	1.4	87
All depths	612	2.1	294	1.5	96	1.3	1002
Within 50- by 50-foot plots of variation study							
0-6 (dry period)		1.0		2.6		1.6	
(wet period)		2.2		1.7		1.7	
6-12 (dry period)		1.7		2.7		1.8	
(wet period)		2.3		1.0		1.6	

of samples within the 50- x 50-foot plots of the variation study are given at the bottom of the table.

Variation in the several layers of the upper 4 feet of soil seems to be of the same general magnitude; standard deviations range from about 1 to somewhat over 2 percent by weight. At Mound, standard deviations were appreciably larger for the second foot, mostly because uneven wetting after summer storms caused several large differences between duplicate samples. Variation within the calibration plots was somewhat less than that within the 50- x 50-foot plots of the variation study.

Point Estimates and Area Estimates

In a study using electrical-resistance units, sampling methods will depend on whether the soil-moisture record is to represent the point at which the unit is installed or a larger surrounding area. With the point concept, an attempt is made to determine the moisture content of the soil in contact with the unit. Since this soil cannot be directly sampled in the field without destroying the installation, laboratory calibration is required. Field calibration with sampling plots as small as possible (4 x 4 feet, for example) would approach the point-concept requirements.

The area concept is probably more generally useful in soil-moisture studies. The resistance of the soil-moisture unit is considered an index of the average soil-moisture content of a given soil layer over a calibration area whose size is determined by the objectives of the study. Field calibration by random gravimetric sampling relates the unit's resistance to the average moisture content of the layer. The calibration area should be homogeneous, or nearly so, with respect to soil and vegetation.

Thames [5] has used the area concept for extending information gained from installations of electrical-resistance units. At Rhinelander, Wisconsin, he related resistances of units installed at one site to moisture contents of gravimetric samples obtained from calibration plots at other sites up to 4 miles distant. Calibration curves were prepared, and, from daily resistance determinations at the first site, a daily record of moisture content at the other sites was secured. Vegetation was similar at sites where fiberglas units had been installed and at corresponding sites without units. Soil texture, however, varied considerably. For

---

[5] Unpublished study on file at Southern Forest Experiment Station.

example, units in one silt loam site were used to
at a sandy site.

Relative to the effort required to secure
accuracies were very encouraging. Except for the ex
loam and sand, a consistent relationship was found be
at the unit installations and moisture content at the cc
without units. Probably the general good agreement
ties of depletion forces and rainfall at the pairs of sit
watershed studies for soil-moisture records, coupled
cilities for installing innumerable soil-moisture stati
necessity of devising short-cut methods such as thi

Another method of securing records over a w
to install electrical-resistance units in as many locat
and then secure frequent (preferably daily) soil-moist
each soil horizon. For each soil-cover complex in th
average depth and the field maximum and field minim
contents of each horizon would also be determined. A
estimates for any soil-cover complex would be made
moisture content to be at the same relative position b
and minimum values as at the most nearly comparabl
which a complete record is available. Soil-moisture
course be adjusted for differences in depth of corresp

Sampling Design

The character of any specific study will dic
procedures applicable. Objectives may vary from
of mean moisture content for a treated or untrea
square feet in area to the estimation of the mean
soil-cover complex many acres in extent. Again, cor
desired between areas at one specific time, or a comp
record covering several days, months, or years ma

Cline (24) discussed the principles of soil
from the standpoint of soil chemistry, and stressed th
is commonly much greater than analytical error. Muc
presentation is equally applicable to sampling for mo
Reed and Rigney (58) presented data on the number of
to attain specified limits of accuracy in sampling for
mostly chemical, in fields of uniform and non-unifc

The sampling design will depend in part on h
variation is expected. This variation is of the same

that associated with soil fertility, timber volume, and numerous other characteristics subject to measurement on an area basis. However, soil-moisture content differs from most other soil or site characteristics in being subject to great change in relatively short periods of time. Samples taken from different parts of an area must ordinarily be secured on the same day in order to be comparable. When samplings are made at different times, it is often hard to determine how much of the moisture content difference is associated with time differences and how much with areal variation.

Soil-moisture content varies vertically and horizontally. Location of samples and determination of mean moisture content should ordinarily be limited to essentially homogeneous soil volumes. Vertically, the soil horizon provides a logical division into separate soil populations. However, especially near the surface, moisture content varies greatly with depth even within horizons. On the Vicksburg Project the surface foot was subdivided into four 3-inch layers for sampling purposes. Because of variation with depth, it is also advisable to obtain sample cores extending from top to bottom of the layer being sampled, especially for layers near the surface.

The dual requirement of determining moisture content through space and time often dictates an intensity and frequency of sampling far above that prescribed by financial limitations. The usual result is that experiments are designed to get the best estimates possible for a given expenditure rather than estimates to any predetermined standard of accuracy. Even so, experimental design should be such that the accuracy of the results can be determined.

In situ methods for determining moisture content enable frequent measurement and avoid difficulties encountered when successive estimates are based upon different samples. Even here, however, areal variation is a problem because of the necessity of obtaining gravimetric samples for field calibration or for checking calibrations.

Random location of samples is the usual procedure when data are to be analyzed statistically. However, according to L. R. Grosenbaugh[6], "Systematic sampling according to a rectangular grid pattern probably affords the most efficient estimate of the mean volume of water in a specified rectangular space, though of course an element of randomness should govern the initial positioning of the grid, if bias from edge effect is to be avoided. DeLury (31) has provided an interesting and elegant

---

6/ Personal communication, December 6, 1954.

method for fitting a surface to such a systematic sample, and obse
deviations from this surface provide a realistic estimate of error.
simplified computational method allows picking out by inspectic
highest degree terms which contribute to a good fit, and residual (
can be obtained by simple addition."

The simplest method for sampling moisture consists of obta
on a given date, duplicate samples from each soil depth under cons
ation. Variation between paired samples will give a measure of rel
Sampling in duplicate was the regular procedure for calibrating fibe
electrical-resistance units on the Vicksburg Project.

Increasing the number of samples will of course incre
reliability of the mean. If an estimate of the standard deviation (
moisture content is obtained, an estimate of the standard deviation
mean, or standard error ($s_{\bar{x}}$), can be made for 'any given numbe
random samples (n):

$$s_{\bar{x}} = \frac{s}{\sqrt{n}}$$

The standard deviation can be estimated if the range betwee
largest and smallest values in a sample of given size is known. Sne
(75), in a table based on the work of Tippett and Pearson, gives
values of range/s. For example, with sample sizes of 2, 5, 10, ar
range/s equals 1.13, 2.33, 3.08, and 3.73 respectively.

Compositing samples helps to attain increased accuracy f
given sampling effort. Two or more samples procured from a giv
soil depth by soil tube or other instrument are placed in one conta
Thereafter, through weighing, drying, and reweighing, they are ha
as one sample. In any series of samplings, enough uncomposited s
should be obtained to give some idea of the magnitude of variation a
reliability of the results from composited samples. Care should b
not to composite samples from too large an area because there is c
of covering up differences between locations.

Plant spacing, as in orchards, may result in a uniform pa
of soil-moisture variation. Jackson and Weldon (42) developed for
for determining the weight of water in a soil or subsoil mass in v
the moisture content increases with distance from a plant or gro
plants. Where plants vary widely in size and are unevenly distribu
at the Escalante site, the only method for determining mean moistu
content is to sample the area as a whole, increasing the sample si
accordance with the amount of variation present.

LITERATURE CITED

Aldous, W. M., Lawton, W. L., and Mainport, R. C.
   1952. The measurement of soil moisture by heat diffusion. Civil
      Aeronautics Admin. Tech. Devel. Rpt. 165.

Allyn, R. B.
   1942. A calibrated soil probe for measuring field soil moisture.
      Soil Sci. 53: 273-285.

_____ and Work, R. A.
   1941. The availameter and its use in soil moisture control
      I.   The instrument and its use. Soil Sci. 51: 307-321.
      II. Calibration methods. Soil Sci. 51: 391-406.

Begemann, E.
   1936. Determination of the water content of soil samples by
      means of xylol. Proc. Internatl. Conf. Soil Mech.
      and Found. Engin. 2: 66-67.

Belcher, D. J., Cuykendall, T. R., and Sack, H. S.
   1950. The measurement of soil moisture and density by neutron
      and gamma ray scattering. Civil Aeronautics Admin.
      Tech. Devel. Rpt. 127.

Bethlahmy, N.
   1951. An alinement chart for use with the fiberglas soil-moisture
      instrument. Soil Sci. 71: 377-380.

_____
   1952. A method for approximating the water content of soils.
      Trans. Amer. Geophys. Union 33: 699-706.

Bidwell, G. L., and Sterling, W. F.
   1925. Preliminary notes on the direct determination of moisture.
      Jour. Assoc. Off. Agr. Chemists 8: 295-301.

Bouyoucos, G. J.
   1931. The alcohol method for determining moisture content of
      soils. Soil Sci. 32: 173-179.

_____
   1937. Evaporating the water with burning alcohol as a rapid
      means of determining moisture content of soils. Soil
      Sci. 44: 377-381.

(11) Bouyoucos, G. J.
      1949. Nylon electrical resistance unit for continuous measurement of soil moisture in the field. Soil Sci. 67: 319-330.

(12) _____
      1950. A practical soil moisture meter as a scientific guide to irrigation practices. Agron. Jour. 42: 104-107.

(13) _____
      1952. Improvements in the nylon method of measuring soil moisture in the field. Agron. Jour. 44: 311-314.

(14) _____
      1953. More durable plaster of paris moisture blocks. Soil Sci. 76: 447-451.

(15) _____
      1954. New type electrode for plaster of paris moisture blocks. Soil Sci. 78: 339-342.

(16) _____
      1954. Newly developed nylon units for measuring soil moisture in the field. Highway Res. Abs. 29-32.

(17) _____ and Mick, A. H.
      1940. An electrical resistance method for the continuous measurement of soil moisture under field conditions. Mich. Agr. Expt. Sta. Tech. Bul. 172.

(18) _____ and Mick, A. H.
      1947. Improvements in the plaster of paris absorption block electrical resistance method for measuring soil moisture under field conditions. Soil Sci. 63: 455-465.

(19) Broadfoot, W. M.
      1954. Procedures and equipment for determining soil bulk density. U. S. Forest Serv. South. Forest Expt. Sta. Occas. Paper 135, pp. 2-11.

(20) Carlson, C. A.
      1953. Moisture equilibration in natural cores during laboratory calibration of fiberglas soil-moisture units. U. S. Forest Serv. South. Forest Expt. Sta. Occas. Paper 128, pp. 31-39.

(21) Carlson, C. A.
1954. Comparison of laboratory and field calibration of fiberglas moisture units. U. S. Forest Serv. South.
Forest Expt. Sta. Occas. Paper 135, pp. 34-42.

(22) Carlton, P. F., Belcher, D. J., Cuykendall, T. R., and Sack, H. S.
1953. Modifications and tests of radioactive probes for
measuring soil moisture and density. Civil Aeronautics
Admin., Tech. Devel. Rpt. 194.

(23) Carter, C. E.
1938. A field method for determining soil moisture. Austral.
Forestry 3: 15-16.

(24) Cline, M. G.
1944. Principles of soil sampling. Soil Sci. 58: 275-288.

(25) Coile, T. S.
1953. Moisture content of small stone in soil. Soil Sci. 75:
203-207.

(26) Colman, E. A.
1946. The place of electrical soil-moisture meters in hydrologic
research. Trans. Amer. Geophys. Union 27: 847-853.

(27) _____ and Hendrix, T. M.
1949. The fiberglas electrical soil-moisture instrument. Soil
Sci. 67: 425-438.

(28) Conrad, J. P., and Veihmeyer, F. J.
1929. Root development and soil moisture. Hilgardia 4: 113-134.

(29) Davis, W. E., and Slater, C. S.
1942. A direct weighing method for sequent measurements of
soil moisture under field conditions. Jour. Amer.
Soc. Agron. 34: 285-287.

(30) Davisson, B. S., and Sivaslian, G. K.
1918. The determination of moisture in soils. Amer. Soc.
Agron. Jour. 10: 198-204.

(31) DeLury, D. B.
1950. Values and integrals of the orthogonal polynomials up
to $n = 26$. Univ. Toronto Press, Toronto, Ont.,
33 pp.

(32) Diebold, C. H.
    1953. Rapid methods for estimating readily available moisture and bulk density of medium- to fine-textured soils. Agron. Jour. 45: 36-37.

(33) Dimbleby, G. W.
    1954. A simple method for the comparative estimation of soil water. Plant and Soil 5: 143-154.

(34) Doss, B. D., and Broadfoot, W. M.
    1954. Terminal panel for electrical soil-moisture instruments. U. S. Forest Serv. South. Forest Expt. Sta. Occas. Paper 135, pp. 30-31.

(35) Emmert, E. M.
    1937. A rapid method for determining soil moisture. Soil Sci. 43: 31-36.

(36) Ewart, G. Y., and Baver, L. D.
    1951. Salinity effects on soil moisture-electrical resistance relationships. Soil Sci. Soc. Amer. Proc. (1950) 15: 56-63.

(37) Ferguson, E. R., and Duke, W. B.
    1954. Devices to facilitate King-tube soil-moisture sampling. U. S. Forest Serv. South. Forest Expt. Sta. Occas. Paper 135, pp. 26-29.

(38) Haise, H. R., and Kelley, O. J.
    1950. Causes of diurnal fluctuations of tensiometers. Soil Sci. 70: 301-313.

(39) Henrie, J. O.
    1950. A direct method of computing and correcting the equivalent depth of water. Agr. Engin. 31: 290-291.

(40) Hoover, M. D., Olson, D. F., Jr., and Metz, L. J.
    1954. Soil sampling for pore space and percolation. U. S. Forest Serv. Southeastern Forest Expt. Sta. Sta. Paper 42.

(41) Horonjeff, R., Goldberg, I., and Trescony, L. J.
    1954. The use of radioactive material for the measurement of water content and density of soil. Presented at the Sixth Ann. Street and Hwy. Conf., Univ. of Cal.

(42) Jackson, M. L., and Weldon, M. D.
  1939. Determination of the weight of water in a soil or sub-
    soil mass in which the moisture content increases
    with distance from a plant or group of plants. Jour.
    Amer. Soc. Agron. 31: 116-127.

(43) Keen, B. A.
  1931. The physical properties of the soil. Longmans, Green,
    and Co., Inc., 380 pp.

(44) Korty, B. L., and Kohnke, H.
  1953. Recording soil moisture automatically. Soil Sci. Soc.
    Amer. Proc. 17: 307-310.

(45) Lane, D. A., Torchinsky, B. B., and Spinks, J. W. T.
  1952. Determining soil moisture and density by nuclear radia-
    tions. Symposium on the use of radioisotopes in soil
    mechanics. Amer. Soc. for Testing Materials, Spec.
    Tech. Pub. 134, pp. 23-34.

(46) LeCompte, S. B., Jr.
  1952. An instrument for measuring soil moisture. New Jersey
    Hort. News.

(47) Livingston, B. E., and Koketsu, R.
  1920. The water-supplying power of the soil as related to the
    wilting of plants. Soil Sci. 9: 469-485.

(48) Lutz, H. J.
  1944. Determination of certain physical properties of forest
    soils. I: Methods utilizing samples collected in metal
    cylinders. Soil Sci. 57: 475-487.

(49) Mackaness, F. G., and Rowse, R. W.
  1954. Equipment for installing gypsum moisture blocks. Agr.
    Engin. 35: 337.

(50) Momim, A. H.
  1947. A new simple method of estimating the moisture content
    of soil in situ. Indian Jour. Agr. Sci. 17: 81-85.

(51) Nikiforoff, C. C.
  1948. Stony soils and their classification. Soil Sci. 66
    347-363.

(52) Noyes, H. A. , and Trost, J. F.
    1920. Determination of moisture in field samples of soil.
        Jour. Assoc. Off. Agr. Chemists 4: 95-97.

(53) Olson, D. F. , Jr. , and Hoover, M. D.
    1954. Methods of soil moisture determination under field
        conditions. U. S. Forest Serv. Southeastern Forest
        Expt. Sta. Sta. Paper 38.

(54) Palpant, E. H.
    1953. An inserter for fiberglas soil-moisture units. U. S.
        Forest Serv. South. Forest Expt. Sta. Occas.
        Paper 128, pp. 16-20.

(55) _____ and Lull, H. W.
    1953. Comparison of four types of electrical resistance in-
        struments for measuring soil moisture. U. S. Forest
        Serv. South.Forest Expt. Sta. Occas. Paper 128,
        pp. 2-15.

(56) _____ , Thames, J. L. , and Helmers, A. E.
    1953. Switch shelters for use with soil-moisture units. U. S.
        Forest Serv. South. Forest Expt. Sta. Occas. Paper
        128, pp. 21-30.

(57) Papadakis, J. S.
    1941. A rapid method for determining soil moisture. Soil Sci.
        51: 279-281.

(58) Reed, J. F. , and Rigney, J. A.
    1947. Soil sampling from fields of uniform and nonuniform
        appearance and soil types. Jour. Amer. Soc. Agron.
        39: 26-40.

(59) Reinhart, K. G.
    1953. Installation and field calibration of fiberglas soil-moisture
        units. U. S. Forest Serv. South. Forest Expt. Sta.
        Occas. Paper 128, pp. 40-48.

(60) _____
    1954. Relation of soil bulk density to moisture content as it
        affects soil-moisture records. U. S. Forest Serv.
        South. Forest Expt. Sta. Occas. Paper 135, pp. 12-21.

(61) Richards, L. A.
    1942. Soil moisture tensiometer materials and construction.
    Soil Sci. 53: 241-248.

(62) _____
    1949. Methods of measuring soil moisture tension. Soil Sci.
    68: 95-112.

(63) _____ and Gardner, W.
    1936. Tensiometers for measuring the capillary tension of
    soil water. Jour. Amer. Soc. Agron. 28: 352-358.

(64) Richards, S. J.
    1938. Soil moisture content calculations from capillary tension
    records. Soil Sci. Soc. Amer. Proc. 3: 57-64.

(65) Rowe, P. B., and Colman, E. A.
    1951. Disposition of rainfall in two mountain areas of
    California. U. S. Dept. Agr. Tech. Bul. 1048, 84 pp.

(66) Rowland, E. F., Fagan, T. D., and Crabb, G. A., Jr.
    1954. A slide rule for soil moisture determinations. Jour.
    Amer. Soc. Agr. Engin. 35: 163-164.

(67) Russell, M. B.
    1950. A simplified air-picnometer for field use. Soil Sci.
    Soc. Amer. Proc. (1949) 14: 73-76.

(68) Russell, Sir E. J.
    1950. Soil conditions and plant growth. Longmans, Green and
    Co., Inc.

(69) Scofield, C. S.
    1945. The measurement of soil water. Jour. Agr. Res. 71:
    375-402.

(70) Sen, A.
    1936. Rapid determination of soil moisture by drying the soil
    over Bunsen burner. Indian Jour. Agr. Sci. 6:
    1076-1080.

(71) Shaw, B., and Baver, L. D.
    1939. Heat conductivity as an index of soil moisture. Jour.
    Amer. Soc. Agron. 31: 886-891.

(72) Slater, C. S.
> 1942. A modified resistance block for soil moisture measurement. Jour. Amer. Soc. Agron. 34: 284-285.

(73) _____ and Bryant, J. C.
> 1946. Comparison of 4 methods of soil moisture measurement. Soil Sci. 61: 131-155.

(74) Smith, A., and Flint, F. W.
> 1930. Soil moisture determination by the alcohol method. Soil Sci. 29: 101-107.

(75) Snedecor, G. W.
> 1946. Statistical methods. (Ed. 4) Iowa State College Press, 485 pp.

(76) Soil Survey Staff.
> 1951. Soil survey manual. U. S. Dept. Agr. Handb. 18, 503 pp.

(77) Stackhouse, J. M., and Youker, R. E.
> 1954. Evaluation of the accuracy of fiberglas-gypsum blocks for measuring soil moisture changes. Agron. Jour. 46: 405-407.

(78) Stoeckeler, J. H.
> 1937. Measuring soil moisture in the forest nursery. U. S. Forest Serv. Lake States Forest Expt. Sta. Mimeo.

(79) Stout, G. J., and Holben, F. J.
> 1946. Simple method for rapid drying of soil moisture samples. Amer. Soc. Hort. Sci. Proc. 47: 238.

(80) Tanner, C. B., and Hanks, R. J.
> 1952. Moisture hysteresis in gypsum moisture blocks. Soil Sci. Soc. Amer. Proc. 16: 48-51.

(81) Uhland, R. E.
> 1951. Rapid method for determining soil moisture. Soil Sci. Soc. Amer. Proc. 15: 391-393.

(82) Underwood, N., van Bavel, C. H. M., and Swanson, R. W.
> 1954. A portable slow neutron flux meter for measuring soil moisture. Soil Sci. 77: 339-340.

(83) Van Bavel, C. H. M., and Gilbert, M. J.
        1954. Discussion of "A method for approximating the water
            content of soils" by N. Bethlahmy. Trans. Amer.
            Geophys. Union 35: 168-169.

(84) Veihmeyer, F. J.
        1929. An improved soil-sampling tube. Soil Sci. 27: 147-152.

(85) Waterways Experiment Station, Corps of Engineers
        1955. Field tests of nuclear instruments for the measurement
            of soil moisture and density. Misc. Paper 3-117,
            26 pp.

(86) Weaver, H. A., and Jamison, V. C.
        1951. Limitations in the use of electrical resistance soil
            moisture units. Agron Jour. 43: 602-605.

(87) White-Stevens, R. H., and Jacob, W. C.
        1940. The rapid detection of soil moisture. Amer. Soc. Hort.
            Sci. Proc. 37: 261-266.

(88) Whitney, M. A.
        1894. Instructions for taking samples of soil for moisture
            determination. U.S. Dept. Agr., Div. Agr. Soils,
            Cir. 2.

(89) _____
            1896. Methods of the mechanical analysis of soils and of the
            determination of the amount of moisture in soils in
            the field. U.S. Dept. Agr., Div. Agr. Soils, Bul. 4.

(90) _____
            1896-1899. Report of the Chief of the Division of Agricultural
            Soils. From annual reports of the Secretary of
            Agriculture.

(91) Wilson, J. D.
        1927. The measurement and interpretation of the water-
            supplying power of the soil with special references to
            lawn grasses and some other plants. Plant Physiol.
            2: 385-440.

(92) Woods, F., and Hopkins, W.
        1954. Phone-jack terminals for soil-moisture units. U. S.
            Forest Serv. South. Forest Expt. Sta. Occas. Paper
            135, pp. 32-33.

(93) Youker, R. E., and Dreibelbis, F. R.
     1951. An improved soil-moisture measuring unit for hydrologic
          studies. Trans. Amer. Geophys. Union 32: 447-449.

CATTLE GRAZING DAMAGE
TO PINE SEEDLINGS

John T. Cassady
Walt Hopkins
L. B. Whitaker

Contents

CATTLE GRAZING DAMAGE TO PINE SEEDLINGS

John T. Cassady, Walt Hopkins, and L. B. Whitaker
Southern Forest Experiment Station

This paper summarizes pertinent information on the damage that grazing cattle have done to pine seedlings in several areas of central Louisiana.

Until recently, grazing damage has been of little or no concern to most forest landowners. Today, however, intensive land management is focussing attention on the problem. There are more cattle, more trees, less grass--and more foresters to observe the effects of grazing. The dry years since 1951 have no doubt intensified the situation.

The studies reported here were installed in areas where grazing damage was noticeably heavy. On only one tract was grazing controlled. The results are therefore not average. Instead, they represent the most serious damage encountered in pine plantations from 1946 to 1955 (see cover photo). The information, however, is reliable and can help guide landowners in deciding when cattle grazing can be permitted on areas where pine is being regenerated.

DAMAGE TO PLANTATIONS

The Chandler Plantation

Longleaf pine saplings on a National Forest plantation near Dry Prong, Louisiana, were heavily browsed by cattle during January and February 1947. The 10-year-old stand contained about 200 longleaf trees per acre; the trees ranged from 2 to 15 feet in height. The area also supported scattered to heavy stands of loblolly and shortleaf pines. These were not grazed.

The plantation consisted of 1,250 acres, all fenced. It had adequate amounts of native grass and was grazed most of the year by a herd of 60 to 90 beef cattle of average grade. When the trouble occurred, the cattle were being fed a winter ration of cottonseed meal that averaged

2.5 pounds per head daily--considerably more than Louisiana range cattle usually get.

The supplemental feeding was begun on December 31, during a severely cold, wet spell. Over 2 inches of rain, sleet, and snow fell during the first 3 days. Temperatures were freezing or below. Grasses and shrubs were coated with snow and ice.

As soon as feeding started, the cattle lost enthusiasm for grazing and spent most of their time waiting at the troughs. In less than a week, heavy grazing damage was noted on longleaf near the troughs. Before long, most of the cattle "got the habit" and spent an hour or more each day grazing solely on longleaf needles. Some cows would ride down saplings 10 feet high to reach the top foliage (fig. 1).

Pine grazing was measured on 48 one-quarter acre plots distributed evenly over 500 acres near the feeding ground. In addition, 50 saplings that had been completely defoliated were examined several times to determine rate and degree of recovery from damage. The main results and conclusions were:

1. Damage was concentrated near the feed troughs, but some pine grazing was evident in all parts of the pasture.

2. An average of 61 longleaf trees per acre were defoliated in some degree that first winter, but damage was negligible in 1948 and 1949.

3. Cattle browsed the needles only, not the buds and twigs.

4. The grazed trees suffered some loss in height growth but otherwise recovered quickly. Two percent of the defoliated seedlings died, while 86 percent recovered to good or excellent condition by the end of the first growing season.

5. The cattle preferred longleaf to loblolly and shortleaf foliage.

6. Evidently, this heavy longleaf browsing was stimulated by the cold spell and ice storm. Then the relatively large daily ration of high-protein feed apparently intensified the cows' appetite for green pine needles.

7. Analyses showed that during the winter the longleaf needles had higher quantities of essential nutrients than the native grass. The percentages of three important nutrients are tabulated on p. 3.

Figure 1.--Cattle often "ride down" sapling pines in order to browse foliage that would otherwise be out of reach.

Item	Crude protein	Fat	Phosphorus
Longleaf needles	7. 8	6. 6	0. 22
Winter forage	5. 4	2. 6	. 06

The Nebo Plantations

The Nebo Oil Company planted 4, 000 acres of cutover land in La Salle parish during the winter of 1948-49. Most of the planting was slash pine.

Grass was sparse that winter, partly because the 1948 drought had kept forage production low and partly because nearly half the area had been burned a few weeks before planting. Then, the fall and winter rains filled the nearby bottoms, and floodwaters drove hundreds of open-range cattle to the newly planted uplands.

Heavy browsing of pine seedlings was reported in March] a grazing damage study was started in April. Very little evidenc trampling was seen, but 30 percent of the pines had their termina nipped off. Grass became abundant in the spring of 1949 and no damage was observed thereafter.

Trees on study plots were re-examined at the end of the f growing seasons. The browsing did not increase seedling mortal survival of grazed and ungrazed seedlings was:

Season	Grazed trees	Ungrazed trees
	Percent surviving	
Fall 1949	86	80
Fall 1950	84	77
Fall 1951	70	61

Although these figures suggest that grazing actually increased se survival, the differences were not statistically significant.

Grazing did, however, reduce the growth of the seedlings to a significant degree. Thus by the fall of 1949, grazed trees measured 6 inches in height and ungrazed trees 8 inches. By the fall of 1951, grazed trees were 30 inches tall, ungrazed ones 40 inches (fig. 2).

Figure 2. --Three consecutive seasons of cattle browsing reduced the height growth of this slash pine. (Photo by Louisiana Forestry Commission)

It was expected that trees that had lost their terminal buds would grow up bushy or forked. This did not happen. A lateral bud nearly always developed and replaced the terminal shoot. Only 2 percent of the grazed trees had more than one leader in the fall of 1951.

In summary, cattle browsing of slash and loblolly pines on the 4,000-acre Nebo plantations caused no serious or permanent damage, even though it was alarming the first winter.

The Claiborne Plantation

About 15 acres of unfenced cutover land near Melder, Louisiana, was planted with longleaf, loblolly, and slash pine seedlings in February 1948. The area supported a heavy stand of scrub oaks--330 stems per acre, averaging 4.5 inches in diameter at breast height.

Hardwoods on half the tract were girdled or poisoned with Ammate after the pines were planted. This reduction of hardwoods stimulated the growth of grass as well as pines. Cattle were attracted to the released areas and grazed them heavily; by May 1949 they had browsed 40 percent of the loblolly and 25 percent of the slash. Where the hardwoods had not been controlled, 15 percent of the loblolly pines and 4 percent of the slash were browsed.(All the longleaf seedlings had been rooted up by hogs before the spring of 1949!)

The heavier browsing of the loblolly seedlings may have been due to the fact that they were of higher quality when planted and grew more vigorously than the slash pine.

In September 1949, at the end of the second growing season, grazed seedlings were 5 to 7 inches shorter than ungrazed ones. Average seedling heights were:

Species and treatment	Seedling height	
	Grazed	Not grazed
	Inches	
Loblolly pine		
Hardwoods controlled	22	28
Hardwoods not controlled	10	17
Slash pine		
Hardwoods controlled	15	22
Hardwoods not controlled	11	12

The browsed seedlings made good recovery and no other damage surveys were carried out. By February 1955, the pines were 15 to 20 feet high and all evidence of grazing damage had disappeared (fig. 3). This case may be representative of many of the older slash and loblolly plantations in this area--some damage occurred when the seedlings were small, but it was overlooked or forgotten, and the trees have recovered.

The Longleaf Tract Grazing Study

Cattle grazing damage to planted pines is being closely studied on the Longleaf Tract of the Palustris Experimental Forest, at Alexandria. Here a controlled study is being made of forage production, utilization, and vegetation changes caused by grazing, burning, and the growth of pine trees.

On February 29, 1952, 75 longleaf and 75 slash pine seedlings were planted in each of 18 fenced 1/3-acre plots. As the 1952 survival was low--40 percent for longleaf and 60 percent for slash--the dead seedlings were replaced in 1953. Therefore, this study is based on seedlings of two ages. Since 1952, the study plots have been grazed

Figure 3. --Slash and loblolly pine seedlings have remarkable ability to recover from damage. Cattle browsed nearly half of these trees when they were a year old. Now, at age 7, the trees have recovered and average 16 feet in height. (Photo by Louisiana Forestry Commission)

from April to August at three intensities: six plots ungrazed, six moderately grazed, and six heavily grazed. Moderate grazing has averaged 40 cow-days per acre each season, and heavy grazing about 75 cow-days.

Damage to slash pine in 1953 and 1954. --Very little grazing damage was noted in 1952, and no damage survey was made. After the 1953 and 1954 seasons, however, each seedling or planting space was examined. Trees that were expected to recover without permanent injury to their form were listed as slightly or moderately damaged. Heavily damaged seedlings were those expected to die, be deformed, or suffer serious reduction in growth.

Most of the 1953 damage to slash seedlings was from trampling (table 1). The most significant measure of the damage is the striking difference in mortality: 18, 11, and 2 percent for heavy, moderate, and no grazing. The 2-percent mortality of the ungrazed plots was normal. Mortality on the grazed plots above 2 percent was caused by the cattle.

Table 1. --Proportion of slash pine seedlings damaged by grazing. Longleaf Tract 1/

Condition of seedlings	At end of 1953 season			At end of 1954 season		
	No grazing	Moderate grazing	Heavy grazing	No grazing	Moderate grazing	Heavy grazing
	- - -Percent- - -			- - - Percent - - -		
Living						
Lightly or moderately browsed	0	14	15	0	36	25
Heavily browsed	0	1	4	0	2	4
Total browsed	0	15	19	0	38	29
Dead or missing						
Killed by browsing	0	9	16	0	10	19
Normal mortality	2	2	2	5	5	5
Total dead	2	11	18	5	15	24

1/ About half of the seedlings were planted early in 1952 and half in 1953.

In 1954, many pines were 3 to 4 feet high, and the cattle ca
more injury by rubbing and browsing than by trampling. By the en
this year, cumulative normal mortality (ungrazed paddocks) was 5
cent, as contrasted with a two-year loss of 15 percent under mode
grazing and 24 percent under heavy grazing (table 1).

The 1954 damage was worst on the moderately grazed area
probable reason is that a very heavy rough (2,092 pounds per acre
left on these plots from 1953. This old rough, mainly wiry flower
of slender bluestem, prevented easy grazing of the new green gras
thus induced browsing of the pines. The heavily grazed plots did n
have these unpalatable flowerstalks mixed with the green grass.

Damage to longleaf pine seedlings. --Damage to longleaf se·
was also surveyed at the end of the 1953 and 1954 growing seasons.
both years most of the loss was from trampling. Very little brows
was noted.

Table 2 sums up the results at the end of 1954. Survival or
ungrazed plots was 86 percent. Some of the surviving ungrazed se
lings were weak, but 60 percent were healthy and, with proper mai
ment, will form an adequate stand. It appeared unlikely that a sati
factory stand could be obtained on the grazed plots, where 20 and 3
cent were destroyed
moderate and heavy ι
ing. In fact, only 42
26 percent of the ori;
stand were classed a
vigorous in 1954.

Table 2.--Proportion of longleaf pine seedlings damaged by grazing. Long-
leaf Tract, November 1954 1/

Condition of seedlings	No grazing	Moderate grazing	Heavy grazing
	- - - Percent - - -		
Killed by grazing	0	20	34
Normal mortality	14	14	14
Total survivors	86	66	52
Weak trees	26	24	26
Vigorous trees	60	42	26

1/ About half of the seedlings were planted
early in 1952 and half in 1953.

The severe m
tality of seedlings fro
grazing was not notic
suspected until the su
tallies were made at
end of the grazing se
(fig. 4). Similar cat
damage has probably
tributed to the failure
many longleaf plantat
in this area.

Figure 4. --Damage to grass-stage longleaf is not detected without close observation. (Photo by Louisiana Forestry
Commission)

Evaluation of
damage on the Longleaf
Tract. --Certain fea-
tures of the Longleaf
Tract study caused
heavier damage to pines
than would occur on
large, moderately
grazed plantations.

First, the
small grazing paddocks
are only 70 by 210 feet,
or not much larger than
an average city lot. The
cattle were confined to
these plots singly or in
pairs for 1 to 3 days at
a time. There was
probably more trampl-
ing in these confined
plots than would occur in
a large pasture (fig. 5).

Secondly, the forage on each one-third acre consisted of a very uniform stand of slender bluestem grass, a few weeds, and 150 small pine trees. The desire for variety may have encouraged pine grazing.

The third condition that increased cattle damage was that plots were stocked more heavily than is recommended for moderate grazing. Because grass production was high in 1953 and 1954, the grazing rates were increased from 30 to 54 cow-days on the moderately stocked paddocks and from 60 to 90 on the heavily grazed paddocks.

The conspicuous browsing of slash pine seedlings in 1954 was unseasonal. Ordinarily, such browsing occurs only in winter and early spring, when there is not much green forage. However, green grass was not readily available during the 1954 grazing season, simply because it was covered by or mixed with the heavy rough left over from 1953.

The Long-Bell Plantation

A new slash pine plantation of the Long-Bell Lumber Company was severely damaged by open-range cattle and goats during the winter of 1953-54.

The planted area, of 160 acres, supported a scattered stand of scrub oaks but very little underbrush or browse. Grass, though abundant, was dry and unpalatable.

About 60 range beef cattle normally graze this area and an adjoining larger tract of open range. A few days after the pines were planted, the cattle began browsing them. Thereupon the cattle owner penned and fed his herd for more than 2 weeks, but turned them out again on January 15. Grazing damage then became serious and the Southern Forest Experiment Station's Alexandria Research Center was asked to examine the plantation.

One of the first things learned was that a herd of 75 goats also had ranged on the planted area. Most of them had been penned up after grazing the plantation for one day only. However, it was evident from fresh tracks and droppings that some goats still had access to the area. Obviously some of the damage reported here was caused by goats.

The plantation was surveyed on February 5, 1954. By that time, 12 percent of the pines were missing or dead; 28 percent had had their main stems grazed off, usually down to a short stub; 32 percent were lightly browsed, with no damage to the main stem or terminal bud; and 28 percent were undamaged.

DAMAGE TO DIRECT-SEEDED AREAS

Studies have been under way since 1947 to find reliable techniques for successfully direct-seeding longleaf pine. Most of the work has been done on areas protected from livestock, but several test sites were grazed by cattle. Three such tests are described here.

Turnage Road

In March 1951, a 210-acre block of 16-year-old slash pine on the Palustris Experimental Forest was clear-cut in a salvage operation (it had been ruined by a summer wildfire and several ice storms). The clearcut area was burned in September and October 1951 for seedbed preparation, and cattle and hogs were excluded.

Longleaf pine seed was sown from an airplane on November 28, 1951. In April 1952, the initial stand averaged 2,260 longleaf seedlings per acre.

A heavy growth of coarse weeds and grasses shot up on the seeded area and by June was waist-high and maturing: a serious fire hazard. It was decided to stock the area with cattle to reduce the rank vegetation, even though some grazing damage would result.

On July 1, seven months after seeding, 50 range beef cattle were turned into the pasture, which included the seeded area plus 350 acres of 17-year-old slash pine. The cattle found choice grazing on the seeded area, concentrated there, and effectively reduced the weeds. They were taken out that fall, after 3 months of grazing.

An inventory in November 1952 showed a first-year survival of 1,672 pine seedlings per acre. This represented a loss of 26 percent from April to November, which is not excessive for the critical first summer after seeding. There must have been some seedling mortality from cattle grazing that first summer, but it was not noticeable.

The cattle were returned to the pasture in January 1953. Again, they concentrated on the seeded area. Late in September of the same year it became evident that the seedlings were being seriously browsed and trampled, so the cattle were permanently fenced out.

An inventory in April 1955 showed only 705 seedlings per acre.
The mortality between November 1952 and April 1955 was therefore 967
seedlings, or a 58-percent loss. Most of this heavy loss must be attri-
buted to trampling and browsing between January and September 1953.

North Pasture

A test of disking for seedbed preparation was conducted in the
North Pasture of the Palustris Experimental Forest in 1951. A 1-1/2-
acre plot was double-disked in November and hand-seeded with longleaf
pine in early December. The initial catch in April 1952 was 6,150 seedlings per acre. First-year survival, measured in November 1952, was
4,850 seedlings per acre.

Cattle were not permitted on the area until 1953. Then, however,
they concentrated on the seeded area and overgrazed it heavily. The pine
seedlings were browsed and trampled along with the grass (fig. 6). By
November 1953 only 900 seedlings per acre were left. Browsing and
trampling had killed most of the rest.

Figure 6. --Browsing of longleaf is common in early spring. Then cows
are apt to crop the succulent needles of seedlings that are ready to start
height growth. This damage is not as serious as trampling. (Photo by
Louisiana Forestry Commission)

The Crosby Seeding Test

In November 1952, Crosby Chemicals, Inc., direct-seeded long-leaf pine on 945 acres of cutover land in the south part of Vernon parish. The seed was planted in twin rows on thoroughly disked strips that were 8 feet wide and separated from each other by 9-foot strips of undisturbed light grass sod.

The seeded area was in a 10,000-acre fenced pasture that was grazed by 350 cattle and horses in 1953. The pasture had no cross fences and all of the animals had access to the seeded area. To draw them away, a 1,000-acre block of range in another part of the pasture was burned, but the cattle preferred to graze the succulent new grass and weeds on the disked and seeded strips.

The initial stocking in April 1953 was 972 seedlings per acre. By October, there were only 482 seedlings per acre, a mortality of 50 percent. Since first-summer losses of longleaf seedlings usually do not exceed 25 percent, it appears that trampling and heavy grazing by cattle and horses had caused additional severe loss.

SUMMARY AND RECOMMENDATIONS

These eight examples of cattle grazing damage to pine seedlings are the outstanding instances that have come to the attention of the Alexandria Research Center since 1946. No doubt there have been many other cases of serious grazing damage that were not observed or reported. On the other hand, thousands of acres of successful pine plantations have been established in central and southwest Louisiana on lands open to cattle.

For the future, it would seem wise for landowners to realize that grazing damage to pine trees, either planted or natural, can be a serious matter. The problem is likely to be intensified by recent increases in range cattle numbers and the expanding reforestation program. With more cattle and more trees, there is bound to be much less grass per cow. This will cause heavier grazing and more damage to pines in some areas.

These studies have uncovered some facts that can help landowners decide how much grazing, if any, to permit on new plantations or pine regeneration areas. The major points to consider in making a decision are these:

1. Cattle rarely graze pine foliage where other green forage is available, but some trampling and browsing damage can always be expected

if grazing is permitted the first few years after an area has been seeded or planted to pine.

2. Seedling losses will be excessive where cattle   · concentrate and over-graze. Such places are watering and feeding grounds, and any part that offers greener, fresher forage than the average surrounding range. Any treatment that removes the old grass or stimulates new grass is apt to attract cattle. Such treatments include burning, disking, scalping, and hardwood control.

3. Browsing of pine seedlings increases when desirable green forage is limited in variety or quantity. This is mostly during late winter and early spring, when both green and dry forage may be sparse. Browsing damage occurs from the early seedling stage until the saplings are too high for cattle to ride them down.

4. Seedling losses from trampling are greatest during the first year after planting and the first two years after seeding. However, losses may remain high for several years if cattle are allowed to concentrate. Rubbing injuries are greatest when the pines are from two to six feet tall.

5. Longleaf suffers heavier losses from trampling than slash or loblolly pine, because the longleaf seedlings remain for several years in the vulnerable grass stage with the large brittle terminal bud close to the ground. Trampling damage to longleaf is hard to see and identify.

6. If grazing is permitted the first few years after planting, it should not start before May 1, and should be moderate--that is, it should be regulated so that not more than half of the green forage is utilized during the entire season. On open, treeless land, with full growth of grass, each adult animal needs about one acre per month. Where trees and brush reduce the grass growth, correspondingly more acreage is required. [1]

7. The only positive way to prevent grazing damage is to keep cattle and all other kinds of livestock out of young plantations until the pines are 6 to 8 feet high. Where this is impractical, grazing should be very carefully controlled and observed.

---

[1] For information on the grazing capacity of forest range, see:
  Bond, W. E. , and Campbell, R. S. Planted pines and cattle grazing-- a profitable use of southwest Louisiana's cut-over pine land. Louisiana Forestry Commission Bul. 4, 28 pp. 1951.

  Campbell, R. S. , and Cassady, J. T. Grazing values for cattle on pine forest ranges in Louisiana. Louisiana Agricultural Experiment Station, Louisiana Bul. 452, 31 pp. 1951.

Lightning Source UK Ltd.
Milton Keynes UK
UKHW021816281118
333125UK00009B/423/P